Compendium of Organic
Synthetic Methods

Compendium of Organic Synthetic Methods

Volume 4

LEROY G. WADE, Jr.

DEPARTMENT OF CHEMISTRY
COLORADO STATE UNIVERSITY
FORT COLLINS, COLORADO

A Wiley-Interscience Publication

JOHN WILEY & SONS, New York ● Chichester ● Brisbane ● Toronto

Library of Congress Catalog Card Number: 71-162800

ISBN 0-471-04923-9

Printed in the United States of America.

10 9 8 7 6 5 4 3 2 1

PREFACE

By their compilation of Volumes 1 and 2 of this *Compendium,* Ian and Shuyen Harrison filled one of the greatest needs of the synthetic community: a method for rapidly retrieving needed information from the literature by reaction type rather than by the author's name or publication date.

Compendium of Organic Synthetic Methods, Volume 4, presents the functional group transformations and difunctional compound preparations of 1977, 1978, and 1979. We have attempted to follow as closely as possible the classification schemes of the first three volumes; the experienced user of the *Compendium* will require no additional instructions on the use of this volume.

Perhaps it is fitting here to echo the Harrisons' request stated in Volume 2 of the *Compendium:* The synthetic literature would become more easily accessible and more useful if chemists could write well-organized, concise papers with charts and diagrams that allow the reader to assess quickly and easily the scope of the published research. In addition, the reporting of actual, isolated yields and detailed experimental conditions will save a great deal of wasted effort on the part of other chemists hoping to apply the reported reactions to their own synthetic problems.

I wish to express my gratitude to the many people who helped to bring this book to completion: To Mrs. JoAnn Barley for her patience and dedication in the typing of the camera-ready copy; to Roy Smith, James McKearin, and Jon Lawson for proofreading the manuscript with great care and offering hundreds of helpful suggestions; and to my wife Betsy for her patience and moral support throughout the arduous preparation of this Compendium.

LEROY G. WADE, JR.

Fort Collins, Colorado
September, 1980

CONTENTS

ABBREVIATIONS

An attempt has been made to use only abbreviations whose meaning will be readily apparent to the reader. Some of those more commonly used are the following:

Ac	Acetyl
Am	Amyl
Ar	Aryl
9-BBN	9-borabicyclo[3.3.1]nonane
Bu	Butyl
Bz	Benzyl
Cp	Cyclopentadienyl
DBU	1,5-diazabicyclo[5.4.0]undecene-5
DCC	Dicyclohexylcarbodiimide
DDQ	2,3-Dichloro-5,6-dicyanobenzoquinone
DEAD	Diethyl azodicarboxylate
DIBAL (DIBAH)	Diisobutylaluminum hydride
DMAD	Dimethyl acetylenedicarboxylate
DME	1,2-Dimethoxyethane
DMF	Dimethylformamide
DMSO	Dimethyl sulfoxide
Et	Ethyl
Hex	Hexyl
HMPA, HMPT	Hexamethylphosphoramide (hexamethylphosphoric triamide)
hν	Irradiation with light
L	Triphenylphosphine ligand (if not specified)
LAH	Lithium aluminum hydride
LDA	Lithium diisopropylamide
MCPBA	*meta*-Chloroperbenzoic acid
Me	Methyl
MEM	β-Methoxyethoxymethyl
Ms	Methanesulfonyl
MVK	Methyl vinyl ketone
NBS	N-bromosuccinimide
Ni	Raney nickel
℗	Polymeric backbone
Ph	Phenyl
PPA	Polyphosphoric acid

PPE	Polyphosphate ester
Pr	Propyl
Py, Pyr	Pyridine
Sia	*secondary*-isoamyl
Tf	Trifluoromethane sulfonate
TFA	Trifluoroacetic acid
TFAA	Trifluoroacetic anhydride
THF	Tetrahydrofuran
THP	Tetrahydropyranyl
TMEDA	Tetramethylethylenediamine
TMP	2,2,6,6-Tetramethylpiperidine
TMS	Trimethylsilyl
Ts	*p*-Toluenesulfonyl
Δ	Heat

INDEX, MONOFUNCTIONAL COMPOUNDS

Sections—heavy type
Pages—light type

PREPARATION OF →

FROM ↓

FROM ↓ \ PREPARATION OF →	Acetylenes	Carboxylic acids, acid halides, anhydrides	Alcohols, phenols	Aldehydes	Alkyls, methylenes, aryls	Amides	Amines	Esters	Ethers, epoxides	Halides, sulfonates, sulfates	Hydrides (RH)	Ketones	Nitriles	Olefins
Acetylenes	1 / 1	16 / 9	31 / 28		61 / 102					136 / 215		166 / 252	181 / 297	196 / 309
Carboxylic acids, acid halides, anhydrides		17 / 9	32 / 28	47 / 73	62 / 103	77 / 135		107 / 176		137 / 217	152 / 235	167 / 253		
Alcohols, phenols		18 / 12		48 / 76	63 / 104		93 / 146	108 / 182	123 / 198	138 / 217	153 / 236	168 / 257		198 / 314
Aldehydes	4 / 3	19 / 13	34 / 29	49 / 82	64 / 105	79 / 137	94 / 148	109 / 186	124 / 203	139 / 221		169 / 264	184 / 298	199 / 317
Alkyls, methylenes, aryls				50 / 83						140 / 222	155 / 239	170 / 266		200 / 320
Amides			36 / 36	51 / 84	66 / 106	81 / 138	96 / 150	111 / 188					186 / 301	
Amines			37 / 37	52 / 84		82 / 140	97 / 154	112 / 189		142 / 223	157 / 239	172 / 268	187 / 303	202 / 323
Esters	8 / 5	23 / 15	38 / 38	53 / 86	68 / 107	83 / 141	98 / 155	113 / 189	128 / 205		158 / 241	173 / 269	188 / 303	203 / 323
Ethers, epoxides			39 / 39	54 / 86	69 / 107	84 / 141	99 / 156	114 / 191	129 / 206	144 / 225	159 / 242	174 / 270		204 / 324
Halides, sulfonates, sulfates	10 / 5	25 / 19	40 / 44	55 / 87	70 / 109	85 / 142	100 / 156	115 / 191	130 / 206	145 / 226	160 / 243	175 / 271	190 / 303	205 / 327
Hydrides (RH)		26 / 19	41 / 46	56 / 90	71 / 113	86 / 142		116 / 192		146 / 227		176 / 274	191 / 305	
Ketones	12 / 6	27 / 20	42 / 48	57 / 90	72 / 115	87 / 143	102 / 158	117 / 194	132 / 208	147 / 230		177 / 275	192 / 306	207 / 331
Nitriles						88 / 144	103 / 160				163 / 250		193 / 306	208 / 335
Olefins	14 / 7	29 / 21	44 / 59	59 / 92	74 / 118		104 / 161	119 / 196	134 / 209	149 / 231		179 / 282	194 / 307	209 / 336
Miscellaneous compounds		30 / 22		60 / 94		90 / 145	105 / 161	120 / 197	135 / 214	150 / 233	165 / 251	180 / 284	195 / 307	210 / 337

PROTECTION

	Sect.	Pg.
Acetylenes	15A	8
Carboxylic acids	30A	22
Alcohols, phenols	45A	63
Aldehydes	60A	95
Amines	90A	145
Esters	105A	165
Ketones	120A	197
	180A	287
Olefins	210A	338

Blanks in the table correspond to sections for which no additional examples were found in the literature.

INDEX, DIFUNCTIONAL COMPOUNDS

Sections—heavy type

Pages—light type

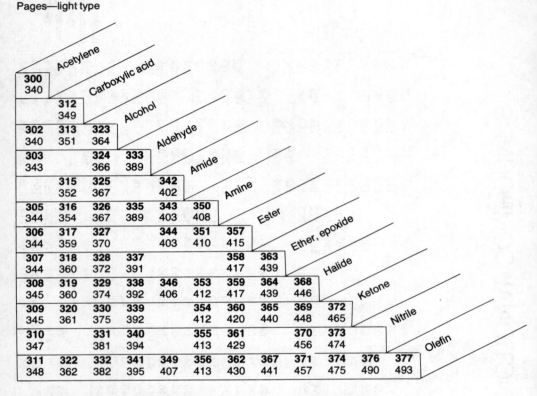

	Acetylene	Carboxylic acid	Alcohol	Aldehyde	Amide	Amine	Ester	Ether, epoxide	Halide	Ketone	Nitrile	Olefin
Acetylene	**300** 340											
Carboxylic acid		**312** 349										
Alcohol	**302** 340	**313** 351	**323** 364									
Aldehyde	**303** 343		**324** 366	**333** 389								
Amide		**315** 352	**325** 367		**342** 402							
Amine	**305** 344	**316** 354	**326** 367	**335** 389	**343** 403	**350** 408						
Ester	**306** 344	**317** 359	**327** 370		**344** 403	**351** 410	**357** 415					
Ether, epoxide	**307** 344	**318** 360	**328** 372	**337** 391			**358** 417	**363** 439				
Halide	**308** 345	**319** 360	**329** 374	**338** 392	**346** 406	**353** 412	**359** 417	**364** 439	**368** 446			
Ketone	**309** 345	**320** 361	**330** 375	**339** 392		**354** 412	**360** 420	**365** 440	**369** 448	**372** 465		
Nitrile	**310** 347		**331** 381	**340** 394		**355** 413	**361** 429		**370** 456	**373** 474		
Olefin	**311** 348	**322** 362	**332** 382	**341** 395	**349** 407	**356** 413	**362** 430	**367** 441	**371** 457	**374** 475	**376** 490	**377** 493

Blanks in the table correspond to sections for which no examples were found in the literature.

INTRODUCTION

Relationship between Volume 4 and Previous Volumes. *Compendium of Organic Synthetic Methods, Volume 4* presents over 1000 examples of published methods for the preparation of monofunctional compounds, updating the 5000 in Volumes 1, 2, and 3. In addition, Volume 4 contains over 1000 additional examples of preparations of difunctional compounds and various functional groups, updating these sections which were initially introduced in Volume 2. The same systems of section and chapter numbering are used in all four volumes.

Classification and Organization of Reactions Forming Monofunctional Compounds. Examples of published chemical transformations are classified according to the reacting functional group of the starting material and the functional group formed. Those reactions that give products with the same functional group form a chapter. The reactions in each chapter are further classified into sections on the basis of the functional group of the starting material. Within each section reactions are listed in a somewhat arbitrary order, although an effort has been made to put similar reactions together when possible.

The classification is unaffected by allylic, vinylic, or acetylenic unsaturation, which appears in both starting material and product, or increases or decreases in the length of carbon chains; for example, the reactions t-BuOH → t-BuCOOH, PhCH$_2$OH → PhCOOH and PhCH=CHCH$_2$OH → PhCH=CHCOOH would all be considered as preparations of carboxylic acids from alcohols. Entries in which conjugate reduction or alkylation of an unsaturated ketone, aldehyde, ester, acid, or nitrile occurs have generally been placed in category 74, Alkyls from Olefins.

The terms hydrides, alkyls, and aryls classify compounds containing reacting hydrogens, alkyl groups, and aryl groups, respectively; for example, RCH$_2$-H → RCH$_2$COOH (carboxylic acids from hydrides), RMe → RCOOH (carboxylic acids from alkyls), RPh → RCOOH (carboxylic acids from aryls). Note the distinction between R$_2$CO → R$_2$CH$_2$ (methylenes from ketones) and RCOR′ → RH (hydrides from ketones). Alkylations which involve additions across a double bond are found in section 74, Alkyls from Olefins.

The following examples illustrate the application of the classification scheme to some potentially confusing cases:

$RCH=CHCOOH \rightarrow RCH=CH_2$	(hydrides from carboxylic acids)
$RCH=CH_2 \rightarrow RCH=CHCOOH$	(carboxylic acids from hydrides)
$ArH \rightarrow ArCOOH$	(carboxylic acids from hydrides)
$ArH \rightarrow ArOAc$	(esters from hydrides)
$RCHO \rightarrow RH$	(hydrides from aldehydes)
$RCH=CHCHO \rightarrow RCH=CH_2$	(hydrides from aldehydes)
$RCHO \rightarrow RCH_3$	(alkyls from aldehydes)
$R_2CH_2 \rightarrow R_2CO$	(ketones from methylenes)
$RCH_2COR \rightarrow R_2CHCOR$	(ketones from ketones)
$RCH=CH_2 \rightarrow RCH_2CH_3$	(alkyls from olefins)
$RBr + RC{\equiv}CH \rightarrow RC{\equiv}CR$	(acetylenes from halides; also acetylenes from acetylenes)
$ROH + RCOOH \rightarrow RCOOR$	(esters from alcohols; also esters from carboxylic acids)
$RCH=CHCHO \rightarrow R_2CHCH_2CHO$	(alkyls from olefins)
$RCH=CHCN \rightarrow RCH_2CH_2CN$	(alkyls from olefins)

Reactions are included even when full experimental details are lacking in the given reference. In some cases the quoted reaction is a minor part of a paper or may have been investigated from a purely mechanistic aspect.

How to Use the Book to Locate Examples of the Preparation or Protection of Monofunctional Compounds. Examples of the preparation of one functional group from another are located via the monofunctional index on p. xi, which lists the corresponding section and page. Thus Section 1 contains examples of the preparation of acetylenes from other acetylenes; Section 2, acetylenes from carboxylic acids; and so forth.

Sections that contain examples of the reactions of a functional group are found in the horizontal rows of the index. Thus Section 1 gives examples of the reactions of acetylenes that form other acetylenes; Section 16, reactions of acetylenes that form carboxylic acids; and Section 31, reactions of acetylenes that form alcohols.

Examples of alkylation, dealkylation, homologation, isomerization, and transposition are found in Sections 1, 17, 33, and so on, which lie close to a diagonal of the index. These sections correspond to such topics as the preparation of acetylenes from acetylenes, carboxylic acids from carboxylic acids, and alcohols and phenols from alcohols and phenols. Alkylations which involve conjugate additions across a double bond are found in section 74, Alkyls from Olefins.

Examples of name reactions can be found by first considering the nature of the starting material and product. The Wittig reaction, for instance, is in Section 199 on olefins from aldehydes and Section 207 on olefins from ketones.

Examples of the protection of acetylenes, carboxylic acids, alcohols, phenols, aldehydes, amides, amines, esters, ketones, and olefins are also indexed on p. xi.

The pairs of functional groups alcohol, ester; carboxylic acid, ester; amine, amide; carboxylic acid, amide can be interconverted by quite trivial reactions. When a member of these groups is the desired product or starting material, the other member should, of course, also be consulted in the text.

The original literature must be used to determine the generality of reactions. A reaction given in this book for a primary aliphatic substrate may also be applicable to tertiary or aromatic compounds.

The references usually yield a further set of references to previous work. Subsequent publications can be found by consulting the Science Citation Index.

Classification and Organization of Reactions forming Difunctional Compounds. This chapter considers all possible difunctional compounds formed from the groups acetylene, carboxylic acid, alcohol, aldehyde, amide, amine, ester, ether, epoxide, halide, ketone, nitrile, and olefin. Reactions that form difunctional compounds are classified into sections on the basis of the two functional groups of the product. The relative positions of the groups do not affect the classification. Thus preparations of 1,2-aminoalcohols, 1,3-aminoalcohols and 1,4-aminoalcohols are included in a single section. The following examples illustrate the application of this classification system:

Difunctional Product	Section Title
$RC{\equiv}C\text{-}C{\equiv}CR$	Acetylene—Acetylene
$RCH(OH)COOH$	Carboxylic Acid—Alcohol
$RCH{=}CHOMe$	Ether—Olefin
$RCHF_2$	Halide—Halide
$RCH(Br)CH_2F$	Halide—Halide
$RCH(OAc)CH_2OH$	Alcohol—Ester
$RCH(OH)COOMe$	Alcohol—Ester
$RCH{=}CHCH_2COOMe$	Ester—Olefin
$RCH{=}CHOAc$	Ester—Olefin

How to Use the Book to Locate Examples of the Preparation of Difunctional Compounds. The difunctional index on p. xii gives the section and page corresponding to each difunctional product. Thus Section 327

(Alcohol—Ester) contains examples of the preparation of hydroxyesters; Section 323 (Alcohol—Alcohol) contains examples of the preparation of diols.

Some preparations of olefinic and acetylenic compounds from olefinic and acetylenic starting materials can, in principle, be classified in either the monofunctional or difunctional sections; for example, RCH=CHBr → RCH=CHCOOH, Carboxylic acids from Halides (monofunctional sections) or Carboxylic acid—Olefin (difunctional sections). In such cases both sections should be consulted.

Reactions applicable to both aldehyde and ketone starting materials are in many cases illustrated by an example that uses only one of them.

Many literature preparations of difunctional compounds are extensions of the methods applicable to monofunctional compounds. Thus the reaction RCl → ROH can clearly be extended to the preparation of diols by using the corresponding dichloro compound as a starting material. Such methods are not fully covered in the difunctional sections.

The user should bear in mind that the pairs of functional groups alcohol, ester; carboxylic acid, ester; amine, amide; and carboxylic acid, amide can be interconverted by quite trivial reactions. Compounds of the type $RCH(OAc)CH_2OAc$ (Ester—Ester) would thus be of interest to anyone preparing the diol $RCH(OH)CH_2OH$ (Alcohol—Alcohol).

Compendium of Organic Synthetic Methods

CHAPTER 1
PREPARATION OF ACETYLENES

Section 1 <u>Acetylenes from Acetylenes</u>

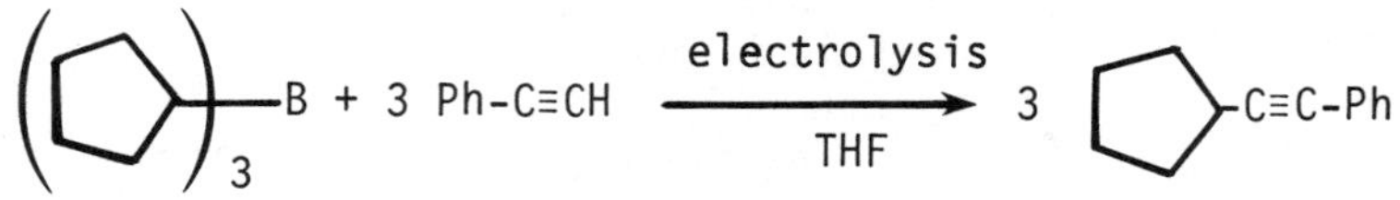

70%

Chem Lett, 999 (1977)

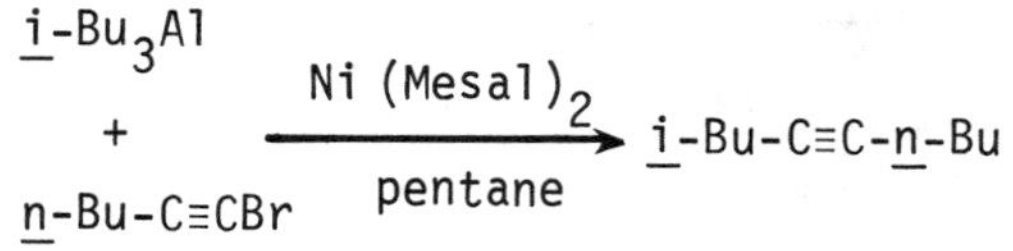

70%

Tetr Lett, 2831 (1978)

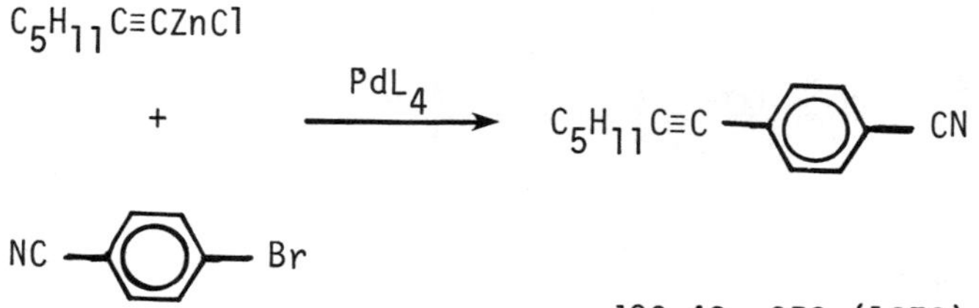

93%

JOC <u>43</u>, 358 (1978)

1

$$\text{LiCH}_2\text{C}\equiv\text{C-Li} \xrightarrow{\begin{array}{c}\text{1) BuBr}\\\hline\\\text{2) BuBr/HMPT}\end{array}} \text{BuCH}_2\text{C}\equiv\text{CBu} \qquad 30\%$$

JCS Perkin I, 1218 (1979)

$$\text{Bu-C}\equiv\text{C-CH}_2\text{CH}_2\text{OH} \xrightarrow[\text{H}_2\text{N(CH}_2)_3\text{NH}_2]{\text{NaNH}_2} \text{HC}\equiv\text{C(CH}_2)_6\text{OH} \qquad 91\%$$

Rec Trav Chim 96, 160 (1977)

$$\xrightarrow[\text{CuI, Et}_3\text{N}]{\text{PdCl}_2\text{L}_2}$$

95%

+

Ph-C≡CH

Chem Pharm Bull 26, 3843 (1978)

Section 2   Acetylenes from Carboxylic Acids

No additional examples

Section 3   Acetylenes from Alcohols

No additional examples

Section 4 Acetylenes from Aldehydes

$$Ar-CHO \xrightarrow{\begin{array}{c} 1) \\ \\ 2) \quad \Delta \end{array}} Ar-C\equiv CH$$

up to 90%

Ar = pyrrole, furan, azulene, etc.

Angew Int Ed 17, 609 (1978)

$$Ph-CH_2CHO \xrightarrow[\text{THF}]{\underline{t}-BuOK, \ (MeO)_2\overset{O}{\overset{\|}{P}}CHN_2} Ph-CH_2-C\equiv CH$$

80%

JOC 44, 4997 (1979)

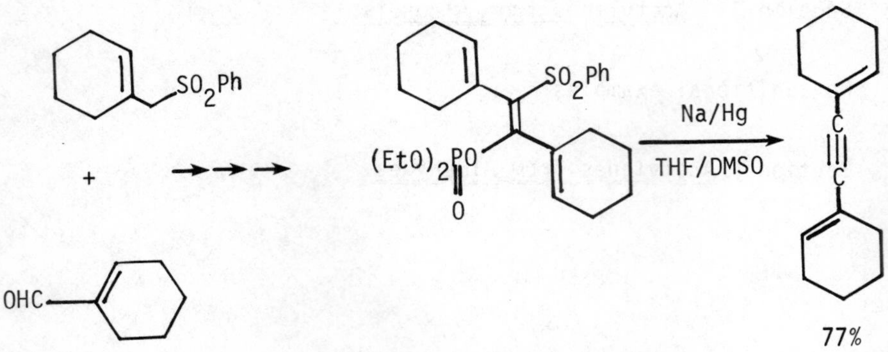

Tetr Lett, 2625 (1978)

Section 5   Acetylenes from Alkyls, Methylenes and Aryls

No Examples

Section 6   Acetylenes from Amides

No additional examples

Section 7   Acetylenes from Amines

No additional examples

Section 8   Acetylenes from Esters

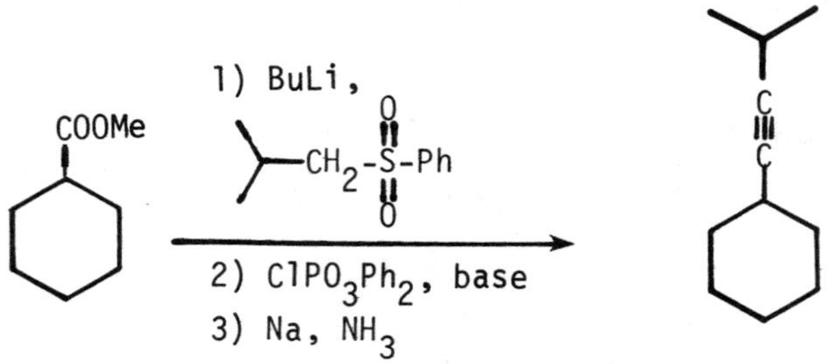

1) BuLi,

$$-CH_2-\overset{O}{\underset{O}{\overset{\|}{\underset{\|}{S}}}}-Ph$$

2) ClPO$_3$Ph$_2$, base
3) Na, NH$_3$

54%

JACS 100, 4852 (1978)

Section 9   Acetylenes from Ethers

No examples

Section 10   Acetylenes from Halides

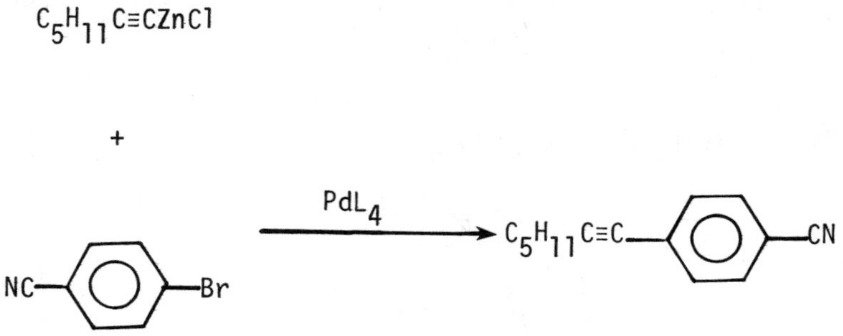

$C_5H_{11}C\equiv CZnCl$

+

NC——◯——Br

PdL$_4$

$C_5H_{11}C\equiv C$——◯——CN   93%

JOC 43, 358 (1978)

1) BuLi

Ph-CH=CH-CH₂CHCl₂ $\xrightarrow{\text{1) BuLi}}$ Ph-CH=CH-C≡CH   70%

2) H₃O⊕

↑ LiCHCl₂

Ph-CH=CH-CH₂Br

Synthesis, 502 (1979)

BuLi

Et-CH₂CF₃ $\xrightarrow{\text{BuLi}}$ Et-C≡C-Bu   73%

Tetr Lett, 3103 (1978)

Section 11   Acetylenes from Hydrides

No examples

For examples of the reaction RC≡CH → RC≡C-C≡CR' see section 300
(Acetylene - Acetylene)

Section 12   Acetylenes from Ketones

O
‖
Ph-C-CH₂Et $\xrightarrow[\text{Et}_3\text{N, CH}_2\text{Cl}_2]{}$ Ph-C≡C-Et   50%

Chem Lett, 481 (1979)

$$Ph-C \equiv C-Ph \qquad 80\%$$

Me$_3$SiCHN$_2$ / BuLi

JCS Perkin I, 869 (1977)

$$\text{Ac}_2\text{O} / \text{pyr, reflux}$$

65%

JOC 44, 4116 (1979)

1) H$_2$NNHTs

2) MeLi

$$Ph-C \equiv C-Me \qquad 90\%$$

Synthesis, 305 (1978)

Section 13   Acetylenes from Nitriles

No examples

Section 14   Acetylenes from Olefins

1) LiAlH$_4$, TiCl$_4$

2) CH$_2$=C=CHBr, CuCl

$$HC \equiv C-$$

54%

Chem Lett, 789 (1978)

Section 15   <u>Acetylenes from Miscellaneous Compounds</u>

No additional examples

Section 15A   <u>Protection of Acetylenes</u>

Use of the Trimethylsilyl group to terminate polyeneynes of the form $-(CH=CH)_n-C\equiv C-TMS$. Stable to Grignard reagents etc., but removed by aqueous base.

Tetrahedron <u>34</u>, 1037 (1978)

CHAPTER 2
PREPARATION OF CARBOXYLIC ACIDS, ACID HALIDES, AND ANHYDRIDES

Section 16   Carboxylic Acids from Acetylenes

$$BuC\equiv C-H \xrightarrow[\text{CH}_2\text{Cl}_2, \text{ Adogen}]{\text{KMnO}_4, \text{ H}_2\text{O}} Bu-COOH \qquad 61\%$$

Synthesis, 462 (1978)

Section 17   Carboxylic Acids and Acid Halides from
             Carboxylic Acids

1)BuLi
2)$H_3O^{\oplus}$

55%

96% ee

JOC 44, 2250 (1979)

modified Arndt-Eistert

94%

Tetr Lett, 2667 (1979)

R-CH$_2$COOH   1)LDA,THF   R-CH-COOH
fatty acid |
              2)O$_2$,HMPA   OH
                                              CrO$_3$,NaIO$_4$

                                              AcOH/H$_2$O

                              R-COOH   ~58%

Synth Comm 9, 63 (1979)

Review: "Synthesis of Aldehydes, Ketones, and Carboxylic Acids
        from Lower Carbonyl Compounds by C-C Coupling Reactions"

Synthesis, 633 (1979)

Carboxylic Acids may be alkylated and homologated via ketoacid,
ketoester and olefinic acid intermediates. See section 320
(Carboxylic Acid - Ketone), section 360 (Ester - Ketone) and
Section 322 (Carboxylic Acid - Olefin). Conjugate reductions of
unsaturated acids are listed in Section 74 (Alkyls from Olefins).

$$Cl-N \overset{N}{\underset{Cl}{\bigcirc}} N-Cl$$

R-COOH $\xrightarrow[\text{Et}_3\text{N}]{}$ $R-\overset{\overset{\displaystyle O}{\|}}{C}-Cl$

Tetr Lett, 3037 (1979)

R-COOH $\xrightarrow{\overset{\displaystyle Cl}{\underset{\displaystyle Me_2C=CNMe_2}{|}}}$ $R-\overset{\overset{\displaystyle O}{\|}}{C}-Cl$ 100%

R = Cl_3C, t-Bu, HCO, furan-2-yl, etc.

Chem Comm, 1180 (1979)

$R-\overset{\overset{\displaystyle O}{\|}}{C}-OH$ $\xrightarrow{CF_3CF_2CF=O}$ $R-\overset{\overset{\displaystyle O}{\|}}{C}-F$ 80-90%

Chem Lett, 483 (1977)

$Ph-\overset{\overset{\displaystyle O}{\|}}{C}-Cl$ $\xrightarrow[\text{KF, CH}_3\text{CN}]{\text{polyethylene glycol}}$ $Ph-\overset{\overset{\displaystyle O}{\|}}{C}-F$ 98%

Chem Lett, 283 (1978)

Review: "Activation and Protection of the Carboxyl Group"

Chem & Ind, 610 (1979)

Section 18   Carboxylic Acids from Alcohols

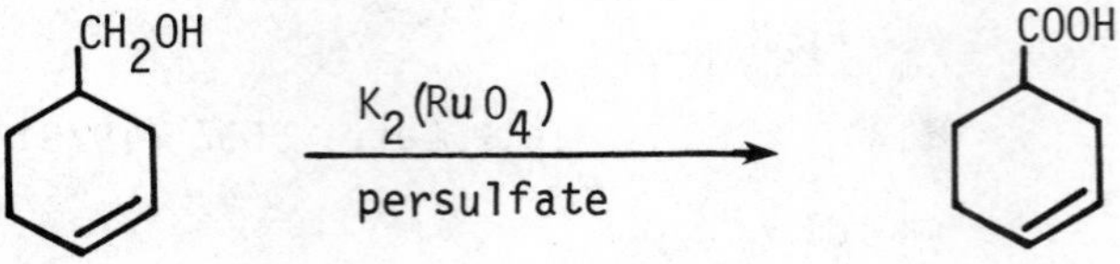

JCS Chem Comm, 58 (1979)

$Pr-CH_2OH$ $\xrightarrow[\text{Ni(OH)}_2 \text{ anode}]{\text{electrolysis}}$ $Pr-COOH$ 92%

Synthesis, 513 (1979)

JCS Chem Comm, 253 (1978)

Section 19   Carboxylic Acids and Acid Halides from Aldehydes

$$C_6H_{13}CHO \xrightarrow[\substack{2)ZnCl_2,Ac_2O \\ 3) \ \overset{\ominus}{OH}}]{\substack{1)NaH,(EtO)_2\overset{\overset{O}{\|}}{P}CH(CN)O-\underline{t}-Bu}} C_6H_{13}CH_2COOH \qquad 95\%$$

JACS 99, 182 (1977)

$$\underline{n}-C_5H_{11}CHO \xrightarrow[CH_2Cl_2]{BzEt_3\overset{\oplus}{N}MnO_4^{\ominus}} C_5H_{11}COOH \qquad 89\%$$

Monatsh Chem 110, 1471 (1979)

$$\xrightarrow{Bu_4\overset{\oplus}{N}MnO_4^{\ominus}}$$

99%

JCS Chem Comm, 253 (1978)

$$\xrightarrow[H_2O,NaOH]{Nickel\ peroxide}$$

96%

Chem Pharm Bull 26, 299 (1978)

$$C_6H_{13}-\underset{\underset{H}{|}}{C}\overset{O}{\underset{O}{\diagup}} \quad \xrightarrow[\text{2)Zn/ZnCl}_2]{\text{1)NBS}} \quad C_6H_{13}-COOH \qquad 60\%$$

JOC <u>43</u>, 3417 (1978)

$\xrightarrow[\text{hv}]{\text{NBS}}$

>82%

(converted to ester)

Tetr Lett, 3809 (1979)

PhCHO

+

$PhCCl_3$

$\xrightarrow{FeCl_3}$

$$Ph-\overset{O}{\overset{\|}{C}}-Cl$$

+

$PhCHCl_2$

~80%

J Gen Chem (USSR) <u>47</u>, 1531 (1977)

Related methods: Carboxylic Acids from Ketones (Section 27).
Also via: Esters - Section 109.

Section 20   Carboxylic Acids from Alkyls

No additional examples

Section 21   Carboxylic Acids from Amides

No additional examples

Section 22   Carboxylic Acids from Amines

No additional examples

Section 23   Carboxylic Acids and Acid Halides from Esters

$$R-\overset{O}{\overset{\|}{C}}-OR' \xrightarrow[\text{2) } H_2O]{\text{1) }Me_3SiI,CCl_4} RCOOH + R'I \qquad 80\text{-}95\%$$

R = alkyl, aryl, heterocyclic

R' = Me, Et

JACS 99, 968 (1977)

$$\text{(cyclohexyl-COOMe)} \xrightarrow[\text{2) } H_2O]{\text{1)ClSiMe}_3, \text{ NaI}} \text{(cyclohexyl-COOH)} \quad 86\%$$

JCS Chem Comm, 874 (1978)

$$Me_3C\text{-}\overset{O}{\underset{\|}{C}}\text{-}OMe \xrightarrow[\text{NaI}]{\text{ClSiMe}_3} Me_3C\text{-}COOH \quad 79\%$$

JOC **44**, 1247 (1979)

$$PhCH_2\text{-}COOEt \xrightarrow[\text{2) } H_2O]{\text{1)Me}_3SiSiMe_3/I_2} PhCH_2COOH \quad 95\%$$

Angew Int Ed **18**, 612 (1979)

$$Bz\text{-}CO_2Et \xrightarrow{I_2, PhSiMe_3} Bz\text{-}COOH \quad 96\%$$

Synthesis, 417 (1977)

$$Ph\text{-}COOMe \xrightarrow[\text{AlBr}_3]{\text{EtSH}} Ph\text{-}COOH \quad 94\%$$

Tetr Lett, 5211 (1978)

$$\text{NaO}_2, \text{Me}_2\text{SO}$$

97%

JOC **44**, 4727 (1979)

$$\text{Ph-C-OMe} \xrightarrow{\;2\ \underline{t}\text{-BuOK, 1 H}_2\text{O}\;} \text{Ph-C-O}^{\ominus} \quad + \quad \text{MeO}^{\ominus}$$

100%

JOC **42**, 918 (1977)

$$\text{R-C-O-CH}_2\text{CH=CH}_2 \xrightarrow{\;\text{Me}_2\text{CuLi}\;} \text{R-C-OLi}$$

75-85%

R = alkyl, aryl

Synth Comm **8**, 15 (1978)

$$\text{R-C-OCH}_2\text{CH=CHPh} \xrightarrow[\text{2) KSCN}]{\text{1) Hg(OAc)}_2} \text{R-COOH}$$

∼90%

Tetr Lett, 2081 (1977)

100%

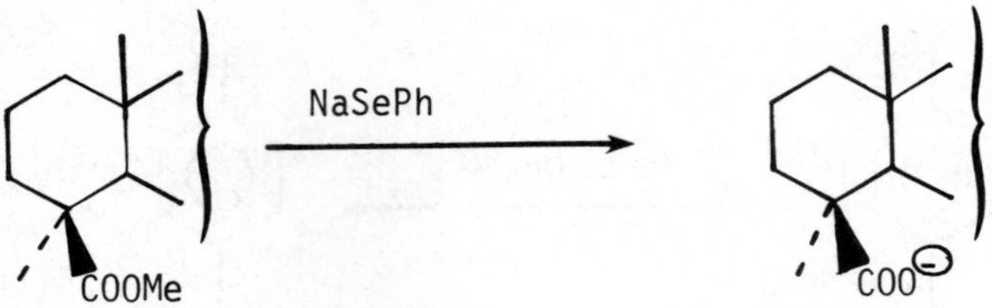

Tetr Lett, 4365 (1977)

Other reactions useful for the hydrolysis of esters may be found in Section 30A (Protection of Carboxylic Acids).

$CH_3(CH_2)_5CO_2Si(Me)_2$-*t*-Bu $\xrightarrow[\text{DMF}]{(COCl)_2}$ $CH_3(CH_2)_5C(=O)$-Cl 92%

JOC 43, 3972 (1978)

Section 24 Carboxylic Acids from Ethers

No additional examples

Section 25   <u>Carboxylic Acids and Acid Halides from Alkyl Halides</u>

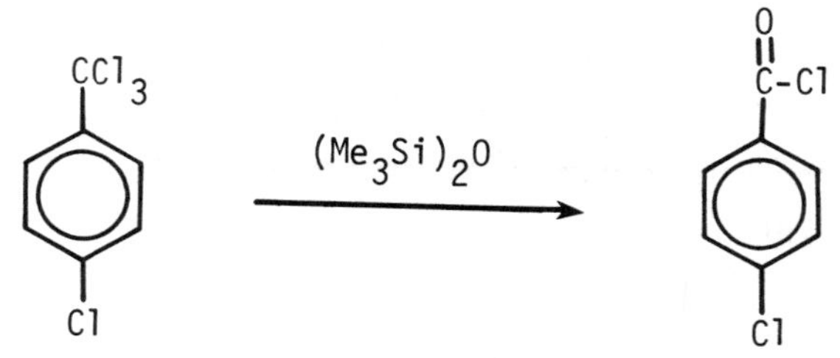

JCS Chem Comm, 808 (1977)

Also via: Esters - Section 115

Section 26   <u>Carboxylic Acids from Hydrides</u>

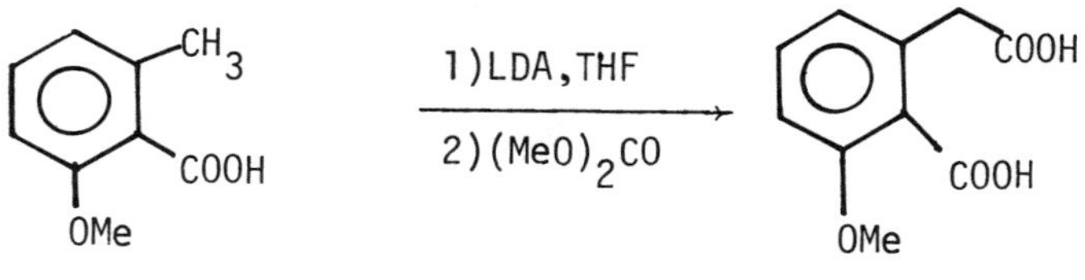

Synthesis, 245 (1977)

SECTION 27   Carboxylic Acids from Ketones

1) NaH, $(EtO)_2\overset{\overset{O}{\|}}{P}CH(CN)O\text{-}\underline{t}\text{-}Bu$

2) $ZnCl_2$, Ac_2O

3) $^\ominus OH$

~70%

JACS 99, 182 (1977)

Review: "Synthesis of Aldehydes, Ketones, and Carboxylic Acids from Lower Carbonyl Compounds by C-C Coupling Reactions"

Synthesis, 633 (1979)

Also via: Esters - Section 117.

Section 28   Carboxylic Acids from Nitriles

No additional examples

Section 29   Carboxylic Acids from Olefins

$$C_8H_{17}-CH=CH_2 \xrightarrow[\text{H}_2\text{O}]{\text{KMnO}_4\text{,HOAc}} C_8H_{17}-COOH \qquad 94\%$$

JOC 42, 3749 (1977)

$$\underline{n}-C_6H_{13}CH=CH_2 \xrightarrow[\text{2) H}_2\text{/Pd, CaCO}_3\text{, PbO}]{\text{1) O}_3} \underline{n}-C_6H_{13}COOH \quad 81\%$$

JOC (USSR) 14, 48 (1978)

1) O$_3$

2) H$_2$, Lindlar

98%

Bull Akad USSR Chem 25, 1790 (1977)

1) $\overset{O}{\diagup}\!\!\diagdown \text{Br}$ PhSe

2) NaIO$_4$,NaHCO$_3$

3) C$_6$H$_{13}$NH$_2$, Δ

4) KOH, H$_2$O

77%

Helv Chim Acta 61, 2286 (1978)

Section 30   Carboxylic Acids from Miscellaneous Compounds

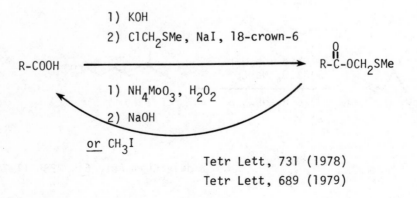

$$Bu_3B \xrightarrow[\text{3) } H_3O^{\oplus}]{\substack{\text{1) } PhO\overset{Li}{C}HCOOLi \\ \text{2) } NaOH, H_2O_2}} Bu\text{-}CH_2COOH \qquad 100\%$$

Tetr Lett, 2891 (1978)

1) BuMgBr,
   NiCl$_2$ THF

2) H$_3$O$^{\oplus}$

92%
99%ee

BCS Japan **51**, 3368 (1978)

Section 30A   Protection of Carboxylic Acids

R-COOH

1) KOH
2) ClCH$_2$SMe, NaI, 18-crown-6

R-C-OCH$_2$SMe

1) NH$_4$MoO$_3$, H$_2$O$_2$
2) NaOH

or CH$_3$I

Tetr Lett, 731 (1978)
Tetr Lett, 689 (1979)

$$R-COO^{\ominus}Et_3\overset{\oplus}{N}H \xrightarrow{Me_2SCl_2} R-\overset{O}{\overset{\|}{C}}-OCH_2SCH_3 \qquad \sim 85\%$$

Synth Comm <u>9</u>, 267 (1979)

88%

Synthesis, 567 (1977)

$$\begin{array}{c} CH_2CH_2SCH_3 \\ | \\ H_2N-CH-COOH \end{array} \xrightarrow[\substack{CCl_4, \text{ azeotropic dist.}}]{TsOH, \ Cl_3C-CH_2OH} \begin{array}{c} CH_2CH_2SCH_3 \\ \overset{\oplus}{H_3N}-\overset{|}{CH}-\overset{O}{\overset{\|}{C}}-OCH_2CCl_3 \\ \ominus OTs \end{array}$$

60%

Synthesis, 24 (1979)

PhCH$_2$CH$_2$COOH
+
EtSH

$\xrightarrow[Et_3N, \ DMF]{DEPC}$

$$PhCH_2CH_2\overset{O}{\overset{\|}{C}}-SEt \qquad 85\%$$

Chem Pharm Bull <u>25</u>, 2423 (1977)

$$\underset{\text{MeCH-COOH}}{\overset{\text{OPh}}{|}} \quad \xrightarrow[\text{EtSH,DME}]{\text{DMPADC,Et}_3\text{N}} \quad \underset{\text{Me-CH-C-SEt}}{\overset{\text{OPh} \quad \text{O}}{| \qquad \|}} \qquad 100\%$$

Synth Comm <u>9</u>, 91 (1979)

$$\text{Z-gyl-OH} \quad \xrightarrow[\text{2) PhSH}]{\overset{\text{O}}{\overset{\|}{\text{1) Ph}_2\text{PCl, Et}_3\text{N}}}} \quad \text{Z-gly-SPh} \qquad 98\%$$

Synth Comm <u>7</u>, 251 (1977)

1,4-cyclohexadiene/Pd-C removes benzyl ester protecting groups.

JOC <u>43</u>, 4194 (1978)

$$\underset{\text{R-C-OBz}}{\overset{\text{O}}{\overset{\|}{}}} \quad \xrightarrow[\text{PhOCH}_3]{\text{AlCl}_3} \quad \text{R-COOH}$$

Tetr Lett, 2793 (1979)

peptide - Bzl $\xrightarrow[\text{Pd/C}]{}$ peptide-OH

JCS perkin I, 490 (1977)

$R-\overset{O}{\overset{\|}{C}}-O-CH_2-\langle\text{ring}\rangle-NO_2 \xrightarrow[\text{THF}]{Na_2S}$ R-COOH

R = Azetidinone

JOC $\underline{43}$, 1243 (1978)

$R-\overset{O}{\overset{\|}{C}}-OCH_2CH_2X \xrightarrow[\substack{\text{or 1) } Me_3SnLi \\ \text{2) } Bu_4NF}]{NaHSe}$ $RCOO^{\ominus}$

X = Cl, Br

Synth Comm $\underline{8}$, 301 and 359 (1978)

peptide-$\overset{O}{\overset{\|}{C}}$-O-CH$_2$-CH$_2$Br $\xrightarrow[\text{2) Zn}]{\text{1) NaI, DMF}}$ peptide-COOH

Chem Ber $\underline{112}$, 2145 (1979)

Use of the p-methoxyphenacyl acid protecting group in gibberellin synthesis. Stable to Koenigs-Knorr conditions, MnO_2 oxidation, etc.

Removed by photolysis in ethanol or by Zn/HOAc.

Tetrahedron 34, 345 (1978)

H-Phe-OH
⟶
1) tetramethylguanidine,
   CH_3COCH_2COOEt, DMF
2) Nbzl-Br
3) HCl, MeOH

H-Phe-ONbzl·HCl

88%

Aust J Chem 31, 1865 (1978)

Z-Phe-OH ⟶ Z-Phe-OCH$_2$-N

KF

77%

Synth Comm 8, 515 (1978)

Use of immobilized carboxypeptidase Y (at pH 8.5) to remove ethyl ester blocking groups in peptide synthesis.

JACS 101, 3394 (1979)

Use of Maq esters

in peptide synthesis:

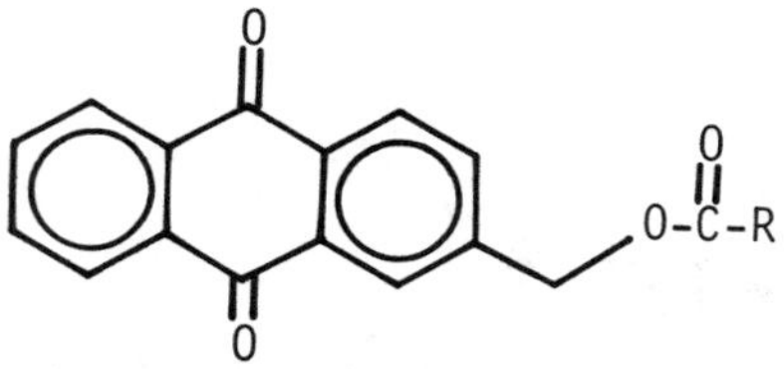

Prepared using Maq-OH and DCC; removed by sodium dithionite, photolysis, or polymer-bound 9,10-dihydroxyanthracene. Soluble in organic solvents, and UV-active allowing facile detection on TLC.

Tetr Lett, 1031 (1977)

Review: "Activation and Protection of the Carboxyl Group"

Chem & Ind, 610 (1979)

Review: "Protecting Groups in Peptide Synthesis"

Chem & Ind, 617 (1979)

Other reactions useful for the protection of carboxylic acids are included in Section 107 (Esters from Carboxylic Acids and Acid Halides) and Section 23 (Carboxylic Acids from Esters).

CHAPTER 3
PREPARATION OF ALCOHOLS
AND PHENOLS

Section 31   <u>Alcohols from Acetylenes</u>

JACS <u>99</u>, 5192 (1977)

Section 32   <u>Alcohols from Carboxylic Acids and Acid Halides</u>

$C_9H_{19}-COOH \longrightarrow C_9H_{19}-\overset{O}{\underset{}{C}}-S-\!\!\!<\text{thiazoline}> \xrightarrow{NaBH_4} C_9H_{19}CH_2OH$ 98%

JCS Chem Comm, 330 (1978)

$CH_3(CH_2)_{14}-\overset{O}{\underset{}{C}}-O-N\!\!<\text{succinimide}> \xrightarrow{NaBH_4} CH_3(CH_2)_{14}-CH_2OH$

Chem Lett, 981 (1979)

28

$$\underset{\text{MeO-CCH}_2\text{CH}_2\text{CCl}}{\overset{\text{O}\qquad\text{O}}{\quad\|\qquad\quad\|}} \xrightarrow{\text{NaBH}_4-\text{Al}_2\text{O}_3} \underset{\text{MeO-C-CH}_2\text{CH}_2\text{CH}_2\text{OH}}{\overset{\text{O}}{\|}} \qquad 80\%$$

Synthesis, 912 (1979)

Also via: Esters (Section 38)

Section 33   Alcohols from Alcohols

No additional examples

Section 34   Alcohols from Aldehydes

$$\underline{n}\text{-C}_5\text{H}_{11}\text{-CHO} \xrightarrow[\text{H}_2\text{O/dioxane}]{\text{Na}_2\text{S}_2\text{O}_4} n\text{-C}_5\text{H}_{11}\text{CH}_2\text{OH} \qquad 63\%$$

Synthesis, 246 (1977)

$$\underline{n}\text{-C}_9\text{H}_{19}\text{CHO} \xrightarrow[\text{alumina}]{\text{2-propanol}} \underline{n}\text{-C}_9\text{H}_{19}\text{CH}_2\text{OH} \qquad 84\%$$

JOC 42, 1202 (1977)

CHO → CH$_2$OH

$$\xrightarrow[\text{Ni(OAc)}_2, \text{ MgBr}_2]{\text{NaH/}\underline{t}\text{-AmONa}}$$

98%

JOC <u>43</u>, 4804 (1978)

$$CH_3(CH_2)_6CHO \xrightarrow[\text{silica gel}]{Bu_3SnH} CH_3(CH_2)_6CH_2OH \qquad 90\%$$

JOC <u>43</u>, 3977 (1978)

$$CH_3(CH_2)_7CHO \xrightarrow{Cp_2Zr(Cl)BH_4} CH_3(CH_2)_8-OH \qquad 90\%$$

Tetr Lett, 4985 (1978)

$$Ph-CH_2CH_2CHO \xrightarrow[\text{ether}]{NaBH_4/Al_2O_3} PhCH_2CH_2CH_2OH \qquad 87\%$$

Synthesis, 891 (1978)

$$PhCH=CHCHO \xrightarrow{BH_3 \cdot SMe_2} PhCH=CH-CH_2OH \quad 82\%$$

JOC **43**, 1829 (1978)

$$C_5H_{11}-CHO \xrightarrow{\text{9-BBN} \cdot \text{pyridine}} C_5H_{11}-CH_2OH \qquad 100\%$$

JOC **42**, 4169 (1977)

$$C_6H_{13}CHO \xrightarrow{(HAlN-i-Pr)_6} C_6H_{13}CH_2OH \qquad 98\%$$

Z Chem **17**, 18 (1977)

$$Ph-CHO \xrightarrow{MgH_2} Ph-CH_2OH \qquad 100\%$$

JOC **43**, 1557 (1978)

$$PhCH=CHCHO \xrightarrow[\text{RhCl}_3]{H_2, CO} PhCH=CH-CH_2OH \qquad 83\%$$

BCS Japan **50**, 2148 (1977)

$$\text{Ph-CH=CH-CHO} \xrightarrow{\text{9-BBN}} \text{Ph-CH=CH-CH}_2\text{OH} \qquad 86\%$$

JOC $\underline{42}$, 1197 (1977)

100%

Chem Lett, 1085 (1977)

$$\text{C}_6\text{H}_{13}\text{-CHO} \xrightarrow[\text{RuCl}_2(\text{CO})_2\text{L}_2]{\text{H}_2} \text{C}_6\text{H}_{13}\text{CH}_2\text{OH} \qquad 88\%$$

J Organometal Chem $\underline{145}$, 189 (1978)

$$\text{R-CHO} \xrightarrow[\text{IrH}_3\text{L}_3,\text{HOAc}]{\text{H}_2} \text{R-CH}_2\text{OH}$$

Ketones are not reduced.

J Organometal Chem $\underline{129}$, C43 (1977)

Aldehydes are selectively reduced in the presence of ketones by LiBH$_4$ on molecular seives.

JOC 44, 3969 (1979)

$$Ph-\overset{O}{\overset{\|}{C}}-D \longrightarrow Ph-\overset{OH}{\underset{H}{C}}-D$$

82%

88%ee

JACS 99, 5211 (1977)

$$R-CDO \xrightarrow[\text{2)}HOCH_2CH_2NH_2]{\text{1)}3-pinanyl-9-BBN} R-\overset{OH}{\underset{H}{C}}-D$$

~80%

58-86%ee

R = alkyl, aryl, allyl

JACS 101, 2352 (1979)

$$Bu-CHO \xrightarrow[\text{acetone}]{BuMnI} Bu_2CHOH$$

89%

Tetr Lett, 3383 (1977)

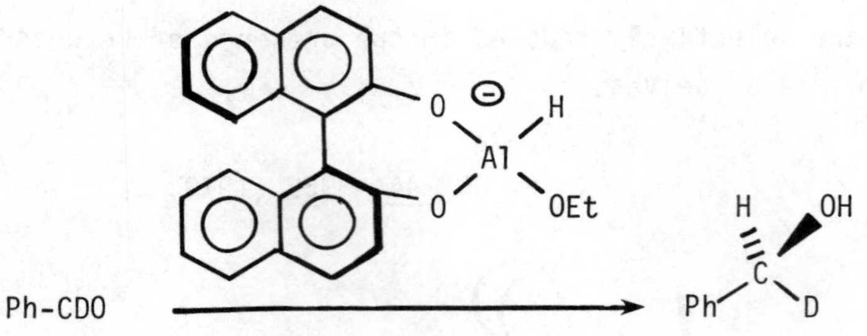

O
Al
H
O
OEt
Ph-CDO
H
OH
C
Ph
D

75%

82%ee

JACS 101, 3129 (1979)

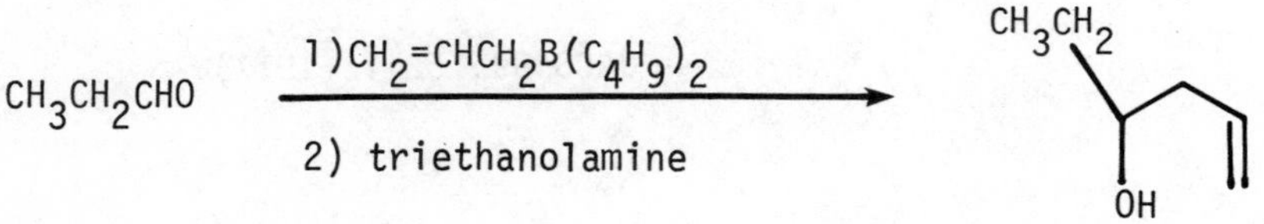

CH_3CH_2CHO
1) $CH_2=CHCH_2B(C_4H_9)_2$
2) triethanolamine
CH_3CH_2
OH

Bull Acad USSR Chem 27, 1663 (1979)

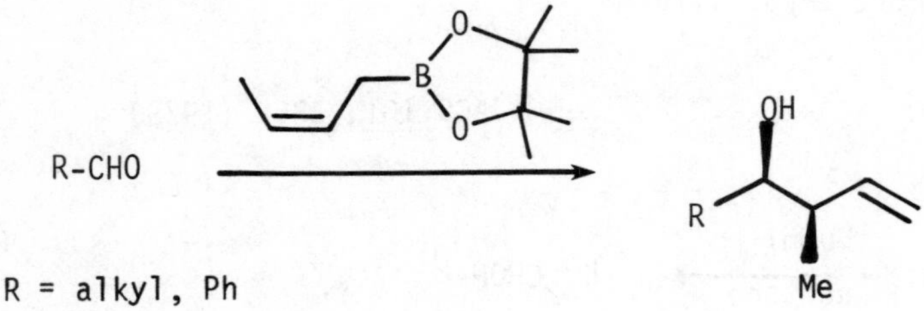

O
B
O
R-CHO
OH
R
Me
R = alkyl, Ph

≤92%

Angew Int Ed 18, 306 (1979)

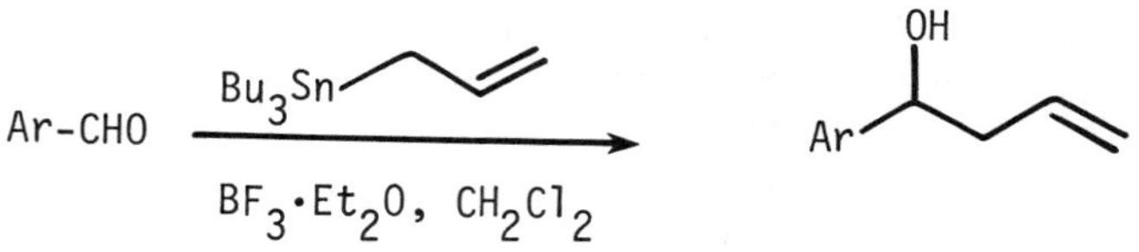

Ar = Subst. Ph, furyl, thiophenyl, etc.

Chem Lett, 919 (1979)

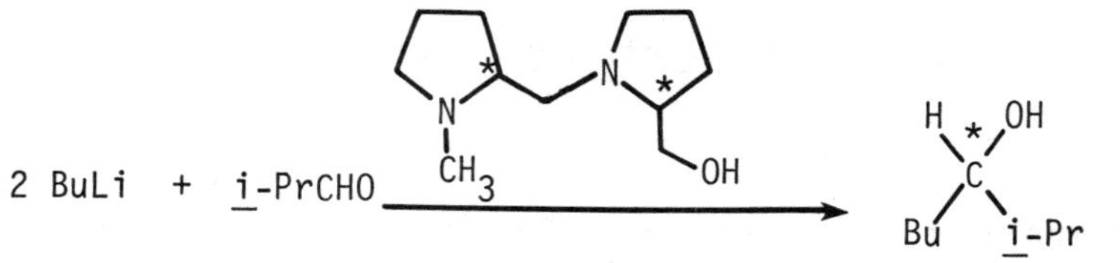

Chem Lett, 219 (1978)

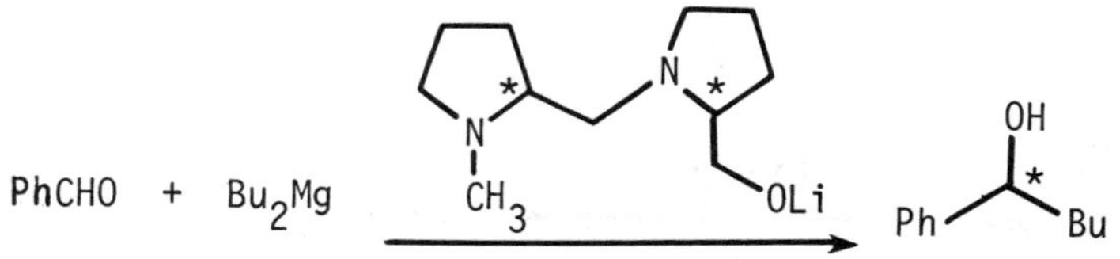

Chem Lett, 601 (1978)

59%

Tetr Lett, 2847 (1978)

Related methods: Alcohols from Ketones (Section 42)

Section 35   Alcohols and Phenols from Alkyls, Methylenes
              and Aryls

No examples of the reaction RR' → ROH (R'=alkyl, aryl, etc.)
occur in the literature. For reactions of the type RH → ROH
(R=alkyl or aryl) see Section 41 (Alcohols and Phenols from
Hydrides).

Section 36   Alcohols from Amides

95%

Synthesis, 635 (1977)

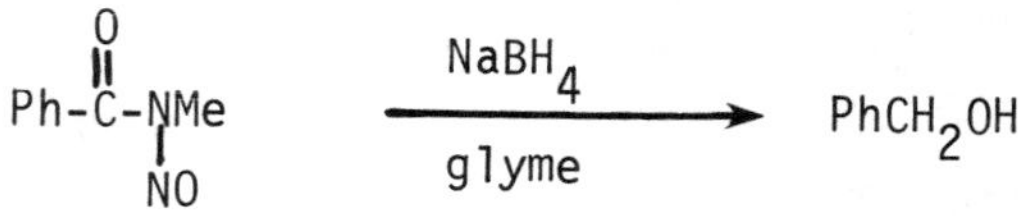

$$\underset{\underset{NO}{|}}{\overset{\overset{O}{||}}{Ph-C-NMe}} \quad \xrightarrow[\text{glyme}]{NaBH_4} \quad PhCH_2OH \qquad 84\%$$

JOC **44**, 860 (1979)

Section 37   Alcohols and Phenols from Amines

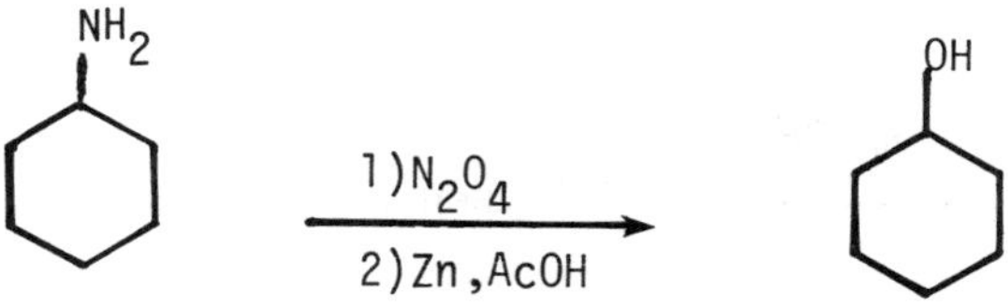

$$\xrightarrow[\text{2) Zn,AcOH}]{\text{1) } N_2O_4} \qquad 68\%$$

JCS Perkin I, 1114 (1977)

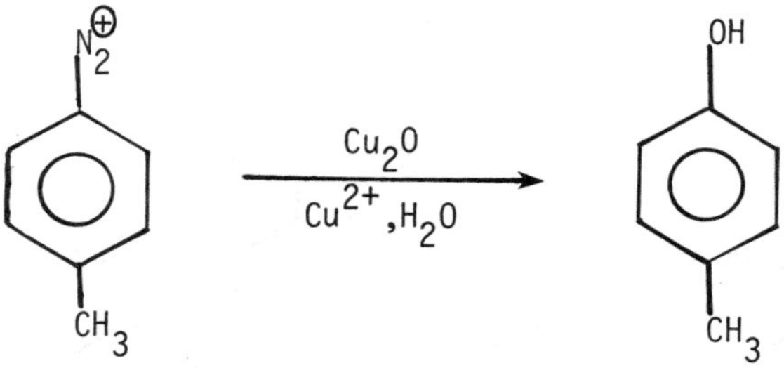

$$\xrightarrow[\text{Cu}^{2+},H_2O]{Cu_2O} \qquad 93\%$$

JOC **42**, 2053 (1977)

Section 38 Alcohols and Phenols from Esters

O
OEt
B-H
THF
OH

71%

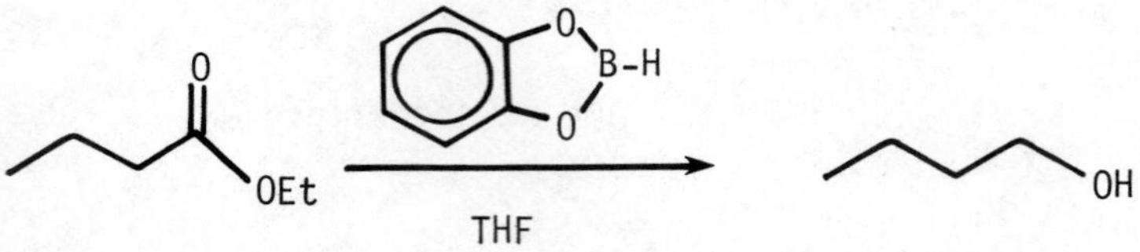

JOC 42, 512 (1977)

Ph-COOEt
$BH_3OH^{\ominus}$
$PhCH_2OH$

89%

JOC 42, 3963 (1977)

C-OMEM
C-OMEM
$LiEt_3BH$
CH_2OH
COOMEM
H

90%

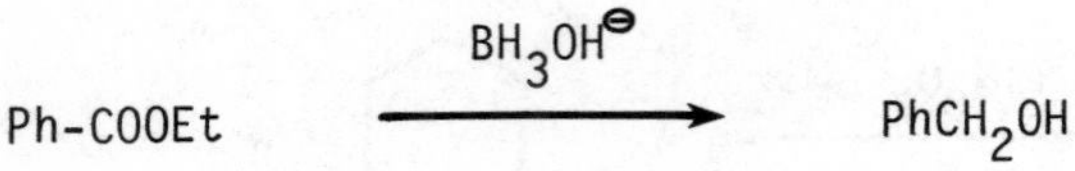

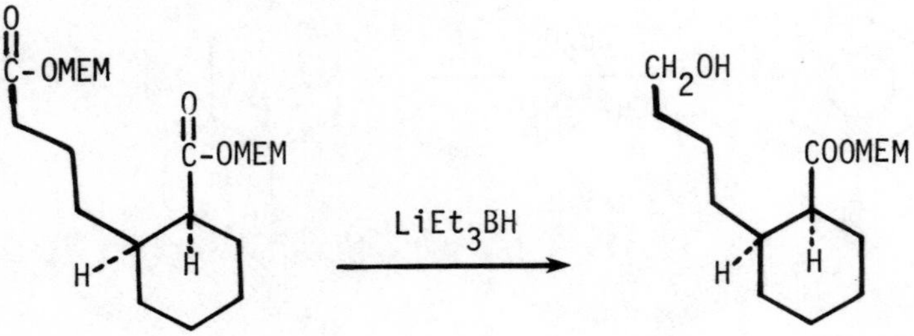

Tetr Lett, 4705 (1979)

$$R-O-\overset{\overset{\text{O}}{\|}}{C}-R' \xrightarrow[NH_3]{Li} ROH \qquad 71-97\%$$

R' = Ph or t-Bu

JOC 44, 2810 (1979)

$$\xrightarrow[DME]{NaBH_4} \qquad 87\%$$

Benzyl esters, benzoates, and cinnamates are unaffected.

JOC 43, 155 (1978)

Related Methods: Carboxylic Acids from Esters - Section 23,
                    Protection of Alcohols - Section 45A

Section 39   Alcohols and Phenols from Ethers and Epoxides

$$\xrightarrow[NaI]{Me_3SiCl} \qquad 90\%$$

JOC 44, 1247 (1979)

$$C_6H_{11}-OMe \xrightarrow[\text{2) } H_2O]{\text{1) } Me_3SiI} C_6H_{11}OH \qquad 95\%$$

JOC <u>42</u>, 3761 (1977)

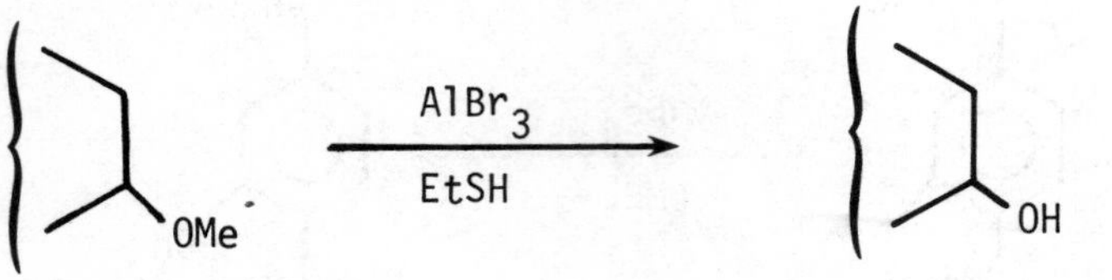

$$\xrightarrow[\text{EtSH}]{AlBr_3} \qquad 98\%$$

$$\xrightarrow[\text{EtSH}]{AlBr_3} \qquad 78\%$$

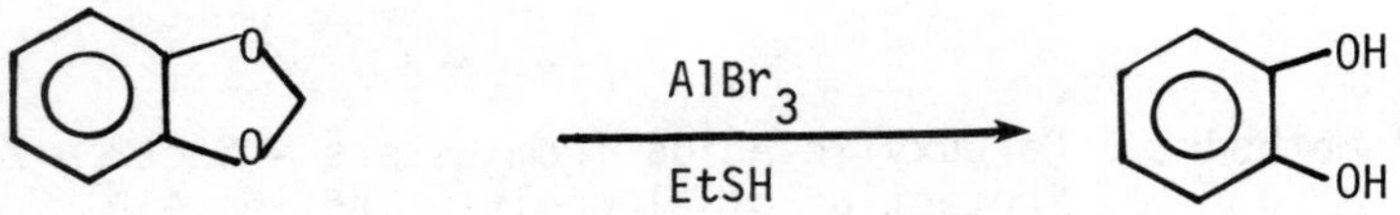

Chem Lett, 97 (1979)

$$Ph-OMe \xrightarrow{I_2, PhSiMe_3} Ph-OH \qquad 90\%$$

Synthesis, 417 (1977)

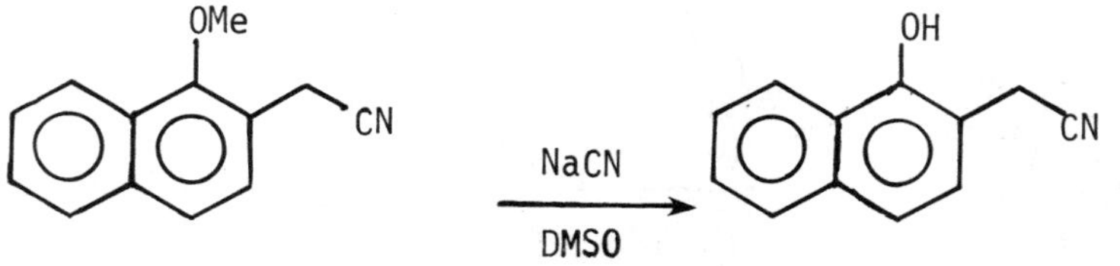

Tetr Lett, 5183 (1978)

$$Ar-OEt \xrightarrow[CH_2Cl_2]{BBr_3} Ar-OH \qquad 64-91\%$$

Synth Comm $\underline{9}$, 407 (1979)

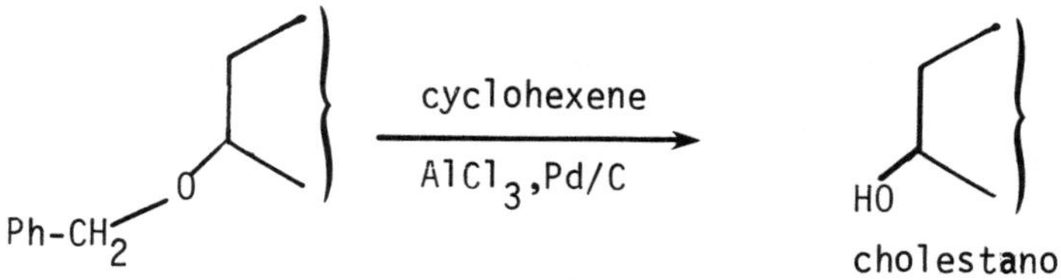

cyclohexene

AlCl$_3$,Pd/C

cholestanol

Synthesis, 825 (1978)

$$Ph-O-CH_2Ph \xrightarrow[\text{2) } H_2O]{\text{1) } Me_3SiCl, \text{ NaI}} \begin{array}{c} PhOH \\ + \\ PhCH_2I \end{array} \qquad 81\%$$

JCS Chem Comm, 874 (1978)

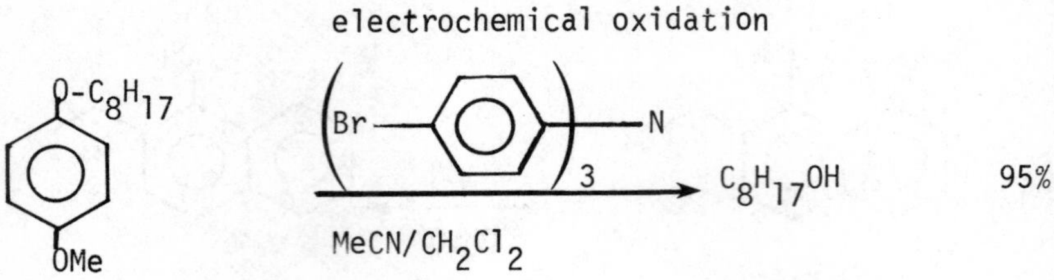

electrochemical oxidation
$O-C_8H_{17}$
Br
N
3
$C_8H_{17}OH$
95%
$MeCN/CH_2Cl_2$
OMe

Angew Int Ed 17. 673 (1978)

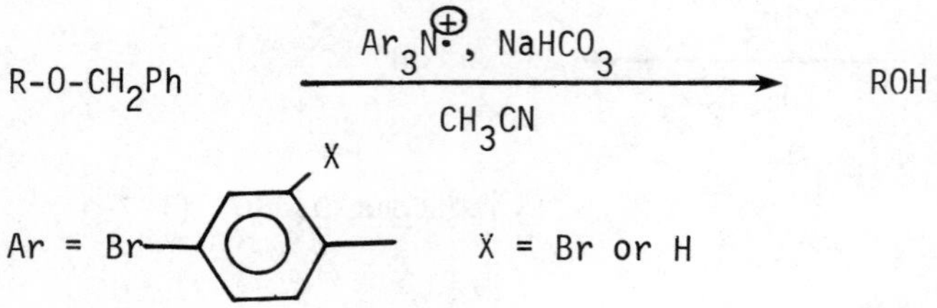

$R-O-CH_2Ph$
$Ar_3N^{\oplus\cdot}$, $NaHCO_3$
CH_3CN
ROH
X
Ar = Br
X = Br or H

Angew Int Ed 18, 801 and 802 (1979)

Additional examples of ether cleavages may be found in Section 45A (Protection of Alcohols and Phenols).

NO_2
sugar
O
$NaBH_4$
EtOH
OH
~60%

Can J Chem 56, 1177 (1978)

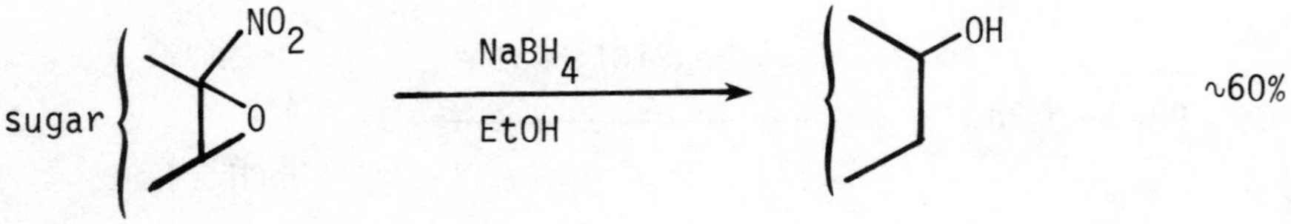

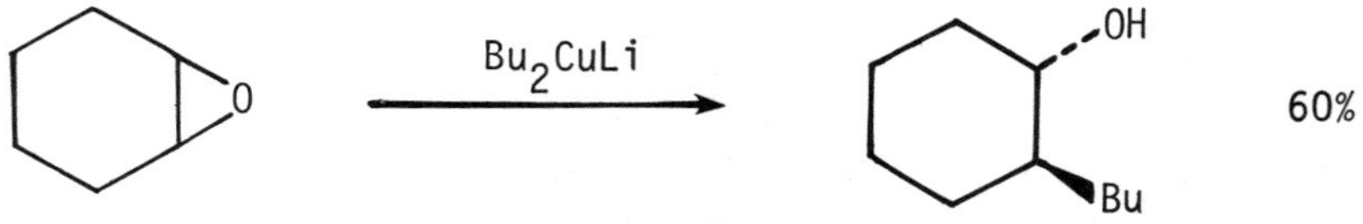

Tetr Lett, 3407 (1977)

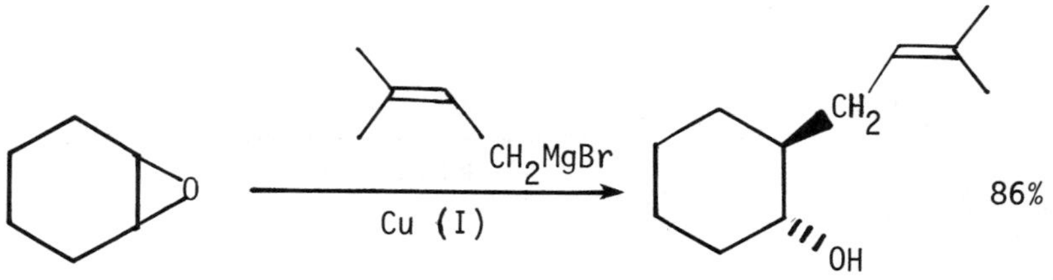

Tetr Lett, 1503 (1979)

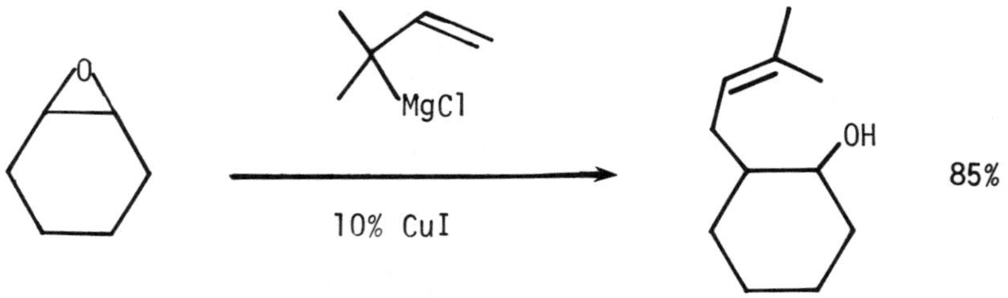

Tetr Lett, 4069 (1978)

Section 40   Alcohols and Phenols from Halides

1) Mg, THF

2) MoO$_5$, pyr, HMPA

70%

Also works with aromatic halides.

Aust J Chem 31, 2091 (1978)

Ph-Br   1) Mg, THF
         ─────────────────→   Ph-OH   89%
         2) MoO$_4$, Pyr, HMPA

JOC 42, 1479 (1977)

1)

Bu-I   ─────────────────────→   Bu-CHCH$_3$ 38%
2)   LiAlH$_4$ (OH)

Ar =

JOC 43, 4255 (1978)

BzCl

$\xrightarrow{\begin{array}{l}\text{1) NaH, } CH_2(COSEt)_2\\[10pt]\text{2) Raney Ni}\end{array}}$

$BzCH_2CH_2OH$ 45%

Can J Chem <u>57</u>, 2522 (1979)

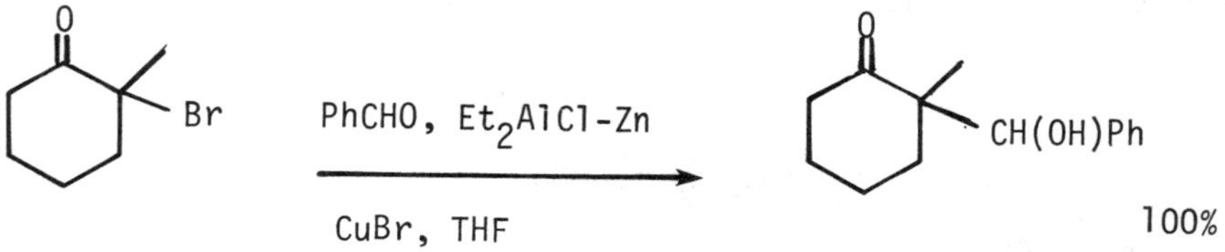

PhCHO, Et_2AlCl-Zn

CuBr, THF

$CH(OH)Ph$ 100%

JACS <u>99</u>, 7705 (1977)

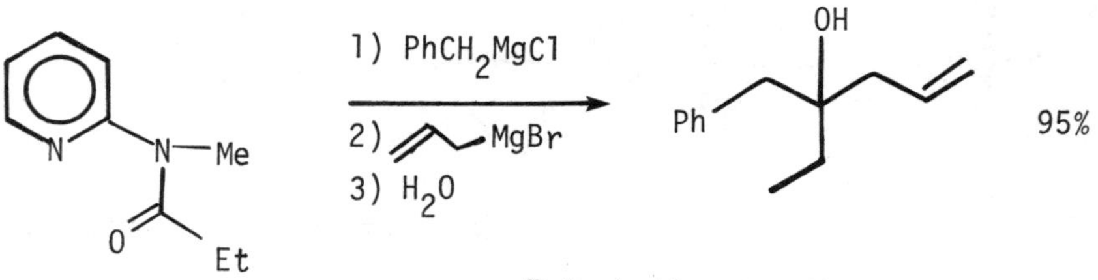

1) $PhCH_2MgCl$

2) ⌁MgBr

3) H_2O

Ph   OH   95%

Tetr Lett, 5179 (1978)

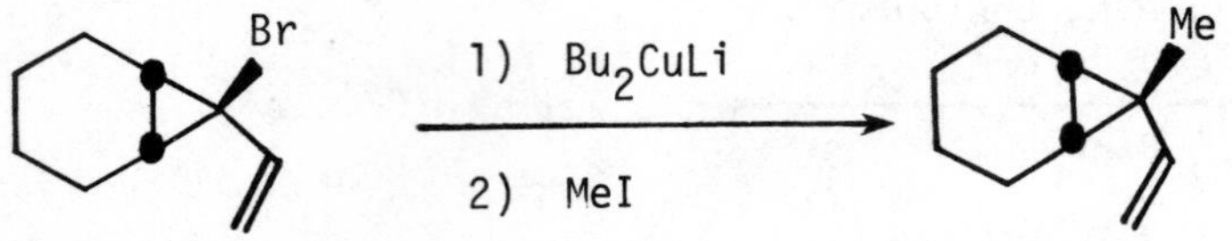

96%

JACS 99, 5816 (1977)

Section 41   Alcohols and Phenols from Hydrides

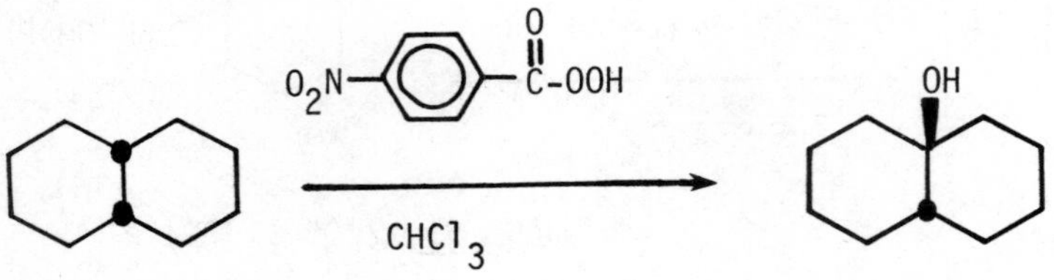

69%

Angew Int Ed 18, 407 (1979)

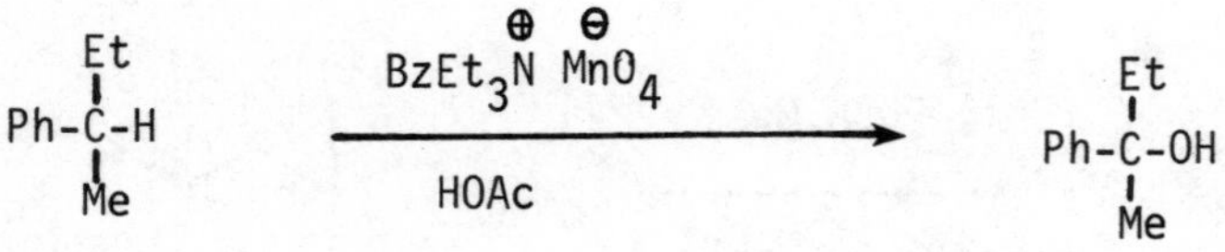

26%

Angew Int Ed 18, 68 and 69 (1979)

Review: "Superacid-Catalyzed Oxygenation of Alkanes"

Angew Int Ed 17, 909 (1978)

$$\text{Ph-H} \xrightarrow[\text{electrocatalysis}]{H_2O_2,\ Fe^{II}} \text{Ph-OH} \qquad 64\%$$

Chem Ber 110, 3561 (1977)

SeO2
pyridine
HOCH2

42%

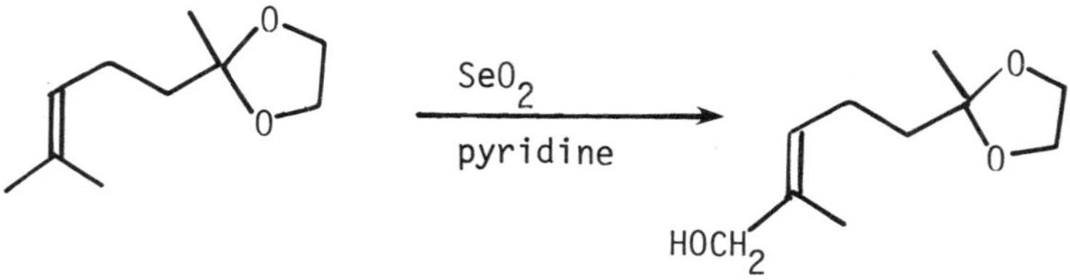

Synthesis, 215 (1978)

Benzeneselenenic anhydride
PhCH3
H
O
Me
OAc
OH

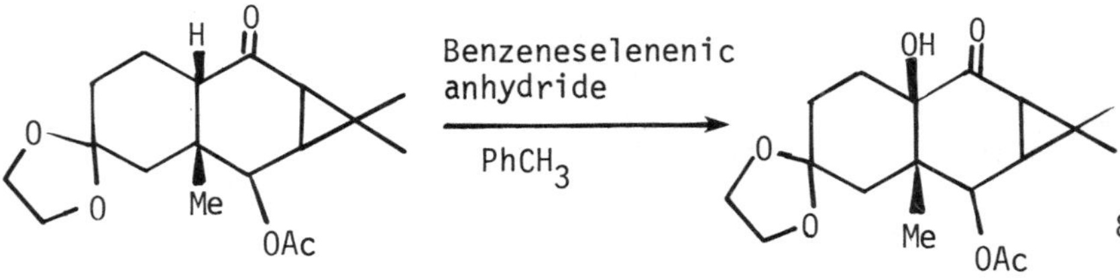

80%

Chem Lett, 763 (1979)

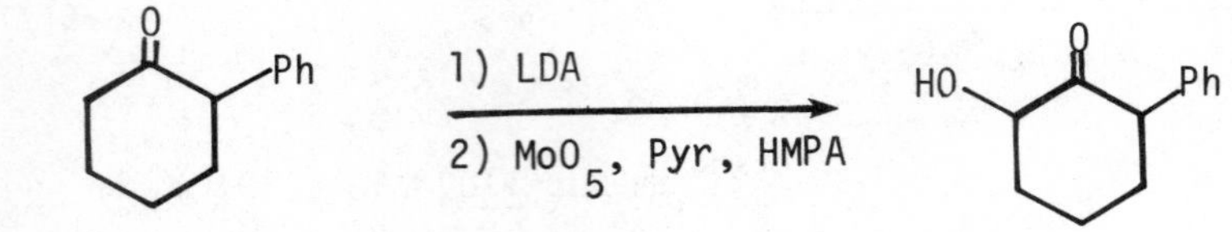

70%

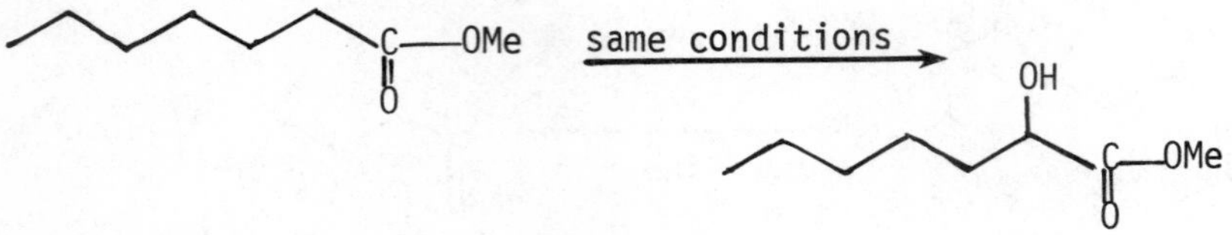

74%

JOC <u>43</u>, 188 (1978)

Section 42   <u>Alcohols from Ketones</u>

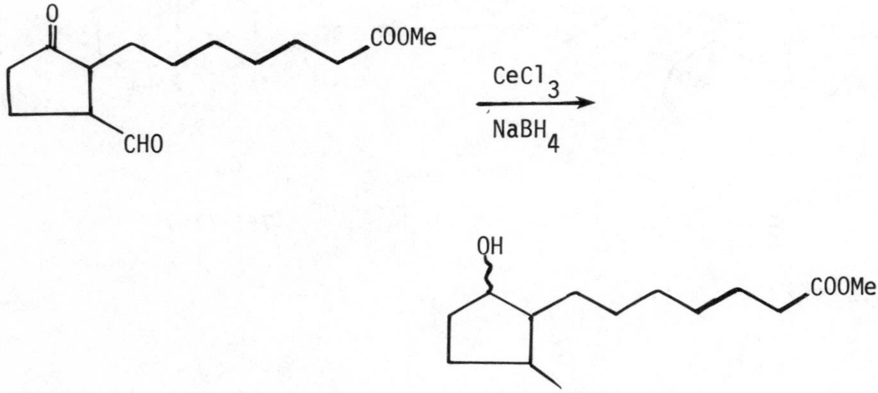

75%

JACS <u>101</u>, 5848 (1979)

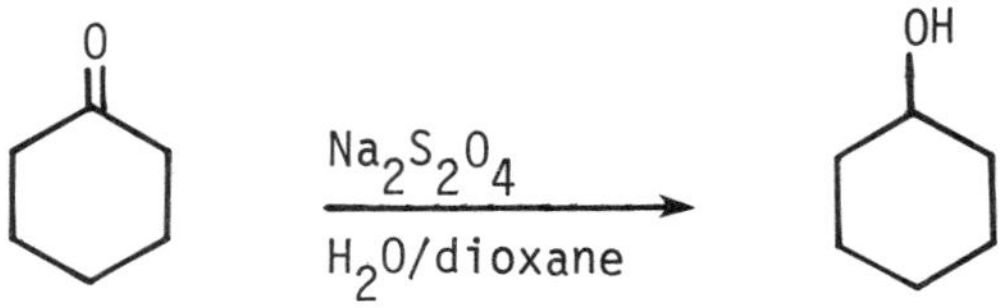

$$\text{Na}_2\text{S}_2\text{O}_4 \quad / \quad \text{H}_2\text{O/dioxane}$$

80%

Synthesis, 246 (1977)

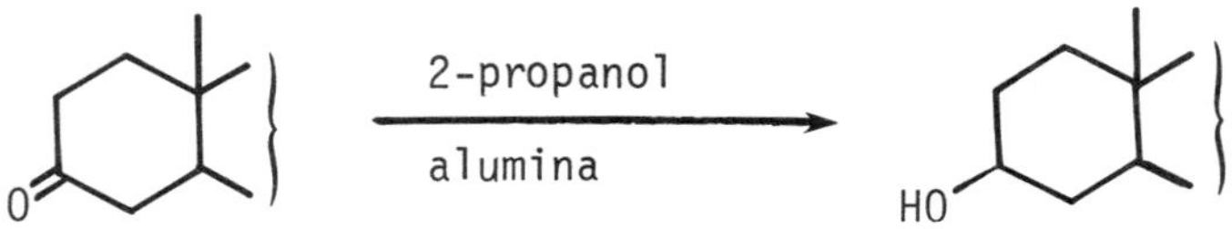

2-propanol / alumina

90%

JOC 42, 1202 (1977)

NaH, t-AmONa / Ni(OAc)$_2$

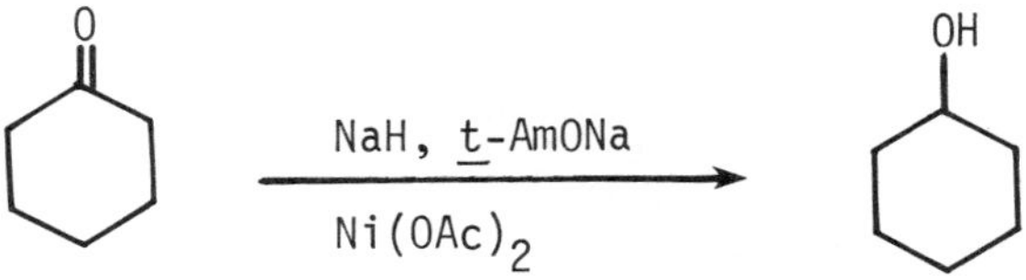

90%

JOC 43, 4804 (1978)
Tetr Lett, 1069 (1977)

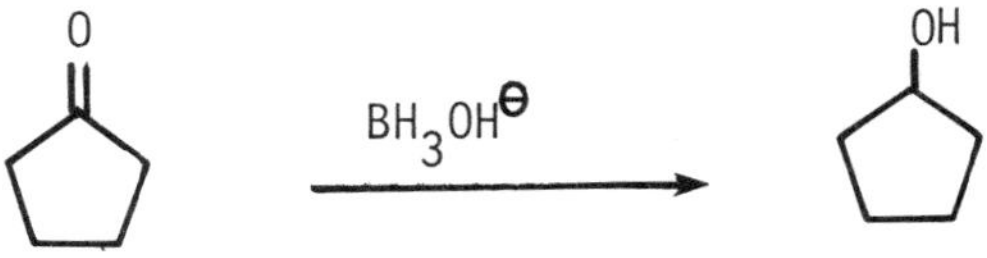

BH$_3$OH$^{\ominus}$

86%

JOC 42, 3963 (1977)

$$n\text{-}C_6H_{13}\overset{\overset{O}{\|}}{-}C\text{-}Me \quad \xrightarrow[\text{ether}]{NaBH_4, \ Al_2O_3} \quad n\text{-}C_6H_{13}\overset{\overset{OH}{|}}{-}CH\text{-}Me \qquad 90\%$$

Synthesis, 891 (1978)

$$\xrightarrow{(HAlN\text{-}\underline{i}\text{-}Pr)_6} \qquad 95\%$$

Z Chem 17, 18 (1977)

$$\xrightarrow{Cp_2Zr(Cl)BH_4} \qquad 90\%$$

Tetr Lett, 4985 (1978)

$$\xrightarrow[\text{silica gel}]{Bu_3SnH} \qquad 89\%$$

JOC 43, 3977 (1978)

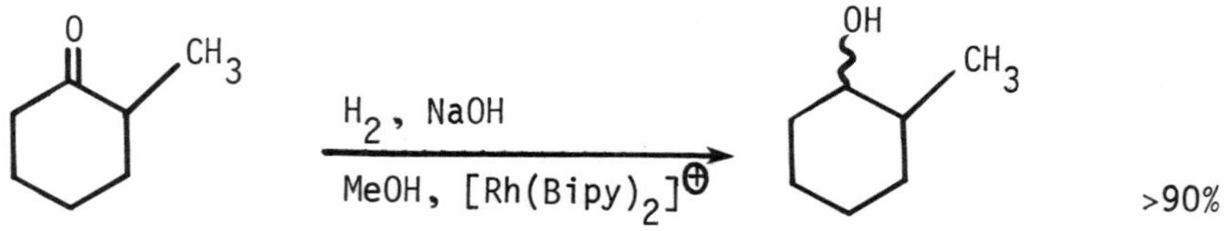

O
CH3
H2, NaOH
MeOH, [Rh(Bipy)2]⊕
OH
CH3
>90%

Can be accomplished in the presence of olefins.

J Organometal Chem 157, 345(1978)

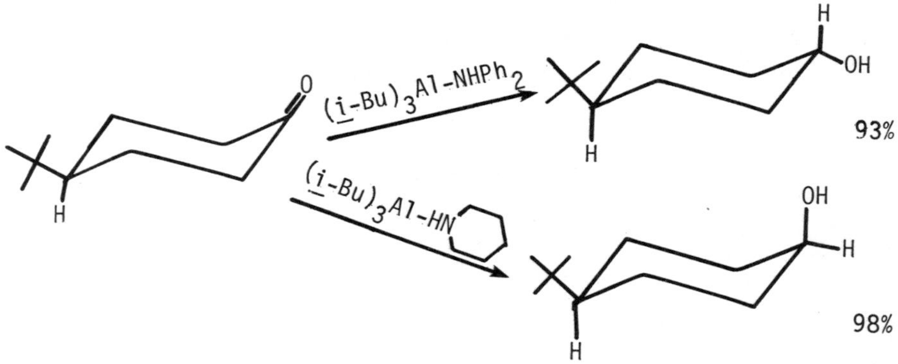

O
H
(i-Bu)3Al-NHPh2
(i-Bu)3Al-HN
H
OH
H
93%
OH
H
H
98%

BCS Japan 51, 2664 (1978)

O
MeMgO-2,6-i-Pr2C6H3
OH
Me
Me
Me
Me Me
H
89%

JOC 43, 4094 (1978)

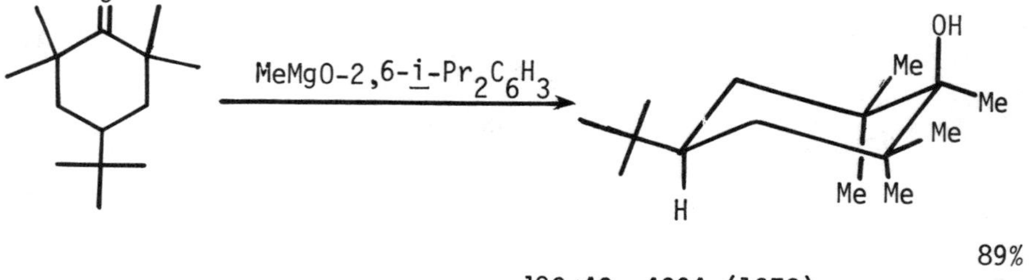

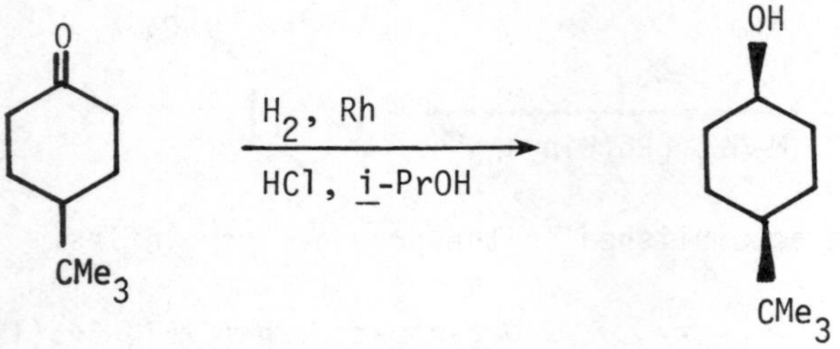

O
OH
H_2, Rh
HCl, i-PrOH
CMe_3
CMe_3

99%

Chem Lett, 963 (1977)

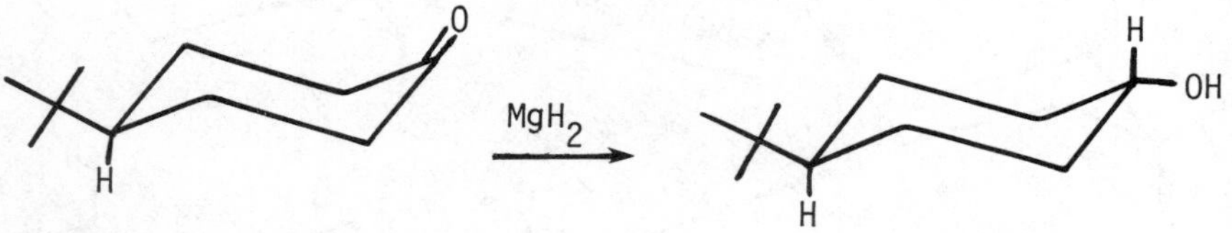

O
H
MgH_2
H
OH
H

76%

JOC 43, 1564 (1978)

CH_3
O
Li(HB-IPC-9-BBN)
THF, -78°
CH_3
OH

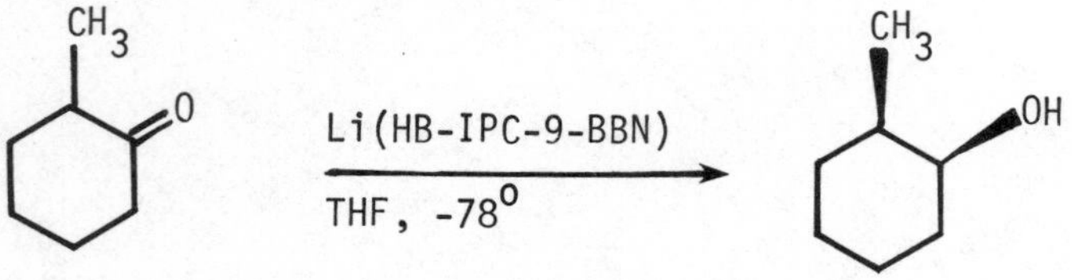

99%

JOC 42, 2534 (1977)

JACS **101**, 3129 (1979)
JACS **101**, 5843 (1979)

JOC **44**, 1363 (1979)

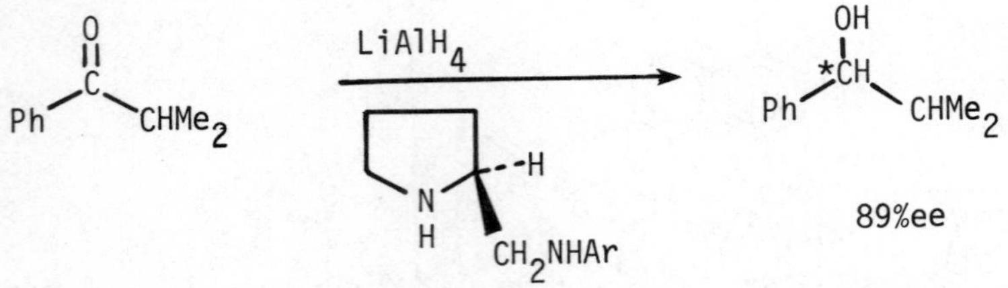

Heterocycles 12, 499 (1979)

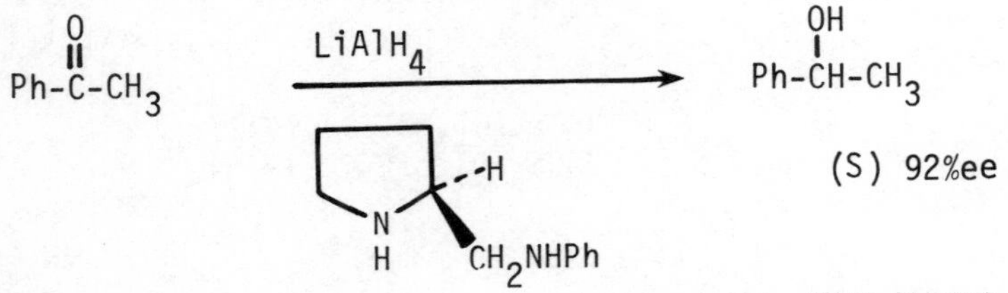

Chem Lett, 783 (1977)

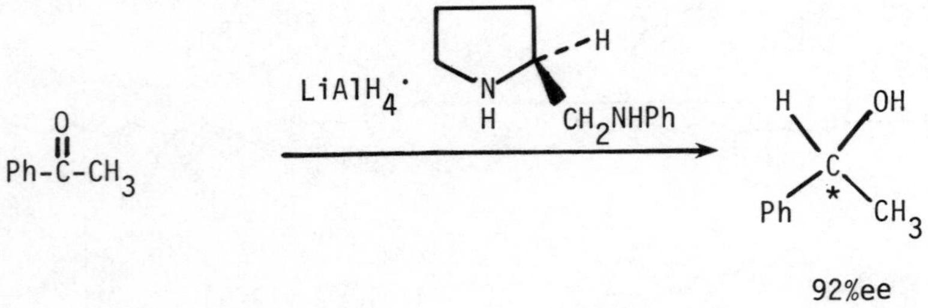

BCS Japan 51, 1869 (1978)

$$Ph-\overset{O}{\overset{\|}{C}}-CH_3 \xrightarrow[\text{THF}]{\begin{array}{c}\text{NH}{\to}\text{BH}_3 \\ \text{COONa}\end{array}} \overset{H \;\; OH}{Ph-\underset{*}{\overset{|}{C}}-CH_3} \quad 92\%$$

32%ee

Chem Pharm Bull <u>27</u>, 1479 (1979)

$$CH_3-\overset{O}{\overset{\|}{C}}-COOEt \xrightarrow{H_2,\; CPPM-Rh} \overset{OH}{CH_3-\underset{*}{\overset{|}{CH}}-COOEt} \quad 99\%$$

65%ee

CPPM =

Tetr Lett, 3735 (1977)

$$\overset{O}{\underset{Ph}{\overset{\|}{C}}}{}_{COOEt} \xrightarrow[\substack{\text{chiral 1,4-dihydropyridine} \\ \text{crown ether}}]{Mg(ClO_4)_2,\; MeCN} \overset{H \qquad OH}{\underset{Ph \qquad COOEt}{C}}$$

61%
86%ee

JACS <u>101</u>, 2759 (1979)
JACs <u>101</u>, 7036 (1979)

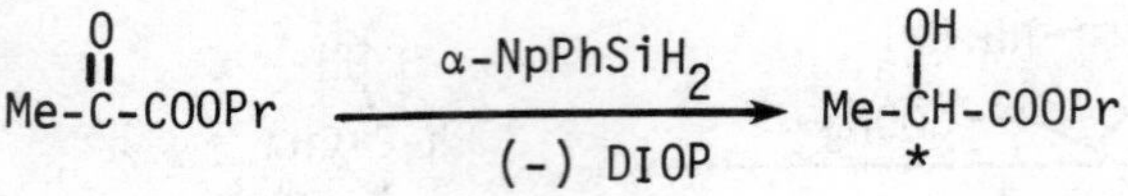

$$\underset{Me-C-COOPr}{\overset{O}{\overset{\|}{}}} \xrightarrow[\text{(-) DIOP}]{\alpha\text{-NpPhSiH}_2} \underset{Me-CH-COOPr}{\overset{OH}{\overset{|}{}}} \quad\quad 90\%$$
$$* \quad\quad\quad\quad\quad 85\%ee$$

JOC <u>42</u>, 1671 (1977)

9-BBN   85%

JOC <u>42</u>, 1197 (1977)

$(HAlN-\underline{i}-Pr)_6$

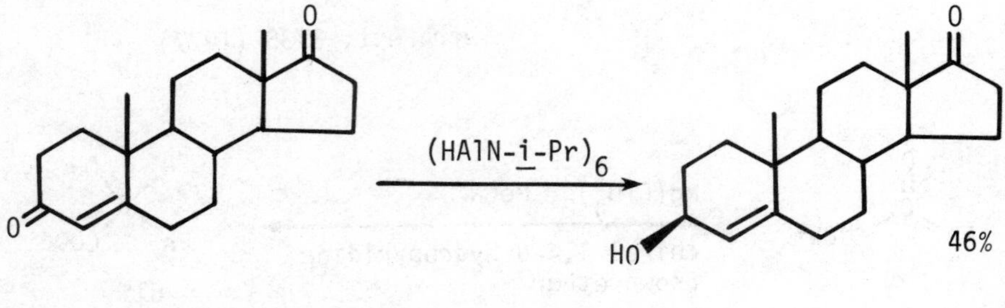

46%

Tetr Lett, 2369 (1977)

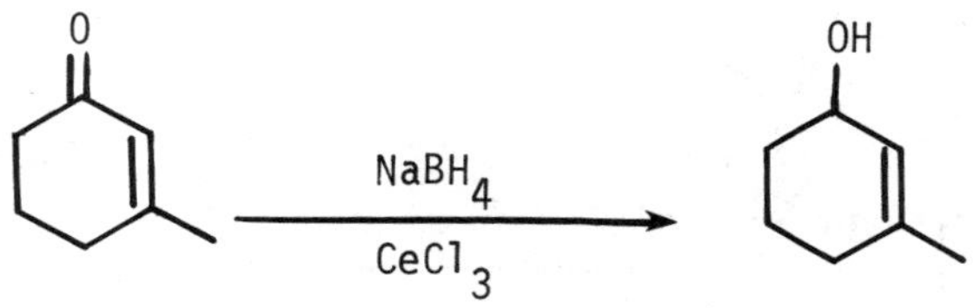

100%

JACS 100, 2226 (1978)

The catalyst Rh(bipy)$_2^{\oplus}$ selectively hydrogenates ketones to alcohols in the presence of olefins.

J Organometal Chem 140, 63 (1977)

94%

JOC 42, 2797 (1977)

$Et_2C=O$　$\xrightarrow[\text{acetone}]{\text{BuMnI}}$　$Bu(Et)_2C\text{-OH}$　　86%

Tetr Lett, 3383 (1977)

81%

JACS **99**, 5317 (1977)

electroreduction

MeOH/dioxane

98%

JACS **100**, 545 (1978)

Review: "Stereochemistry and Mechanism of Ketone Reductions by Hydride Reagents"

Tetrahedron **35**, 449 (1979)

Related methods: Alcohols from Aldehydes (Section 34)

Section 43   Alcohols and Phenols from Nitriles

No additional examples

Section 44   <u>Alcohols from Olefins</u>

For the preparation of diols from olefins see Section 323
(Alcohol-Alcohol)

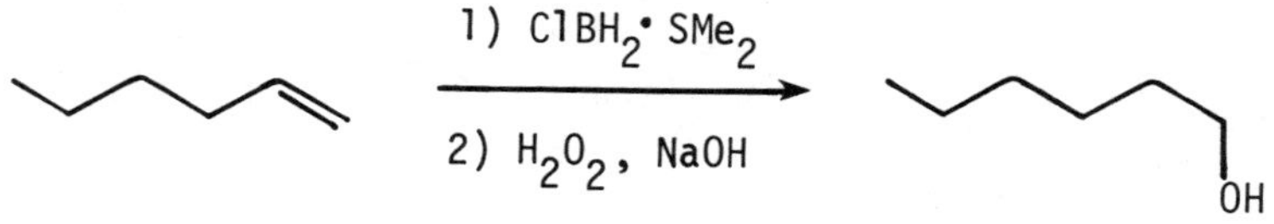

1) $ClBH_2 \cdot SMe_2$

2) H_2O_2, NaOH

99%

JOC <u>42</u>, 2533 (1977)

1) $SnCl_4$, $NaBH_4$

2) H_2O

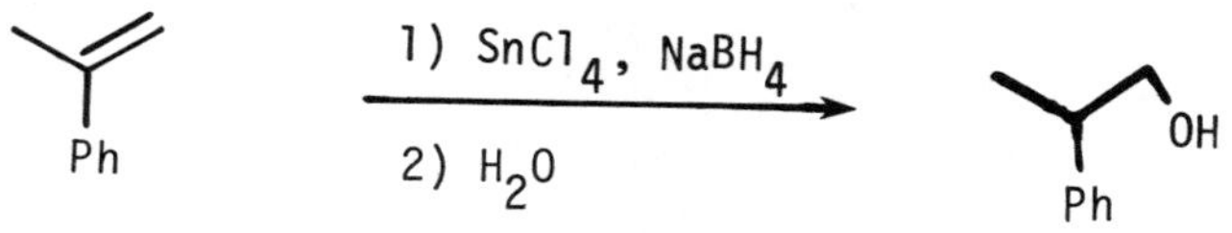

58%

JCS Chem Comm, 796 (1979)

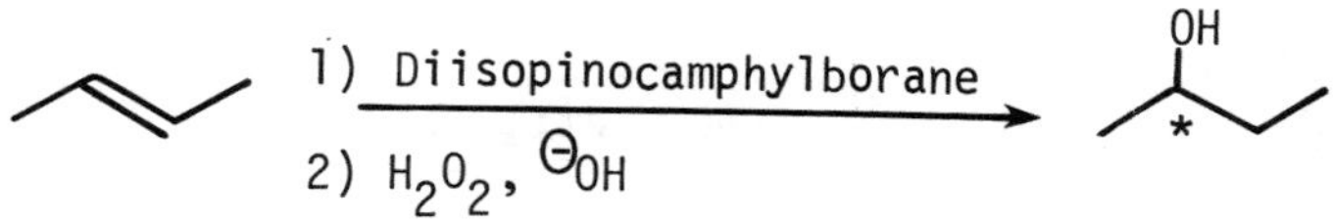

1) Diisopinocamphylborane

2) H_2O_2, $^\ominus OH$

72%

98%ee

Israel J Chem <u>15</u>, 12 (1977)

Ph–C(Et)=CH₂ $\xrightarrow[\text{(-) DIOP-PtCl}_2/\text{SnCl}_2]{\text{H}_2}$ HO–C(Ph)(Et)–CH₃   90%

37%ee

Israel J Chem 15, 221 (1977)

$$Ph-CH=CH_2 \xrightarrow[\text{3) NaOH, H}_2\text{O}_2]{\substack{\text{1) LiEt}_3\text{BH} \\ \text{2) CH}_3\text{SO}_3\text{H}}} Ph-\underset{\underset{OH}{|}}{CH}-CH_3 \quad 90\%$$

JOC 42, 1482 (1977)

$$\text{(cyclohexenyl)}-CH=CH_2 \xrightarrow[\text{2. O}_2,\text{HCl}]{\text{1. LiAlH}_4,\text{Cp}_2\text{Ti(AlH}_3)_2} \text{(cyclohexenyl)}-CH_2CH_2OH$$

86%

Chem Lett, 1117 (1977)

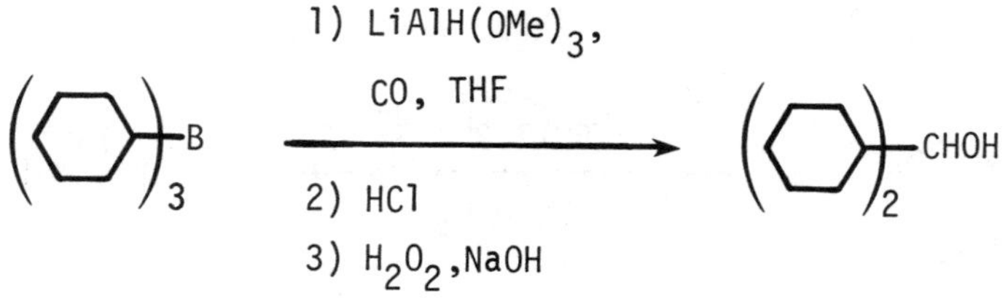

1) LiAlH(OMe)$_3$,
   CO, THF

2) HCl

3) H$_2$O$_2$, NaOH

86%

Synthesis, 676 (1978)

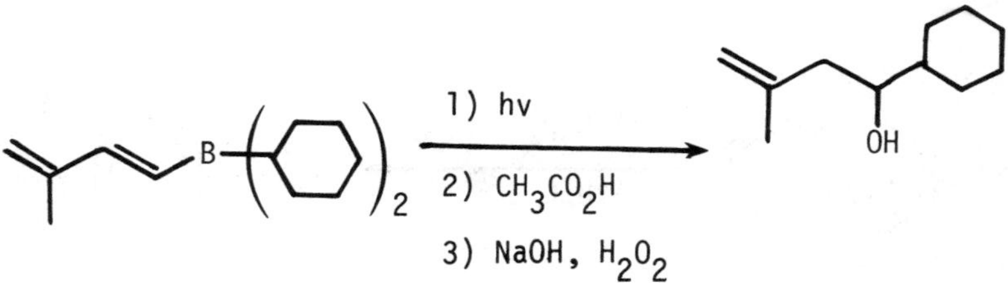

1) hv

2) CH$_3$CO$_2$H

3) NaOH, H$_2$O$_2$

69%

JACS 99, 5192 (1977)

$(C_6H_{13})_3B$

2) HgCl$_2$

3) H$_2$O$_2$/NaOH

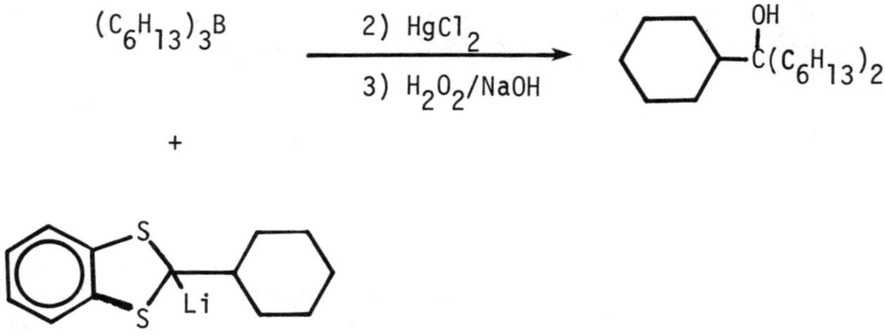

85%

+

Tetr Lett, 1895 (1979)

$$\text{1) }(C_8H_{17})_3B, \text{ HgCl}_2$$
$$\text{2) NaOH, H}_2O_2$$

$(C_8H_{17})_2\overset{\displaystyle C_3H_7}{\underset{}{C}}\text{-OH}$ 96%

JCS Perkin I. 1172 (1977)

$$\text{1) B}_2H_6$$
$$\text{2) H}_2O_2, \ominus\text{OH}$$

30%

Comptes Rendus 289, 227 (1979)

$$\text{1) CH}_3\text{MgI}$$
$$\text{2) CH}_3\text{CHO}$$

61%

CHOH
CH₃

Bull Acad USSR Chem 27, 1364
                                    (1979)

Section 45   Alcohols from Miscellaneous Compounds

No additional examples

For conversions of boranes to alcohols, see Section 44

Section 45A   Protection of Alcohols and Phenols

R-OH $\xrightarrow{\text{TsOH·pyridine}}$ R—O

This catalyst is milder than most others, and is useful with
acid-sensitive alcohols.

JOC 42, 3772 (1977)

R-OH , Amberlyst H-15 / CH₃OH Amberlyst H-15

>90%

Synthesis, 618 (1979)

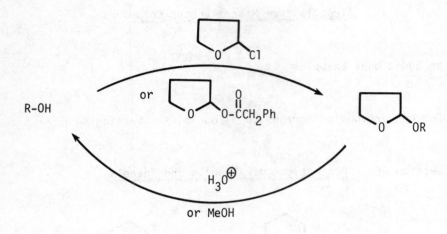

Rec Trav Chim 98, 371 (1979)

R-O-THP $\xrightarrow[\text{MeOH}]{\text{acid-washed Dowex resin}}$ R-OH   >97%

Synth Comm 9, 271 (1979)

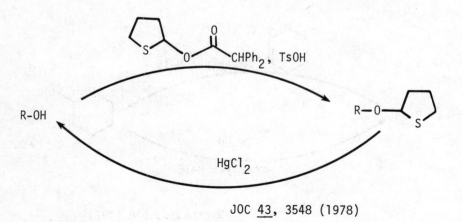

JOC 43, 3548 (1978)

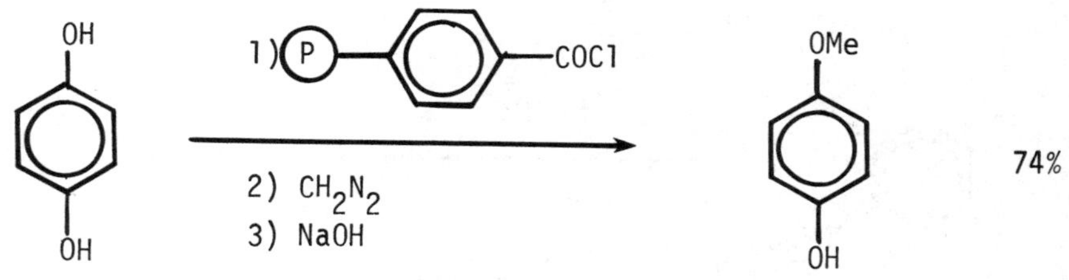

Can J Chem <u>55</u>, 3351 (1977)

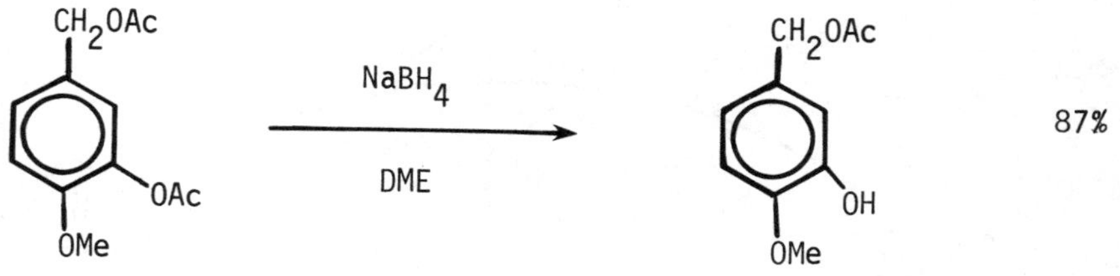

Benzyl esters, benzoates, and cinnamates are unaffected.

JOC <u>43</u>, 155 (1978)

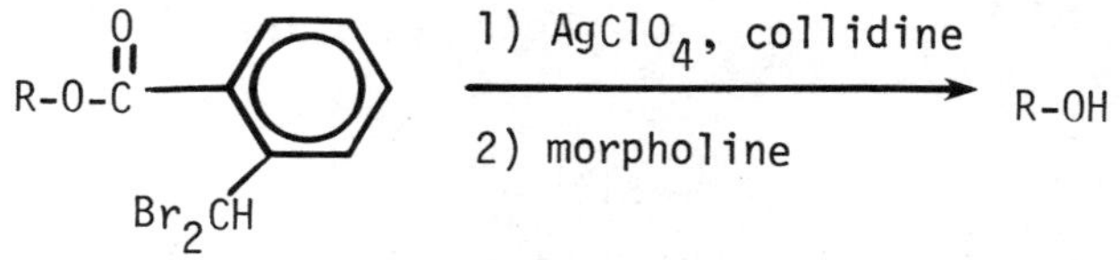

JCS Chem Comm, 987 (1979)

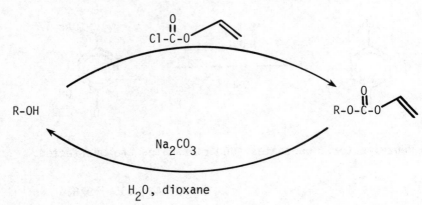

Tetr Lett, 555 (1979)

Tetr Lett, 1571 (1977)

Boc-Val-Tyr(Bzl)·OMe $\xrightarrow{\text{Me}_3\text{SiI}}$ Boc-Val-Tyr·OMe   ~100%

JCS Chem Comm, 495 (1979)

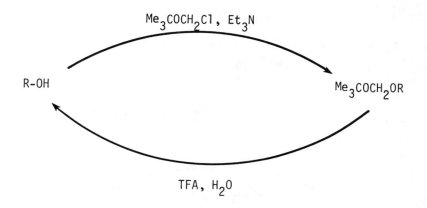

$$R-OH \xrightarrow{\quad Me_3COCH_2Cl, \ Et_3N \quad} Me_3COCH_2OR$$

TFA, H_2O

JOC **43**, 3964 (1978)

$$R-OH \xrightarrow{\quad Me_2SO + Me_3SiCl \quad} R-O-CH_2-O-R$$

JOC **44**, 3727 (1979)

$$Ph-OH \xrightarrow[\text{2) } CH_2(OMe)_2]{\text{1) } BrZnCH_2COOEt} PhOCH_2OCH_3 \qquad 72\%$$

Synthesis, 567 (1977)

$$Ar-OH \xrightarrow[\text{H}_2\text{O, NaOH, Adogen 464}]{\text{CH}_2\text{Cl}_2, \text{ ClCH}_2\text{OMe}} Ar-OCH_2OMe \qquad 79-95\%$$

Tetr Lett, 661 (1978)

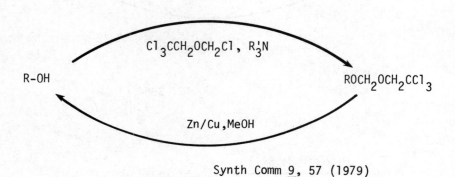

R-OH $Cl_3CCH_2OCH_2Cl$, $R_3'N$ $ROCH_2OCH_2CCl_3$

Zn/Cu, MeOH

Synth Comm <u>9</u>, 57 (1979)

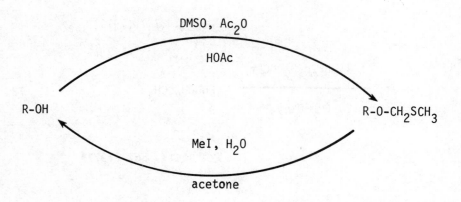

R-OH   DMSO, Ac_2O HOAc $R-O-CH_2SCH_3$

MeI, H_2O

acetone

Aust J Chem <u>31</u>, 1031 (1978)

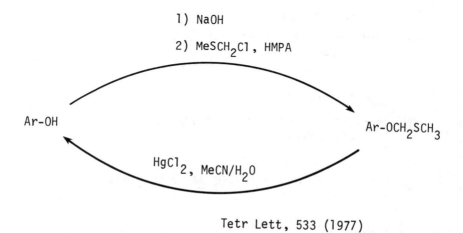

$$\text{R-OH} \xrightarrow[\text{Et}_3\text{N, benzene}]{\text{ClCH}_2\text{SCH}_3,\ \text{AgNO}_3} \text{R-OCH}_2\text{SCH}_3$$

Chem Lett, 1277 (1979)

1) NaOH

2) MeSCH$_2$Cl, HMPA

Ar-OH $\longrightarrow$ Ar-OCH$_2$SCH$_3$

HgCl$_2$, MeCN/H$_2$O

Tetr Lett, 533 (1977)

$$\text{R-OH} \xrightarrow{} \text{R-O-CHS(CH}_2)_3\text{SMe}$$

60%

Synth Comm <u>9</u>, 107 (1979)

$$ROSi(Me)_2-\underline{t}-Bu \xrightarrow[CH_3CN]{Aq.\ HF} R-OH$$

Tetr Lett, 3981 (1979)

$$R-OH \xrightarrow[MeCN]{Me_3SiCl,\ Li_2S} R-OSiMe_3$$

$$R-OH \xrightarrow[Li_2S,\ MeCN]{\underline{t}-BuMe_2SiCl} R-OSiMe_2-\underline{t}-Bu$$

JOC 44, 4272 (1979)

Use of triethylsilyl ethers as -OH protecting groups in the synthesis of prostaglandin D_1.

JCS Chem Comm, 156 (1979)

$$R-O-SiMe_2-\underline{t}-Bu \xrightarrow{BF_3 \cdot Et_2O} R-OH$$

Used with prostaglandin intermediates.

Synth Comm 9, 295 (1979)

Use of polystyrylboronic acid to protect cis-1, 3-diols in glycosides.

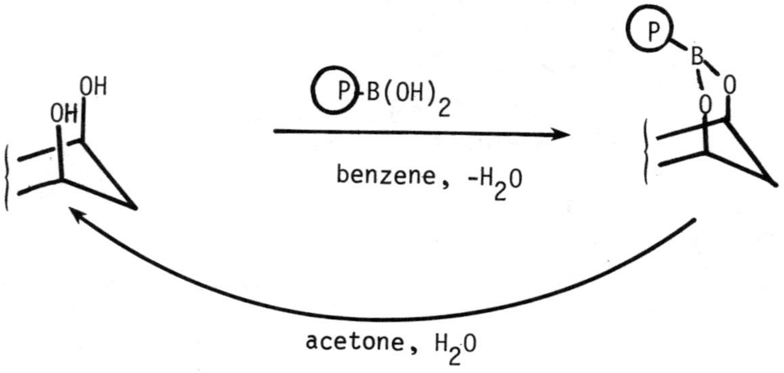

JACS 101, 432 (1979)

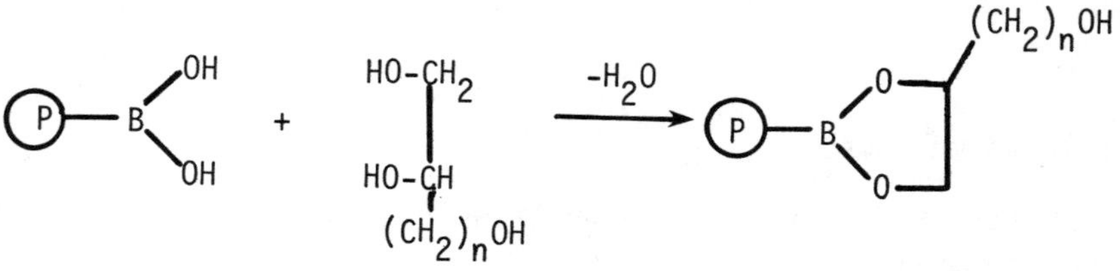

Israel J Chem 17, 253 (1978)

Review: "Carbohydrate Cyclic Acetal Formation and Migration"

Chem Rev 79, 491 (1979)

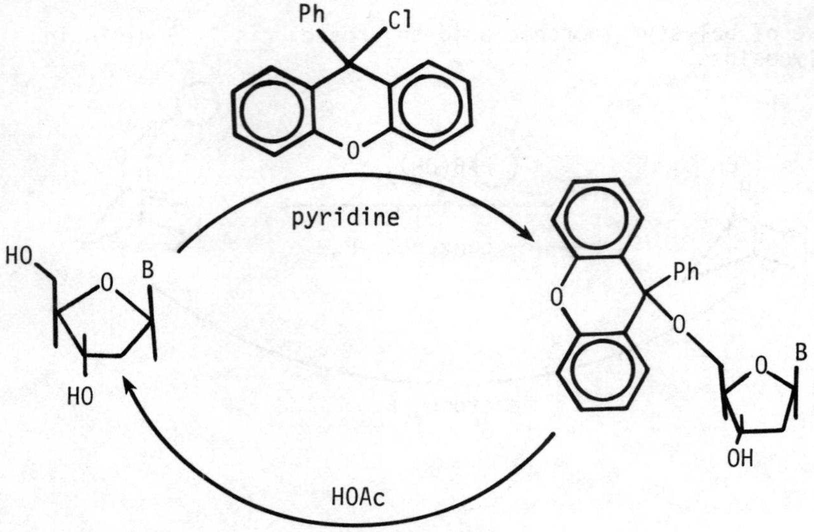

JCS Chem Comm, 639 (1978)

Related methods:

    Ethers from Alcohols - Section 123
    Alcohols from Ethers - Section 39
    Esters from Alcohols - Section 108
    Alcohols from Esters - Section 38

CHAPTER 4
PREPARATION OF ALDEHYDES

Section 46   <u>Aldehydes from Acetylenes</u>

No additional examples

Section 47   <u>Aldehydes from Carboxylic Acids and Acid Halides</u>

$$C_9H_{19}COOH \longrightarrow C_9H_{19}-\overset{\overset{\displaystyle O}{\|}}{C}-S-\left\langle \begin{array}{c} S \\ N \end{array} \right\rangle \xrightarrow{\text{DIBAH}} C_9H_{19}-CHO \quad 72\%$$

JCS Chem Comm, 330 (1978)

COOH → CHO

$$\xrightarrow[\text{2) pyridinium chlorochromate}]{\text{1) Me}_2\text{S:BH}_3} \quad 69\%$$

Synthesis, 704 (1979)

73

$$X-(CH_2)_n COOH \xrightarrow[Et_3N]{} X-(CH_2)_n-\overset{O}{\overset{\|}{C}}-N\overset{S}{\diagdown}\diagup S \xrightarrow[toluene]{DIBAH} X-(CH_2)_n-CHO$$

~70-90%

X = Ph, CN, Br

Bull Chem Soc Japan <u>52</u>, 555 (1979)

1) electrolysis, DMSO

2) NaHCO₃ →   72%

CH₂COOH / OMe → CHO / OMe

JOC <u>42</u>, 1461 (1977)

$$CH_3(CH_2)_5-\underset{SPh}{\overset{|}{CH}}-COOH \xrightarrow[MeOH, LiClO_4]{electrolysis} CH_3(CH_2)_5 CH(OMe)_2 \quad 93\%$$

Tetr Lett, 1045 (1979)

$$C_7H_{15}-CH_2COOH \longrightarrow \longrightarrow C_7H_{15}-\underset{}{\overset{OAc}{CH}}-\boxed{\oplus} \longrightarrow \longrightarrow C_7H_{15}CHO \quad 56\%$$

Tetr Lett, 381 (1977)

$$C_8H_{17}-\overset{\overset{\text{O}}{\|}}{C}-Cl \xrightarrow[\text{DMF}]{\text{NaBH}_4, \text{ CdCl}_2} C_8H_{17}CHO \qquad 56\%$$

JCS Chem Comm, 354 (1978)

$$\xrightarrow{3 \text{ NMe}_4 \bullet \text{HFe(CO)}_4} \qquad 99\%$$

Tetr Lett, 781 (1977)

$$\xrightarrow[\text{Ph}_3\text{P, acetone}]{(\text{Ph}_3\text{P})_2\text{CuBH}_4} \qquad 80\%$$

Tetr Lett, 1437 and 2473 (1978)

$$CH_3(CH_2)_8\overset{\overset{\text{O}}{\|}}{C}-Cl \xrightarrow[\text{PPh}_3, \text{ acetone}]{(\text{Ph}_3\text{P})_2\text{CuBH}_4} CH_3(CH_2)_8CHO \qquad 79\%$$

Tetr Lett, 975 (1979)

Section 48   Aldehydes from Alcohols

$$CH_3-(CH_2)_9-CH_2OH \xrightarrow[\text{Et}_3N]{\text{DMSO, (COCl)}_2} CH_3(CH_2)_9-CHO \qquad 99\%$$

Synthesis, 297 (1978)
Tetrahedron <u>34</u>, 1651 (1978)
JOC <u>43</u>, 2480 (1978)

$$\triangleright\!-CH_2OH \xrightarrow[\text{R}_3N, \ CH_2Cl_2]{\text{DMSO, TFAA}} \triangleright\!-CHO \qquad 78\%$$

Synthesis, 297 (1978)

$$\triangleright\!-CH_2OH \xrightarrow[\text{Et}_3N]{\text{Me}_2SO, \ (COCl)_2} \triangleright\!-CHO \qquad 98\%$$

JOC <u>44</u>, 4148 (1979)

$$CH_3(CH_2)_8CH_2OH \xrightarrow[\text{toluene}]{\text{DMS-NCS}} CH_3(CH_2)_8CHO \qquad 94\%$$

Tetrahedron <u>34</u>, 1651 (1978)

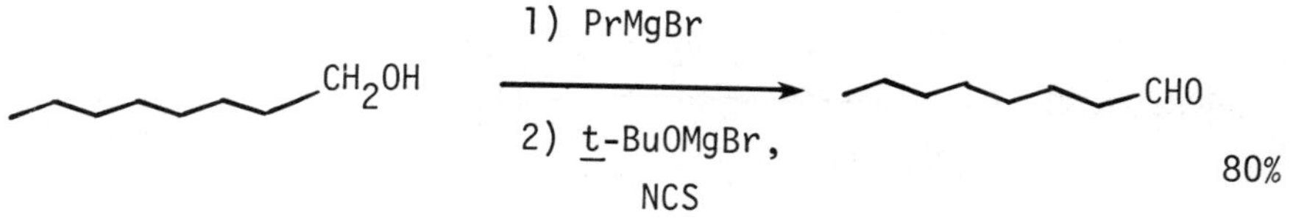

1) PrMgBr
CH2OH
CHO
2) t-BuOMgBr,
NCS
80%

BCS Japan 50, 2773 (1977)

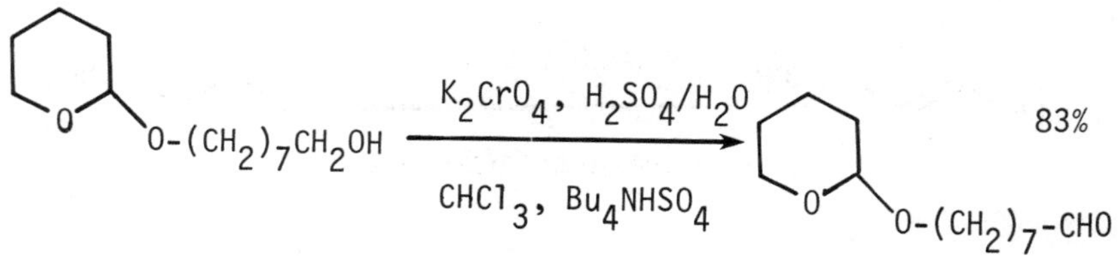

K2CrO4, H2SO4/H2O
O-(CH2)7CH2OH
CHCl3, Bu4NHSO4
O-(CH2)7-CHO
83%

Synthesis, 134 (1979)

$$C_{15}H_{31}CH_2OH \xrightarrow[Bu_4NHSO_4,\ CH_2Cl_2]{Na_2Cr_2O_7,\ H_2SO_4} C_{15}H_{31}CHO \qquad 90\%$$

JCS Perkin II, 788 (1979)

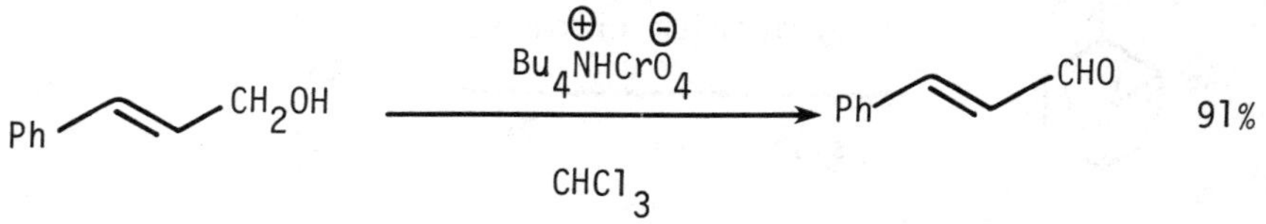

Bu4NHCrO4
Ph
CH2OH
CHCl3
Ph
CHO
91%

Synthesis, 356 (1979)

$$PhCH_2CH_2OH \xrightarrow[\text{CH}_2\text{Cl}_2, \text{ Bu}_4\text{NHSO}_4]{\text{K}_2\text{Cr}_2\text{O}_7, \text{ H}_2\text{SO}_4} PhCH_2CHO \qquad 90\%$$

Tetr Lett, 1601 (1978)

$$Ph\diagup\!\!\!\diagdown CH_2OH \xrightarrow[\text{phase-transfer cat.}]{\text{K}_2\text{Cr}_2\text{O}_7, \text{ benzene}} Ph\diagup\!\!\!\diagdown CHO \qquad 91\%$$

Tetr Lett, 4167 (1977)

$$R-CH_2OH \xrightarrow[\text{CH}_2\text{Cl}_2]{\text{pyridinium dichromate}} R-CHO \qquad 70\text{-}98\%$$

Tetr Lett, 399 (1979)

$$\xrightarrow[\text{silica gel}]{\text{pyridinium chromate}} \qquad 80\%$$

Tetrahadron 35, 1789 (1979)

$\underline{n}\text{-}C_9H_{19}\text{-}CH_2OH$ $\xrightarrow{H_2CrO_4/SiO_2}$ $C_9H_{19}\text{-}CHO$ 86%

Synthesis, 534 (1978)

$Ph\text{-}CH_2OH$ $\xrightarrow[Et_2O/CH_2Cl_2]{CrO_3/celite}$ $Ph\text{-}CHO$ 75%

Synthesis, 815 (1979)

CH_2OH $\xrightarrow[\text{poly (vinylpyridine) resin}]{CrO_3}$ CHO
91%

JOC $\underline{43}$, 2618 (1978)

$C_6H_{13}CH_2OH$ $\xrightarrow[\overset{O}{\underset{\|}{(PhC\text{-}O)_2}}]{NiBr_2}$ $C_6H_{13}CHO$ 92%

JOC $\underline{44}$, 2955 (1979)

$Me_2CH\text{-}CH_2OH$ $\xrightarrow[h\nu]{FeCl_3}$ $Me_2CH\text{-}CHO$ 70%

JOC $\underline{42}$, 171 (1977)

$$PhCH_2OH \xrightarrow[\text{K_2CO_3, phenanthroline}]{\text{CuCl, O_2}} PhCHO \qquad 86\%$$

Tetr Lett, 1215 (1977)

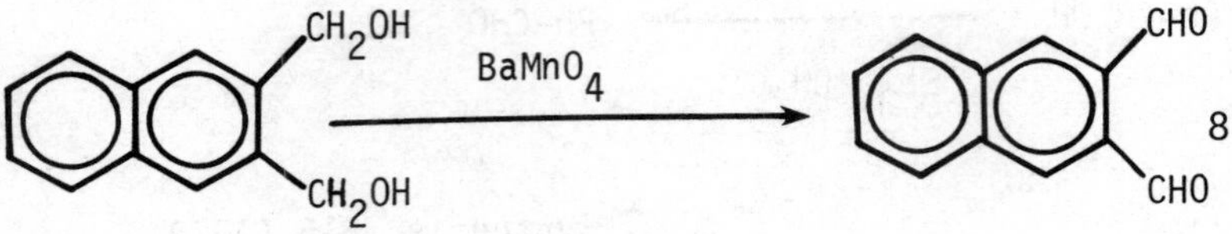

87%

Tetr Lett, 839 (1978)

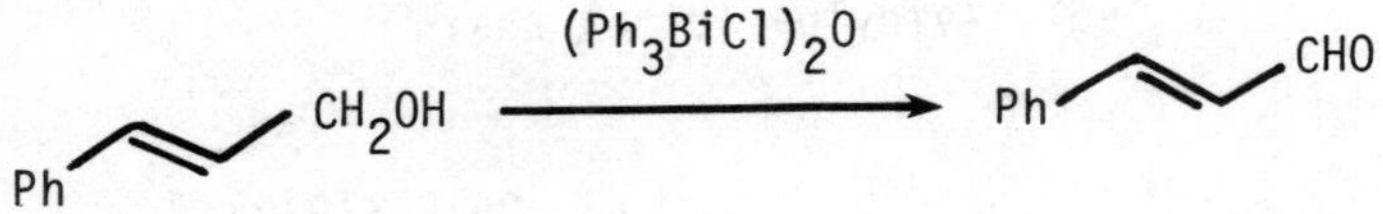

83%

JCS Chem Comm, 1099 (1978)

$$Ph\text{-}CH_2OH \xrightarrow{\quad Ce^{+4} \;/\; BrO_3^{\ominus} \quad} Ph\text{-}CHO \qquad 90\%$$

Synthesis, 936 (1978)

$$CH_3(CH_2)_8CH_2OH \xrightarrow[\underline{t}\text{-BuOOH}]{} CH_3(CH_2)_8CHO \qquad 92\%$$

Tetr Lett, 2801 (1979)

$$Ph\text{-}CH_2OH \xrightarrow{(PhSe)_2O} Ph\text{-}CHO \qquad 99\%$$

JCS Chem Comm, 952 (1978)

$\xrightarrow{I(OAc)_3}$ CHO / CHO   ~70%

JCS Perkin I, 1483 (1978)

$\xrightarrow{DMSO}$ 85%

Chem Lett, 369 (1978)

Related methods: Ketones from Alcohols and Phenols (Section 168)

Section 49   Aldehydes from Aldehydes

Conjugate reductions and Michael alkylations of conjugated
aldehydes are listed in Section 74 (Alkyls from Olefins).

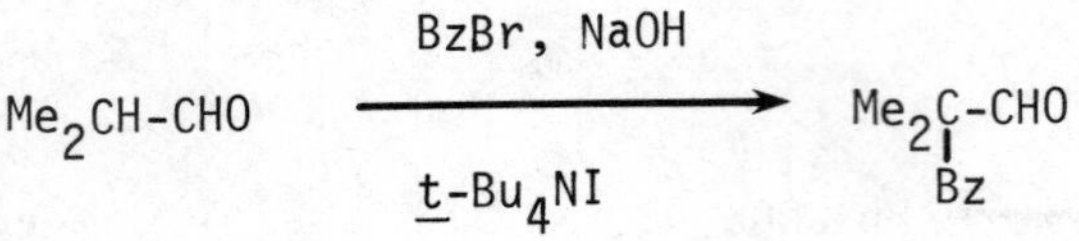

$$Me_2CH-CHO \xrightarrow[\underline{t}-Bu_4NI]{BzBr,\ NaOH} \underset{Bz}{Me_2\overset{|}{C}-CHO}$$ 60%

Chem and Ind, 731 (1978)

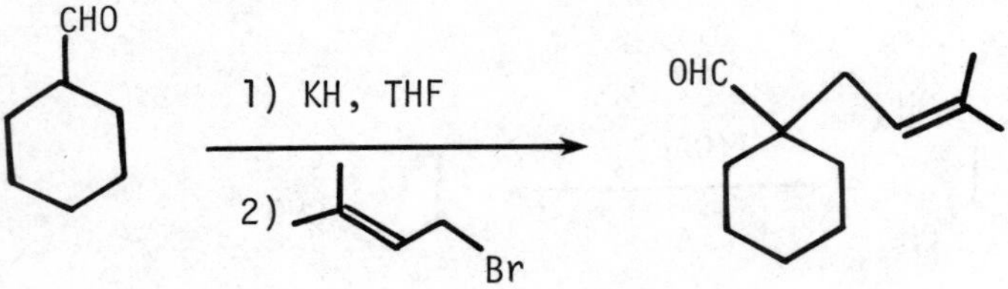

96%

Tetr Lett, 491 (1978)

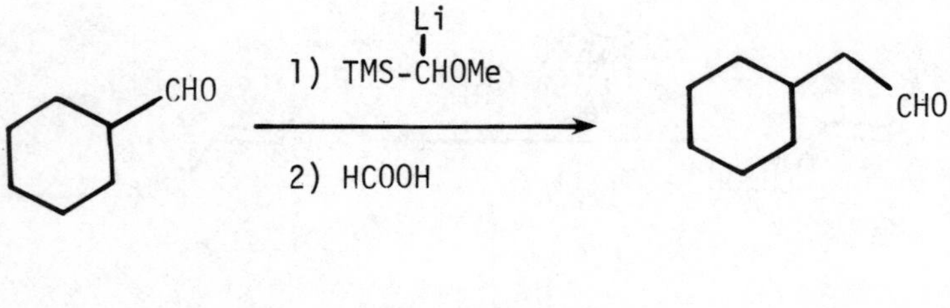

>90%

JCS Chem Comm, 822 (1979)

Review: "Synthesis of Aldehydes, Ketones, and Carboxylic Acids
from Lower Carbonyl Compounds by C-C Coupling Reactions".

Synthesis, 633 (1979)

Related Methods: Aldehydes from Ketones (Section 57), Ketones
from Ketones (Section 177). Also via: Olefinic aldehydes
(Section 341).

Section 50   Aldehydes from Alkyls

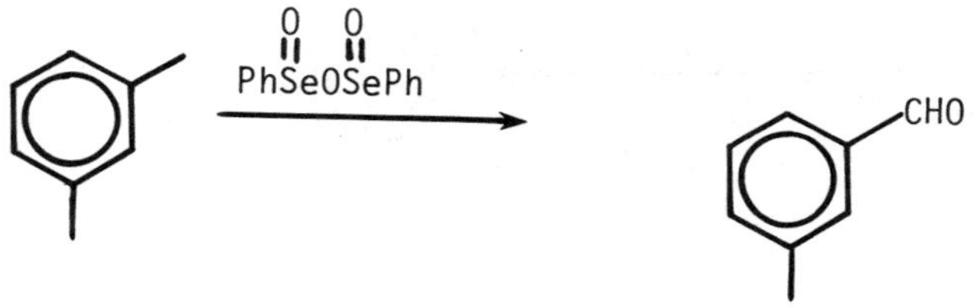

Tetr Lett, 3331 (1979)

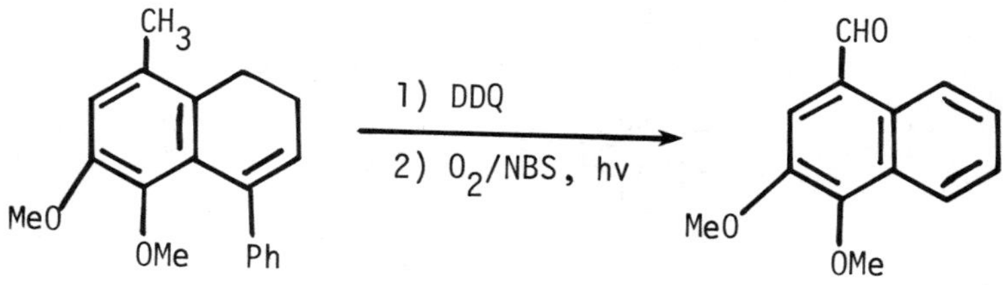

Synthesis, 144 (1979)

Section 51   <u>Aldehydes from Amides</u>

$$\underset{\substack{\text{O}\\\|\\ \text{R-C-NMe}_2}}{} \xrightarrow[\text{toluene}]{\text{LiAlH}_2(\text{OCH}_2\text{CH}_2\text{OMe})_2} \text{R-CHO} \qquad 41\text{-}98\%$$

R = long-chain alkyl

JOC (USSR) <u>13</u>, 1081 (1977)

82%

Tetr Lett, 3875 (1979)

Section 52   <u>Aldehydes from Amines</u>

$$\text{CH}_3(\text{CH}_2)_4\text{-NH-}(\text{CH}_2)_4\text{CH}_3 \xrightarrow[\substack{2)\ \text{KO}_2\\3)\ \text{H}_3\text{O}^{\oplus}}]{1)\ \underline{t}\text{-BuOCl}} \text{CH}_3(\text{CH}_2)_3\text{CHO}$$

+

$$\text{CH}_3(\text{CH}_2)_4\text{NH}_2$$

JOC <u>43</u>, 1467 (1978)

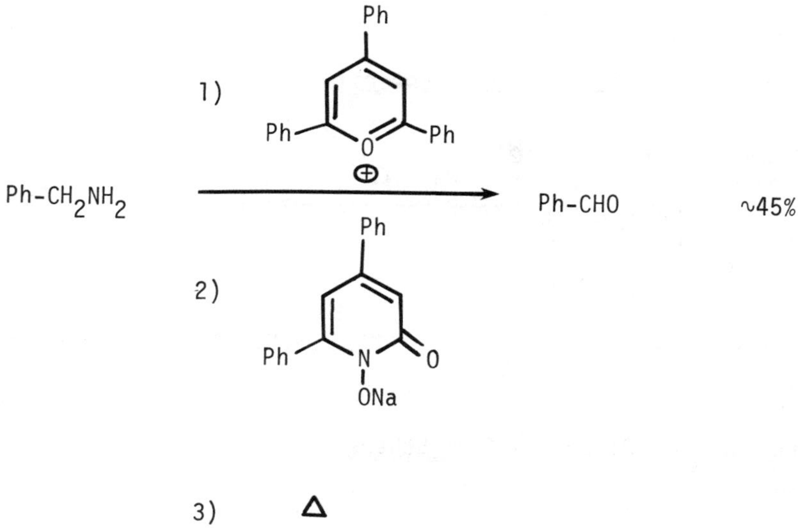

$$Ph-CH_2NH_2 \xrightarrow{\quad\quad} Ph-CHO \quad\quad \sim45\%$$

1)

Ph / Ph–O⁺–Ph / Ph (pyrylium)

2)

Ph / Ph–N(ONa) / O

3)   Δ

JCS Perkin I, 2500 (1979)

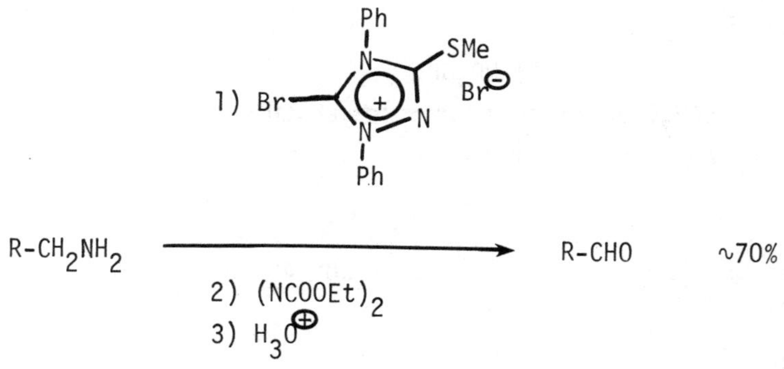

$$R-CH_2NH_2 \xrightarrow{\quad\quad} R-CHO \quad\quad \sim70\%$$

1) Br⁻

Ph / N–SMe / N⁺ / N / Ph   Br⁻

2) (NCOOEt)$_2$
3) H$_3$O⁺

R = alkyl, aryl

Tetr Lett, 2131 (1978)

Section 53   Aldehydes from Esters

$$Ph-CH_2\overset{\overset{Cl}{|}}{C}HOAc \xrightarrow{\;H_2O,\;\Delta\;} PhCH_2CHO \qquad 91\%$$

JOC (USSR) 14, 254 (1978)

Section 54   Aldehydes from Ethers

$$Ph-CH_2OMe \xrightarrow[\text{2) } H_2O]{\text{1) } UF_6} Ph-CHO \qquad 78\%$$

JACS 100, 5396 (1978)

$$PhCH_2OMe \xrightarrow[\text{2) } H_2O]{\text{1) } NO_2BF_4} Ph-CHO \qquad 89\%$$

JOC 42, 3097 (1977)

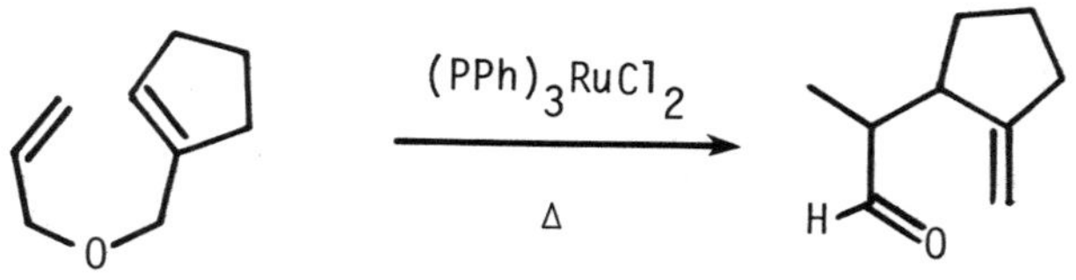

$(PPh)_3RuCl_2$

Δ

84%

JOC <u>42</u>, 3360 (1977)

Section 55   <u>Aldehydes from Halides</u>

$PhCH_2Br$ $\xrightarrow[\text{CHCl}_3]{(Bu_4N)_2Cr_2O_7}$ PhCHO   95%

Chem & Ind, 213 (1979)

$\underline{n}-C_6H_{13}-CH_2Br$ $\xrightarrow[\text{2) }\Delta]{\text{1)}}$ $C_6H_{13}-CHO$ 60%

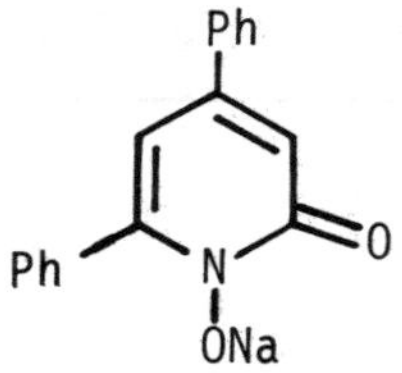

JCS Perkin I, 2493 (1979)

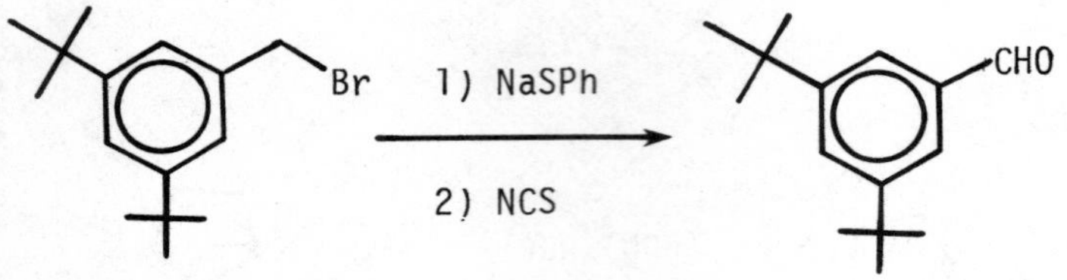

76%

Synth Comm $\underline{6}$, 575 (1976)

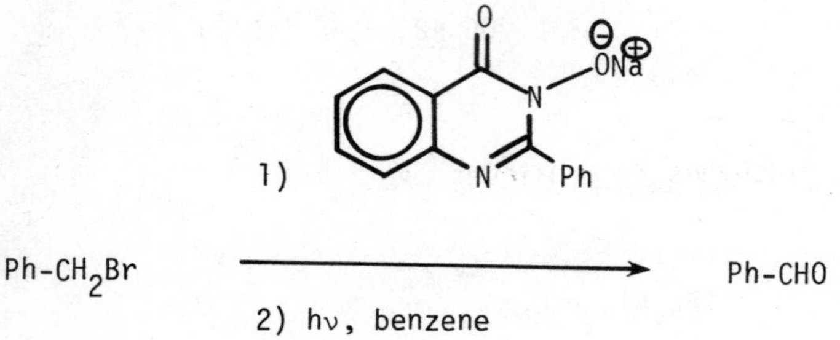

1)

Ph-CH$_2$Br $\longrightarrow$ Ph-CHO   46%

2) hν, benzene

Synthesis, 619 (1978)

PhCH$_2$CH$_2$Br $\xrightarrow{\text{polymer-bound FeH(CO)}_4^{\ominus}}$ PhCH$_2$CH$_2$CHO   80%

JOC $\underline{43}$, 1598 (1978)

$PhCH_2CH_2MgBr$ $\xrightarrow[\text{2) } H_3O^\oplus]{\text{1) }}$ $PhCH_2CH_2CHO$ 75%

1) [pyridine ring with N and N-Me, CHO substituent]

Synthesis, 403 (1978)

$CH_3(CH_2)_8-I$ $\xrightarrow[\substack{\text{2) } (COOH)_2 \\ H_2O, \text{ THF}}]{\text{1) } Et_2NCH_2CN, \text{ LDA}}$ $CH_3(CH_2)_8CHO$ 90%

Tetr Lett, 5175 (1978)

$PhMgBr$ $\xrightarrow[\text{THF}]{\overset{\displaystyle O}{\overset{\|}{Me-S-CH_2SMe}}}$ $Ph-CH(SMe)_2$ 66%

Tetr Lett, 3883 (1977)

$\overset{\displaystyle O}{\overset{\|}{Ph-C}}-CH_2Br$ $\xrightarrow{Et_2NOH}$ $\overset{\displaystyle O}{\overset{\|}{Ph-C}}-CHO$ 78%

JOC $\underline{42}$, 754 (1977)

Section 56   Aldehydes from Hydrides

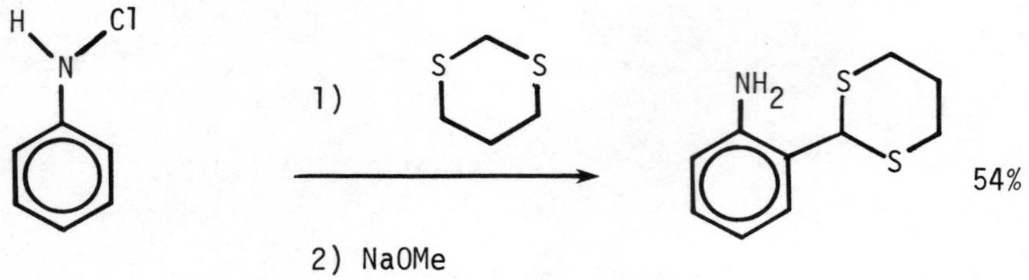

JACS 100, 7600 (1978)

Section 57   Aldehydes from Ketones

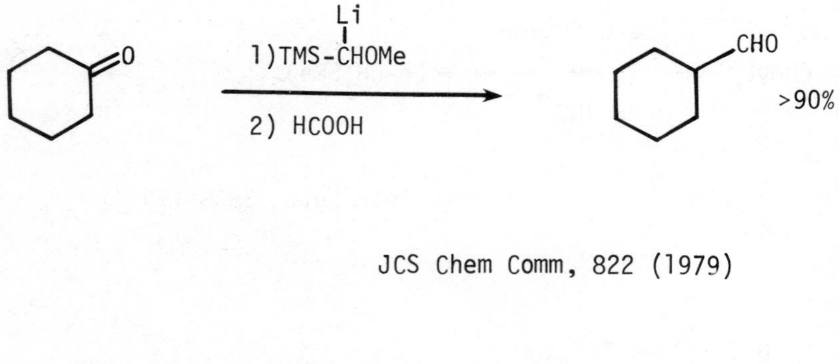

JCS Chem Comm, 822 (1979)

$$\text{1) } Cl\overset{\overset{\displaystyle Li}{|}}{C}HSiMe_3$$

2) $BF_3 \cdot Et_2O$
MeOH

95%

JACS 99, 4536 (1977)

$$\text{1) } (BuO)_2 \overset{\overset{\displaystyle O}{||}}{P} \underset{\underset{\displaystyle Li}{|}}{C}HOTHP$$

2) Δ

3) H_3O^+

~80%

Tetr Lett, 3629 (1978)

$$\text{1) } (RO)_2\overset{\overset{\displaystyle O}{||}}{P}\overset{\ominus}{C}HOR$$

2) Δ

H   OR

$H_3O^{\oplus}$

CHO

84%

JOC 44, 4847 (1979)

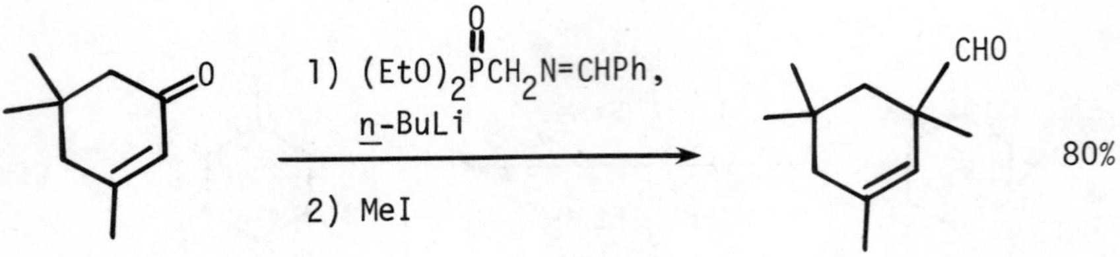

$$\text{1) (EtO)}_2\overset{\overset{\displaystyle O}{\|}}{P}CH_2N=CHPh,\ \underline{n}\text{-BuLi}$$

2) MeI

80%

JOC <u>43</u>, 3792 (1978)

Section 58   Aldehydes from Nitriles

No additional examples

Section 59   Aldehydes from Olefins

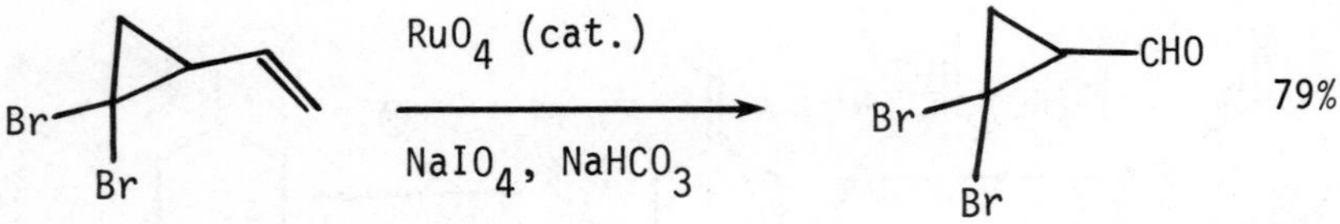

RuO_4 (cat.)

$NaIO_4$, $NaHCO_3$

79%

Acta Chem Scand <u>B32</u>, 693 (1978)

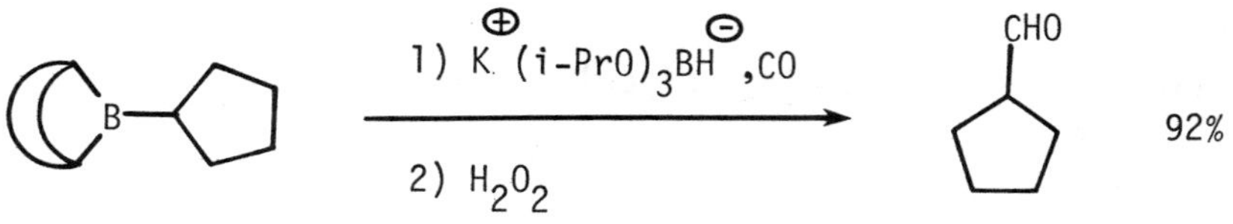

1) K.$\overset{\oplus}{}$ (i-PrO)$_3$BH$\overset{\ominus}{}$,CO

2) H$_2$O$_2$

92%

Synthesis, 701 (1979)

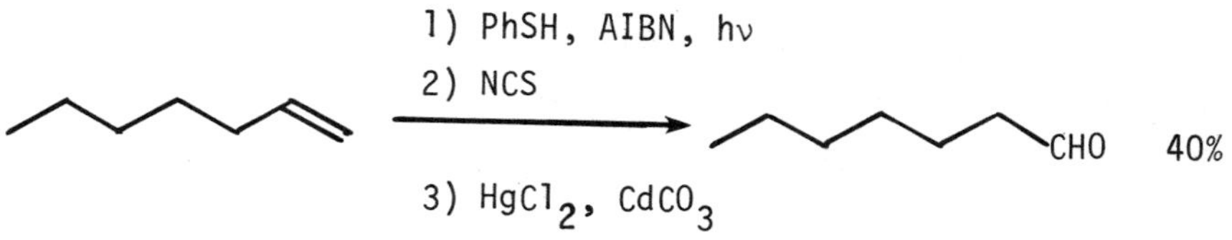

1) PhSH, AIBN, hν

2) NCS

3) HgCl$_2$, CdCO$_3$

40%

Synth Comm <u>6</u>, 575 (1976)

1) LiAlH$_4$, TiCl$_4$

2)

45%

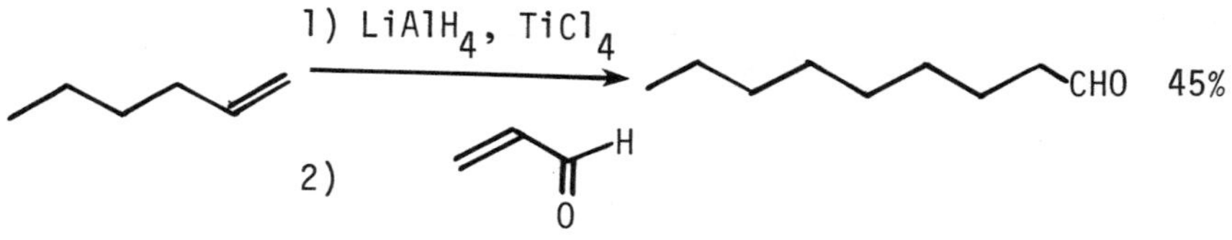

Chem Lett, 167 (1979)

Section 60   <u>Aldehydes from Miscellaneous Compounds</u>

$$Et-CH_2-NO_2 \xrightarrow[\text{CH}_3\text{OH}]{\text{CrCl}_2} Et-CHO \qquad 66\%$$

Synthesis, 792 (1977)

$$CH_3(CH_2)_7NO_2 \xrightarrow[\text{2) }^1O_2]{\text{1) NaOH, MeOH}} CH_3(CH_2)_6CHO \qquad 67\%$$

JOC <u>43</u>, 1271 (1978)

$$\underset{\text{H-C=CH-}\underline{t}\text{-Bu}}{\overset{N_3}{|}} \xrightarrow[\text{MeOH, Et}_3\text{N}]{\text{Na}_2\text{S}} \overset{O}{\underset{\text{H-C-CH}_2\text{-}\underline{t}\text{-Bu}}{||}} \qquad 90\%$$

JOC <u>44</u>, 4712 (1979)

$$\underset{\underline{n}\text{-C}_7\text{H}_{15}\text{-CH}_2\text{-S-Ph}}{\overset{O}{||}} \xrightarrow[\text{2) HgCl}_2\text{, MeCN/H}_2\text{O}]{\text{1) TFAA}} C_7H_{15}\text{-CHO} \qquad 86\%$$

Synthesis, 881 (1978)

Section 60A   <u>Protection of Aldehydes</u>

$$R-CHO \xrightarrow[\text{LnCl}_3]{\text{HC(OMe)}_3,\ \text{MeOH}} \begin{array}{c} \text{MeO} \quad \text{OMe} \\ R-\overset{|}{C}-H \end{array}$$

Ln = any of several lanthamides

Ketones remain unaffected under these reaction conditions.

JOC <u>44</u>, 4187 (1979)

Ph⌒⌒CHO $\xrightarrow[\text{montmorillonite}]{\text{HC(OMe)}_3}$ Ph⌒⌒$\begin{array}{c}\text{MeO} \quad \text{OMe}\\ \text{C}\\ \text{H}\end{array}$

92%

Synthesis, 467 (1977)

$$\text{C}_5\text{H}_{11}\overset{\overset{O}{\|}}{\text{C}}{\sim}\text{H} \xrightarrow[\text{MeOH, ErCl}_3]{\text{HC(OMe)}_3} \begin{array}{c}\text{MeO} \quad \text{OMe}\\ \text{C}\\ \text{C}_5\text{H}_{11} \quad \text{H}\end{array}$$ 96%

JCS Chem Comm, 976 (1978)

MeO   OMe
   \ /
    C
   / \
C$_5$H$_{11}$ H

$\xrightarrow[\text{CHCl}_3/\text{propene}]{\text{Me}_3\text{SiI}}$ C$_5$H$_{11}$-CHO   85%

Tetr Lett, 4175 (1977)

R-CHO $\xrightarrow[\text{Me}_2\text{SO}_4]{\text{HOCH}_2\text{-A-CH}_2\text{OH, DMF}}$

A = -CH$_2$CH$_2$- or -CH=CH-

Synthesis, 975 (1979)

$\xrightarrow[\text{TsOH, benzene}]{\text{Me}_2\text{C(CH}_2\text{OH)}_2}$ 70%

Synthesis, 295 (1978)

A divinylbenzene-styrene copolymer containing 1,3-diol groups can serve as a monoblocking agent for dialdehydes:

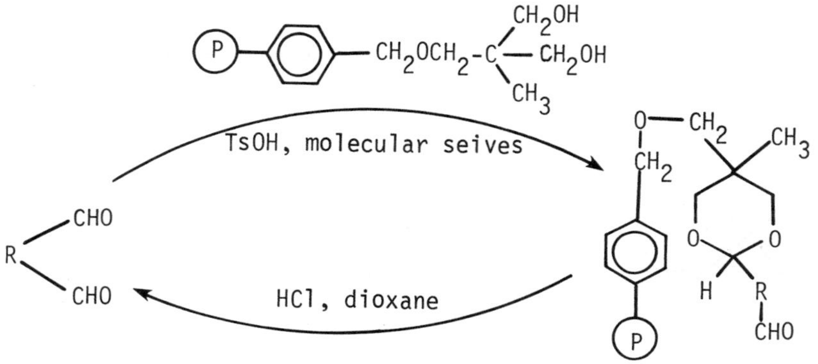

Can J Chem **54**, 3824 (1976)

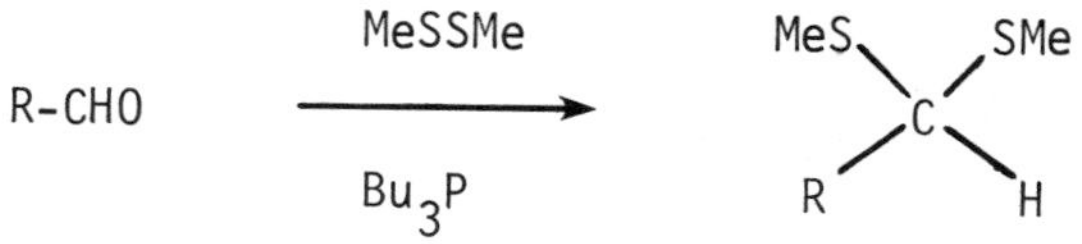

Chem Lett, 767 (1979)

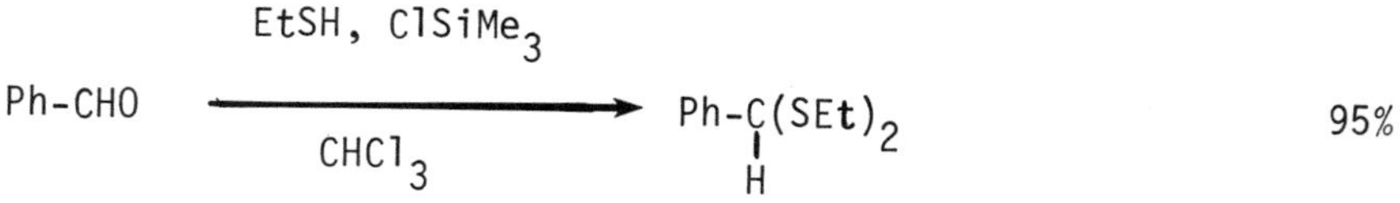

95%

Synth Comm **7**, 283 (1977)

$$CH_3(CH_2)_4CHO \xrightarrow{\text{EtS-SiMe}_3}$$

EtS＼＿＿＿OTMS
   C
CH_3(CH_2)_4   H   82%

JACS 99, 5009 (1977)

Ph   S──
  ＼ ／   electrochemical oxidation
   C   ────────────────────→ PH-CHO   77%
  ／ ＼   MeCN/H_2O, NaClO_4
H   S──

JCS Chem Comm, 255 (1978)

$$\xrightarrow[\text{MeOH/THF}]{\text{Tl(NO}_3)_3}$$

88%

Chem Pharm 26, 3743 (1978)

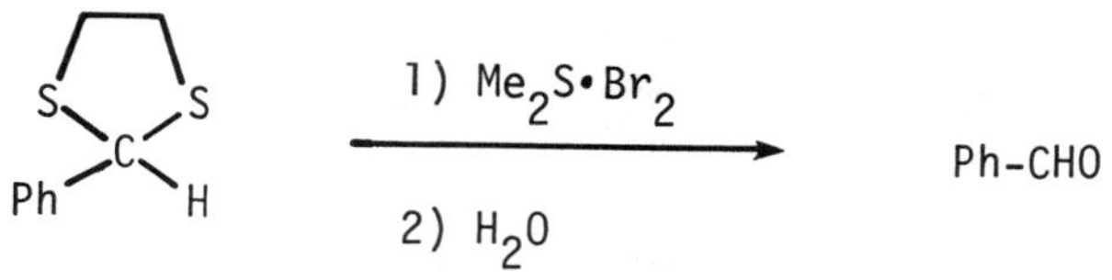

$$\text{1) } Me_2S\cdot Br_2$$
$$\text{2) } H_2O$$

Ph-CHO   75%

Synthesis, 720 (1979)

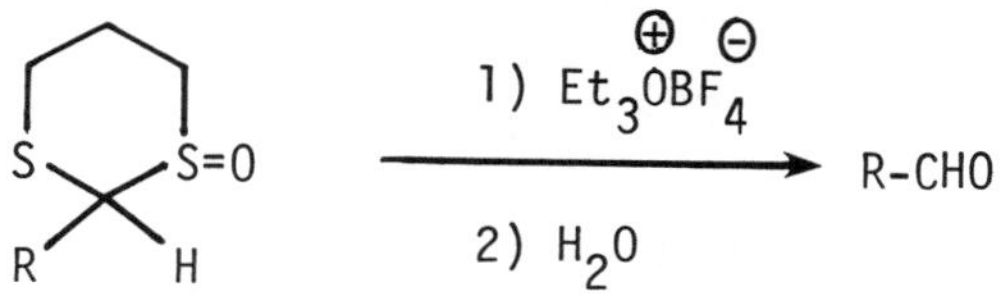

$$\text{1) } Et_3O\overset{\oplus}{}BF_4\overset{\ominus}{}$$
$$\text{2) } H_2O$$

R-CHO   ∿75%

R = alkyl, subst. Ph

Angew Int Ed <u>18</u>, 165 (1979)

$$R-CH=N-NHTs \xrightarrow[\text{MeOH/dioxane}]{H_2O_2, \; K_2CO_3} R-CHO$$

Synthesis, 919 (1978)

NNHTs
‖
C $\xrightarrow{\text{NaNO}_2, \text{ CF}_3\text{COOH}}$ R-CHO
R   H

Synthesis, 207 (1979)

NNTs
‖
C $\xrightarrow[\text{HOAc}]{\text{Tl(OAc)}_3}$ R-CHO
R   H

Tetr Lett, 4583 (1979)

NNHTs
‖
C $\xrightarrow[\text{MeOH, THF}]{\text{CuSO}_4, \text{ H}_2\text{O}}$ O ‖ C   90%
R   H   R   H

Gazz Chim Ital 108, 137 (1978)

Ph-CH=NNHTos $\xrightarrow[\text{NaHCO}_3, \text{ H}_2\text{O/HMPT}]{\text{Br}_2, \text{ CH}_2\text{Cl}_2}$ Ph-CHO   74%

Synthesis, 113 (1979)

$$n\text{-}C_{11}H_{23}\text{-}\overset{\overset{\displaystyle NNMe_2}{\|}}{C}\text{-}H \xrightarrow[\text{methylene blue}]{O_2,\ h\nu} n\text{-}C_{11}H_{23}\text{-}CHO \qquad 52\%$$

Synthesis, 893 (1977)

$$C_{10}H_{21}\text{-}CH(SeMe)_2 \xrightarrow[\hspace{3cm}]{\substack{\text{benzeneselenenic}\\ \text{anhydride}}} C_{10}H_{21}CHO \qquad 84\%$$

Synthesis, 877 (1979)

See Section 367 (Ether - Olefin) for the formation of enol ethers. Many of the methods in Section 180A (Protection of Ketones) are also applicable to aldehydes.

CHAPTER 5
PREPARATION OF ALKYLS, METHYLENES, AND ARYLS

This chapter lists the conversion of functional groups into Me, Et..., CH_2, Ph, etc.

Section 61   <u>Alkyls, Methylenes, and Aryls from Acetylenes</u>

$$CH_3C \equiv C\text{-}C_3H_7 \xrightarrow[\text{FeCl}_3, \text{ THF}]{\text{NaH, } \underline{t}\text{-BuONa}} CH_3\text{-}CH_2\text{-}CH_2\text{-}C_3H_7 \qquad 95\%$$

Tetr Lett, 3947 (1977)

Ph-O-CH-C≡CMgBr
|
Me

+

2   CH_2=CH-CH_2-ZnBr

62%

Synthesis, 838 (1978)

102

TMS

+   CpCo(CO)$_2$ →   TMS / TMS   44%

JACS 99, 4058 (1977)

Review: "Transition - Metal - Catalyzed Acetylene Cyclizations in Organic Synthesis"

Accounts Chem Res 10, 1 (1977)

Review: "Transition Metal Catalyzed Acetylene Cooligomerzations for the Synthesis of Complex Molecules"

Strem Chemiker VI, #2, 1 (1978)

Section 62   Alkyls from Carboxylic Acids

$C_9H_{19}COOH$

1) ⬡ SH / SH, BF$_3$

2) NaBH$_4$

3) Na, NH$_3$

→ $C_9H_{19}-CH_3$ ~80%

JCS Perkin I, 1133 (1978)

77%

Chem Pharm Bull 27, 816 (1979)

Section 63   Alkyls from Alcohols

Reactions in which hydroxyl groups are replaced by alkyl, e.g.,
ROH → RMe, are included in this section. For the conversion
ROH → RH see Section 153 (Hydrides from Alcohols and Phenols)

$$C_5H_{11}-\overset{\overset{O}{\|}}{C}-NHCH_2OH \quad \xrightarrow[\text{benzene}]{AlEt_3} \quad C_5H_{11}-\overset{\overset{O}{\|}}{C}-NHCH_2Et \qquad 71\%$$

Tetr Lett, 1465 (1977)

61%

Tetr Lett, 3003 (1977)

Section 64   Alkyls from Aldehydes

PhCH$_2$CHO

$\xrightarrow{\text{1) } \begin{array}{c}\text{SH}\\\text{SH,}\end{array} \quad \text{HBF}_4}{\text{2) Na, NH}_3}$

PhCH$_2$CH$_3$ ~70%

JCS Perkin I, 1133 (1978)

CHO / OMe (benzene ring)

$\xrightarrow[\text{CH}_2\text{Cl}_2]{\text{Et}_3\text{SiH, BF}_3}$

CH$_3$ / OMe (benzene ring)   100%

JOC 43, 374 (1978)

Ph-CH(OMe)$_2$ $\xrightarrow[\text{AlCl}_3\text{, Pd/C}]{\text{cyclohexene}}$ Ph-CH$_3$ 80%

Synthesis, 825 (1978)

$$PhCH_2CH_2CH=NNHTs \xrightarrow[\text{THF}]{\underline{t}\text{-BuLi}} Ph(CH_2)_3-\underline{t}\text{-Bu} \qquad 61\%$$

Tetr Lett, 135 (1977)

Related methods: Alkyls and Methylenes from Ketones (Section 72)

Also via: Vinyl Ethers (Section 69)

Section 65   Alkyls and Aryls from Alkyls and Aryls

No additional examples

Section 66   Alkyls from Amides

$$Ph-CH_2-N(Tf)_2 \xrightarrow{Me_2CuLi} Ph-CH_2CH_3 \qquad 60\%$$

Tetr Lett, 4727 (1978)

Section 67   Alkyls, Methylenes, and Aryls from Amines

No additional examples

Section 68   Alkyls from Esters

Ph-COOMe $\xrightarrow{\begin{array}{c} 1) \text{ Me}_3\text{SiI/I}_2 \\ 2) \text{ HSiCl}_3\text{-Pr}_3\text{N, MeCN} \\ 3) \ ^{\ominus}\text{OH} \end{array}}$ Ph-CH$_3$ 68%

JOC 44, 2185 (1979)

Bu$_2$CuLi   89%

JACS 101, 4413 (1979)

Section 69   Alkyls and Aryls from Ethers

The conversion ROR → RR' (R' = alkyl, aryl) is included in this section.

PhMgBr   75%

JACS 101. 2246 (1979)

$$\text{(2-pyridyl-OCH}_2\text{CH=CHPr)} \xrightarrow[\text{MgBr}_2, \text{ THF}]{\text{PhCH}_2\text{CH}_2\text{MgBr}} \text{Ph(CH}_2)_3\text{CH=CHPr} \quad 94\%$$

Chem Lett, 689 (1978)

$$\xrightarrow[\text{2) H}_3\text{O}^{\oplus}]{\text{1) BuLi}}$$

91%

JOC 43, 1372 (1978)

$$\text{MeS} \diagup\diagdown \text{C}_6\text{H}_{13} \xrightarrow[\text{NiL}_2\text{Cl}_2]{\text{CH}_3\text{MgBr}} \text{H}_3\text{C} \diagup\diagdown \text{C}_6\text{H}_{13} \quad 71\%$$

JCS Chem Comm, 637 (1979)

PhS⟍＿＿/Ph $\xrightarrow{\text{PhMgBr}}$ Ph⟍＿＿/Ph   97%

$NiCl_2(PPh_3)_2$

Tetr Lett, 43 (1979)

Section 70   Alkyls and Aryls from Halides and Sulfonates

The replacement of halogen by alkyl or aryl Groups is included in this section. For the conversion RX → RH (X=halo) see Section 160 (Hydrides from Halides and Sulfonates).

Ts⟍⟋⟍ $\xrightarrow{\underline{n}-C_6H_{13}MgBr}$ $\underline{n}-C_6H_{13}$ ⟍⟋⟍   100%

10% Cu(AcAc)$_2$

Tetr Lett, 2393 (1979)

Et

I   1) BuLi
     2) Et$_3$B   →   Et   61%
     3) I$_2$

JOC 43, 1279 (1978)

$$PhCH_2Br \xrightarrow[\substack{Pd\ catalyst \\ HMPA}]{Me_4Sn} PhCH_2CH_3 \qquad 70\%$$

JACS 101, 4992 (1979)

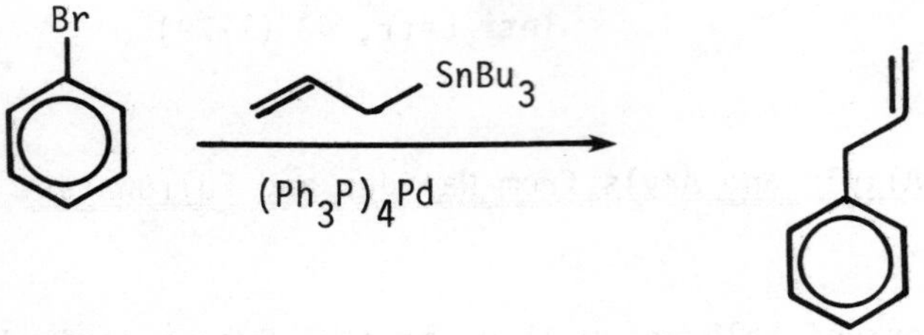

100%

Chem Lett, 301 (1977)

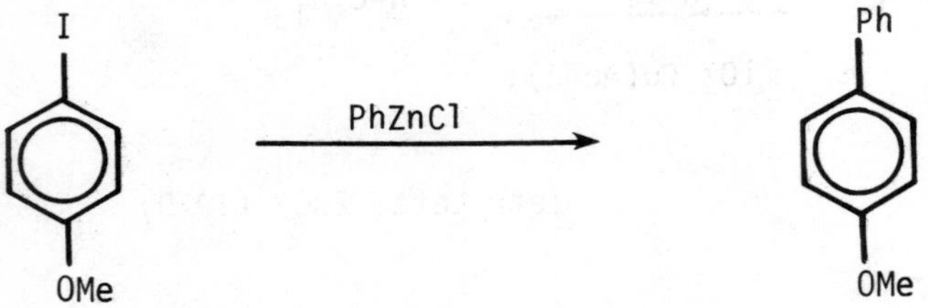

87%

JOC 42, 1821 (1977)

$$2Ph_2CHBr \xrightarrow[\substack{THF}]{VCl_3/LiAlH_4} Ph_2CH-CHPh_2 \qquad 95\%$$

Synthesis, 170 (1977)

$$2 \text{ Ar-I} \xrightarrow[\text{NiBr}_2\text{-Zn}]{\text{Zn, HMPA}} \text{Ar-Ar} \qquad \sim 90\%$$

Ar = subst. Ph

Chem Lett, 917 (1979)

$$C_8H_{17}MgBr$$

+

$$Me_3C-N\overset{O}{\underset{CHPh}{\diagdown|}}$$

$$\xrightarrow[\text{ether}]{} \quad C_8H_{17}-CMe_3 \qquad 90\%$$

JACS 101, 1044 (1979)

1) BuLi, Et$_2$O/HMPA
2) $\underline{n}$-C$_5$H$_{11}$I

3) (COOH)$_2$, H$_2$O

69%

Tetr Lett, 4661 (1978)

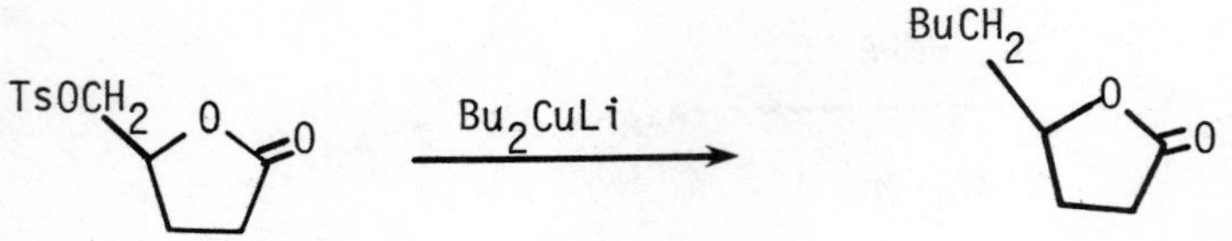

Tetr Lett, 423 (1977)

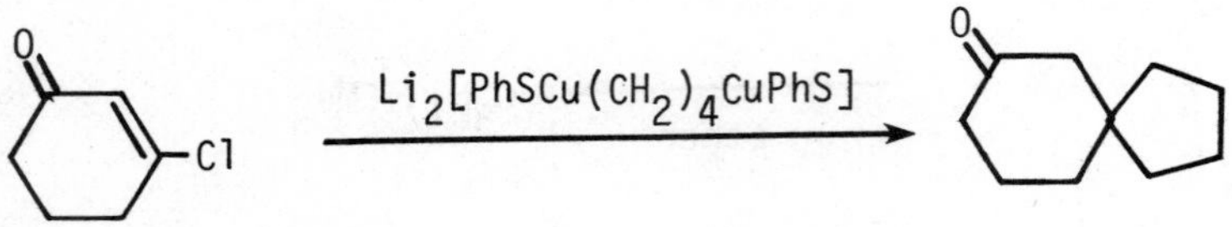

Tetr Lett, 1245 (1977)

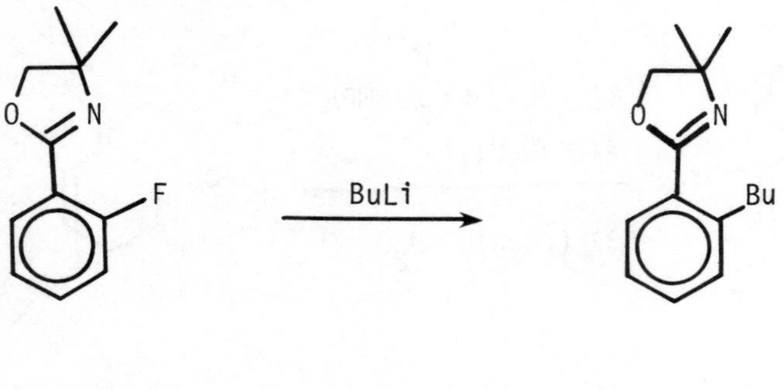

Tetr Lett, 223 (1978)

Ph—CH=CH—CH$_2$Br $\xrightarrow{\text{MeCu·BBu}_3}$ Ph—CH(Me)—CH=CH$_2$ 85%

JACS <u>99</u>, 8068 (1977)

Section 71   <u>Alkyls and Aryls from Hydrides</u>

This section lists examples of the reaction RH → RR' (R, R' =
alkyl or aryl). For the reaction C=CH → C=CR (R = alkyl or aryl)
see Section 209 (Olefins from Olefins). For alkylations of ketones
and esters, see Section 177 (Ketones from Ketones) and Section 113
(Esters from Esters).

PhS-CH=CHMe $\xrightarrow[\substack{\text{Pd(OAc)}_2 \\ \text{TMEDA/MeCN}}]{\text{PhBr, Ph}_3\text{P}}$ PhS-CH=C(Ph)-Me   65%

JACS <u>101</u>, 4743 (1979)

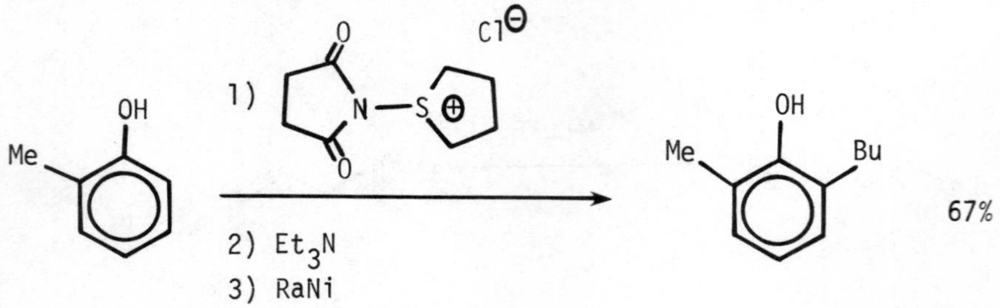

JACS 100, 7611 (1978)

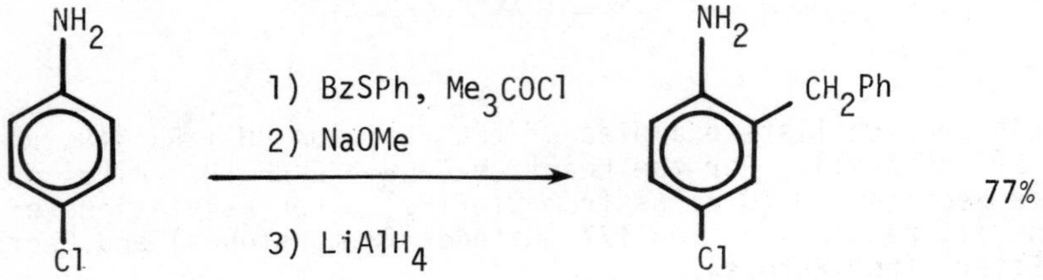

JACS 100, 7600 (1978)

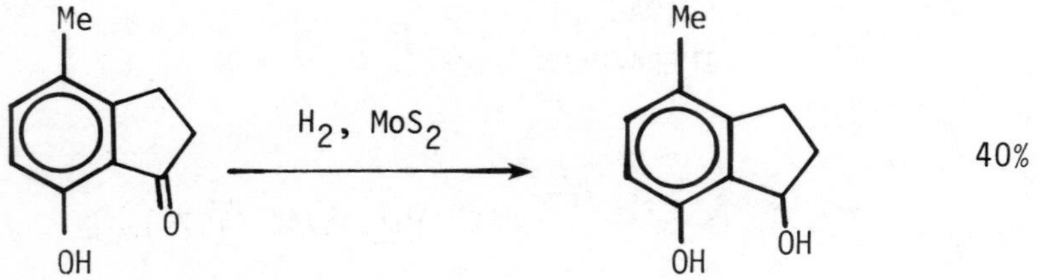

Coll Czech Chem Comm 43, 2174 (1978)

Australian J Chem <u>32</u>, 1531 (1979)

Section 72   Alkyls, Methylenes, and Aryls from Ketones

The conversions $R_2CO \rightarrow RR$, R_2CH_2, R_2CHR', etc. are listed in this section.

JCS Perkin I, 1133 (1978)

Synth Comm <u>9</u>, 275 (1979)

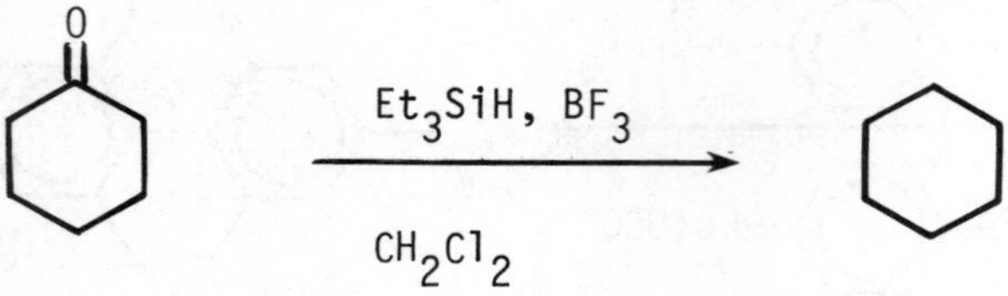

90%

JOC 43, 374 (1978)

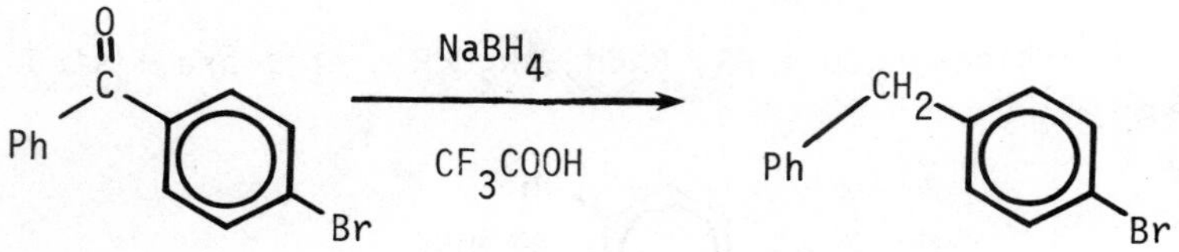

94%

Synthesis, 763 (1978)

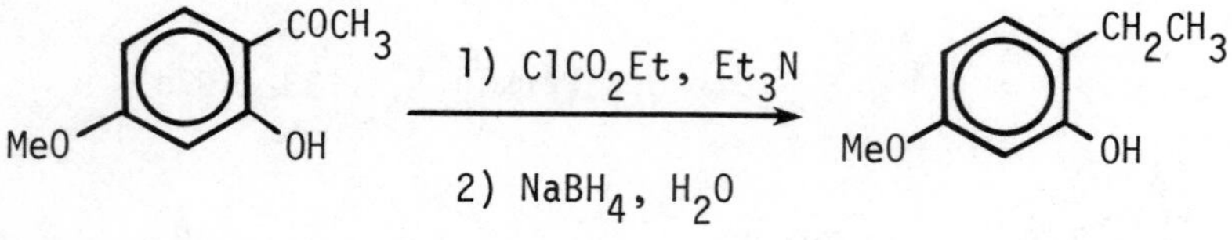

85%

Chem Pharm Bull 27, 1490 (1979)

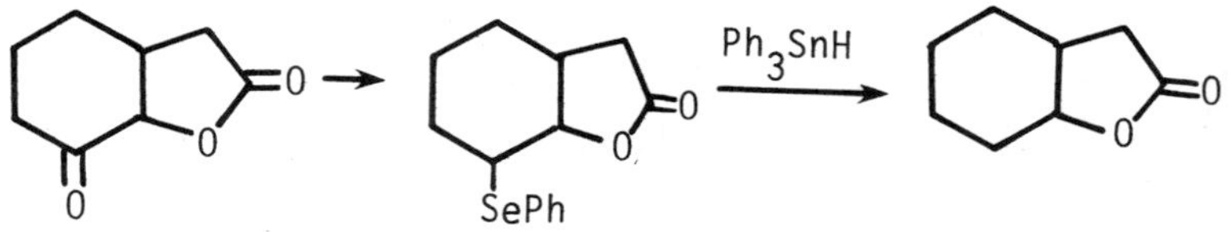

JCS Chem Comm, 41 (1978)

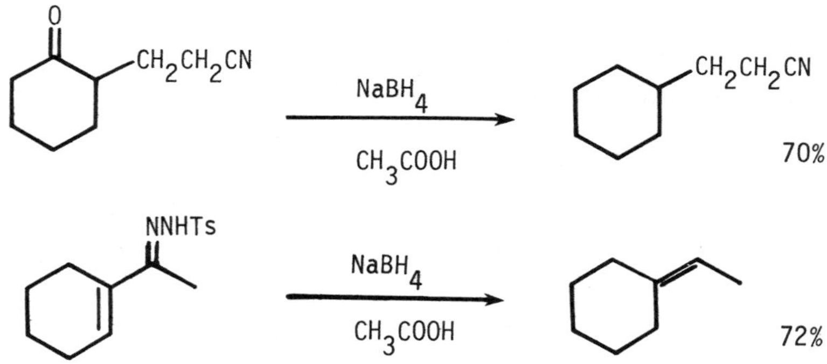

70%

72%

JOC 43, 2299 (1978)

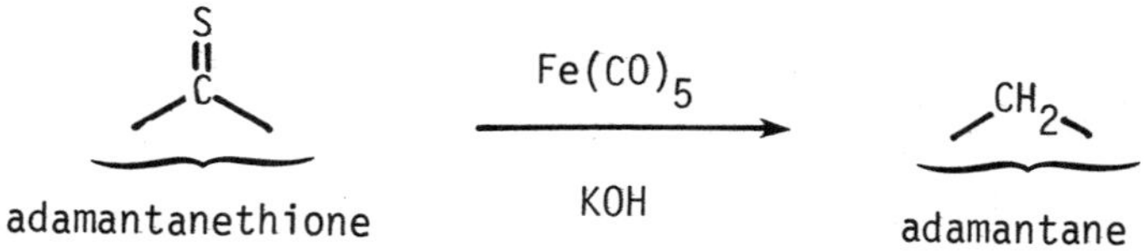

adamantanethione   Fe(CO)$_5$ KOH   adamantane   74%

JOC 42, 3522 (1977)

Synth Comm 7, 161 (1977)

Related methods: Alkyls from Aldehydes (Section 64)

Also via vinyl ethers (Section 69)

Section 73   Alkyls, Methylenes and Aryls from Nitriles

No additional examples

Section 74   Alkyls, Methylenes and Aryls from Olefins

The following reaction types are included in this section:

A.   Hydrogenation of olefins (and aryls).

B.   Dehydrogenations to form aryls.

C.   Alkylations and arylations of olefins.

D.   Conjugate reductions of conjugated aldehydes, ketones, acids, esters and nitriles.

E.   Conjugate alkylations.

F.   Cyclopropanations, including halocyclopropanations.

74A: Hydrogenation of olefins (and aryls)

$$\xrightarrow[\text{Ni(OAc)}_2, \text{ PhOCH}_3]{\text{NaH, } \underline{t}\text{-BuONa}}$$

95%

Tetr Lett, 1069 (1977)

$$\xrightarrow[\text{FeCl}_3, \text{ THF}]{\text{NaH, } \underline{t}\text{-BuONa}}$$

80%

Tetr Lett, 3947 (1977)

$$\xrightarrow[\text{CoCl}_2]{\text{LiAlH}_4}$$

100%

JOC 43, 2567 (1978)

$$\xrightarrow{\text{LiAlH}_4\text{-CoCl}_2}$$

96%

Tetr Lett, 4481 (1977)

$$\text{1) LiAlH}_4, \text{ TiCl}_4$$
$$\text{2) H}_2\text{O}$$

96%

J Organometal Chem 142, 71 (1977)

$(\underline{i}\text{-Pr}_2\text{N})_2\text{AlH}$

cp_2TiCl_2

100%

Tetr Lett, 4579 (1977)

$$\text{H}_2$$
silica-supported L_3RhCl

(Higher rates than for the homogeneous reaction.)

JCS Chem Comm, 510 (1977)

Use of $[\text{Ir(COD)L(py)}]\text{PF}_6$ and $[\text{Ir(COD)L}_2]\text{PF}_6$ as active hydrogenation catalysts for olefins.

J Organometal 141, 205 (1977)

Ir(CO)ClL_2 catalyzes the hydrogenation of cyclopentene, styrene, and ethyl acrylate at 10 atm.

J. Organometal 129, 331 (1977)

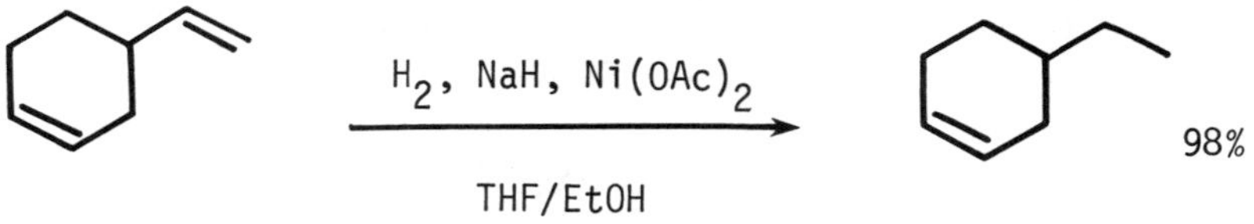

H_2, NaH, Ni(OAc)$_2$

THF/EtOH

98%

Tetr Lett, 3955 (1977)

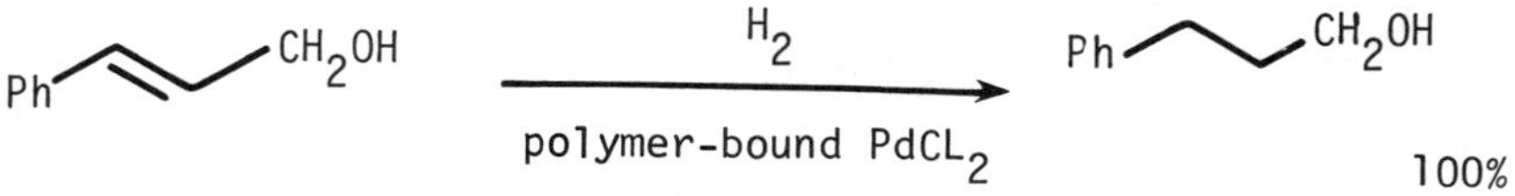

Ph⌇⌇CH$_2$OH

H_2

polymer-bound PdCL$_2$

Ph⌇⌇CH$_2$OH

100%

JOC <u>43</u>, 4686 (1978)

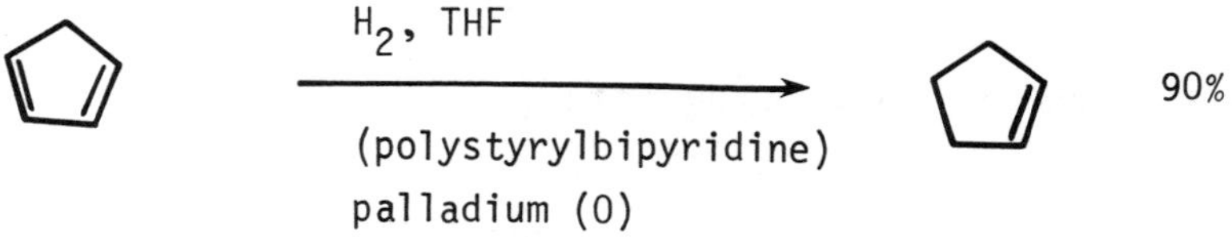

H_2, THF

(polystyrylbipyridine)
palladium (0)

90%

Israel J Chem <u>17</u>, 269 (1978)

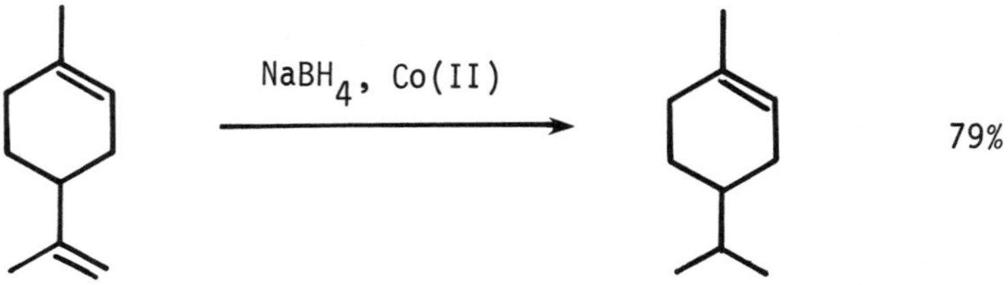

NaBH$_4$, Co(II)

79%

JOC <u>44</u>, 1014 (1979)

$$Ph-CH=CH_2 \xrightarrow[\text{Se, } O_2]{NH_2NH_2} Ph-CH_2CH_3 \qquad 100\%$$

Tetr Lett, 3727 (1977)

$$\xrightarrow[\text{pyridine}]{NaBH_4 \cdot AlCl_3}$$

80%

Chem Pharm <u>26</u>, 108 (1978)

as a chiral ligand for use in rhodium-catatyzed

hydrogenation. Gives 60%ee with α-ethylstyrene.

Tetr Lett, 295 (1977)

$$\xrightarrow[\text{salicylaldehyde·Ni}]{H_2, \ LiAlH_4}$$

96%

Tetr Lett, 2531 (1977)

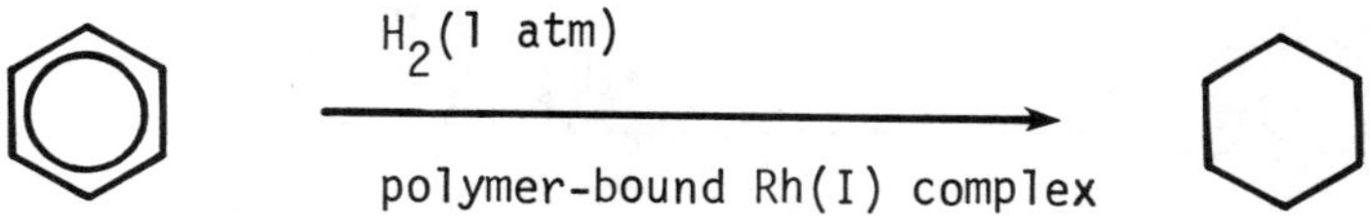

H_2(1 atm)

polymer-bound Rh(I) complex

Tetr Lett, 3703 (1977)

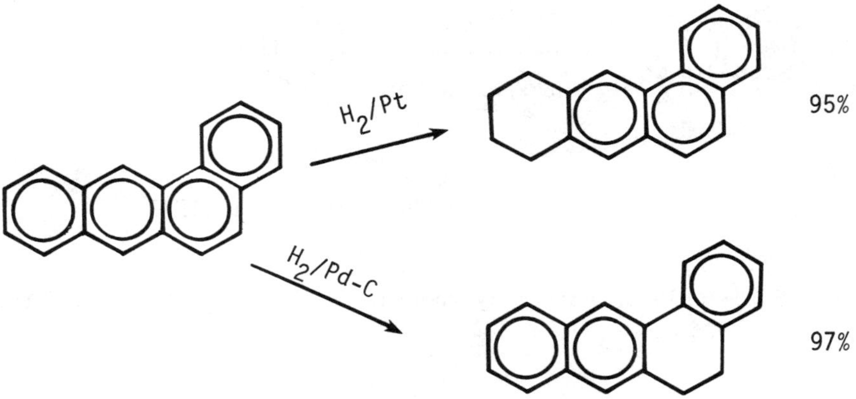

H_2/Pt   95%

H_2/Pd-C   97%

Tetr Lett, 415 (1977)

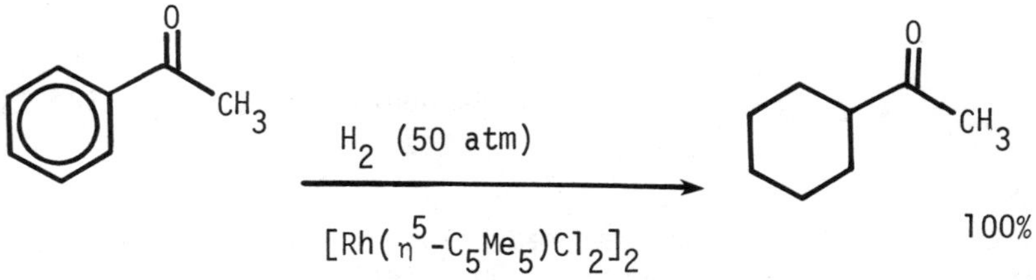

H_2 (50 atm)

$[Rh(\eta^5\text{-}C_5Me_5)Cl_2]_2$ 100%

JCS Chem Comm, 427 (1977)

71%

Synthesis, 447 (1978)

$Na_2S_2O_4$

DMF/H_2O, reflux

73%

Chem Lett, 1091 (1977)

Review: "Highly Selective Hydrogenations over Group VIII Metals"

Strem Chemiker $\underline{VI}$, #2, 7 (1978)

Review: "Catalytic Hydrogenation of Aromatic Hydrocarbons"

Accounts Chem Res $\underline{12}$, 324 (1979)

Reactions involving partial reduction of aromatic rings (to olefins) are found in Section 200 (Olefins from Aryls).

74B: Dehydrogenations to form aryls

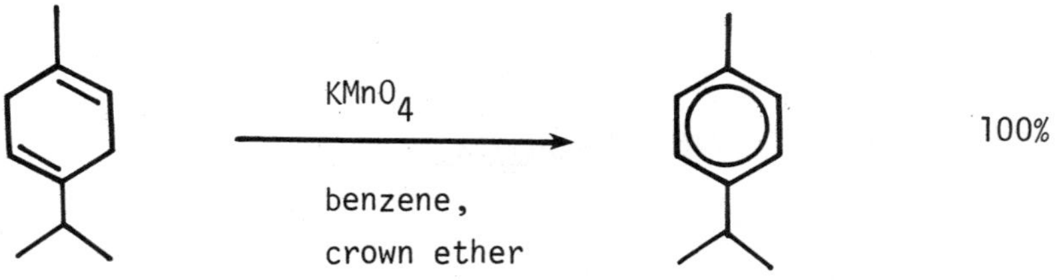

KMnO$_4$

benzene,
crown ether

100%

JCS Chem Comm, 244 (1979)

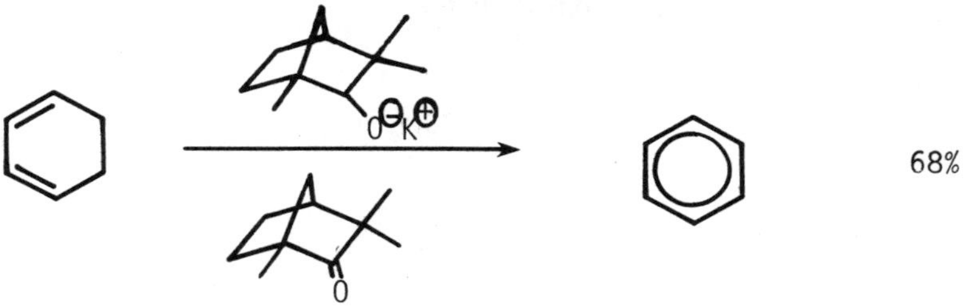

68%

Angew Int Ed 17, 278 (1978)

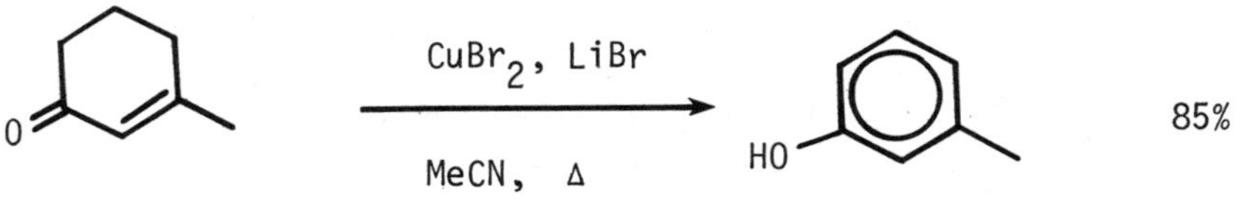

CuBr$_2$, LiBr

MeCN, Δ

85%

Tetr Lett, 821 (1977)

Review: "Dehydrogenation of Polycyclic Hydroaromatic Compounds"

Chem Rev 78, 317 (1978)

74C: Alkylations and arylations of olefins

No additional examples

74D: Conjugate reductions

$$\text{CH}_3\text{CH}=\text{CH}-\text{CHO} \xrightarrow[\text{PdCl}_2/\text{NaBH}_4]{\text{H}_2} \text{CH}_3\text{CH}_2\text{CH}_2\text{CHO}$$

JOC 42, 551 (1977)

$$\xrightarrow[\text{(Li bronze)}]{\text{Li/NH}_3}$$

90%

JOC 43, 4647 (1978)

$$\xrightarrow[\text{CuBr, THF}]{\text{LiAlH(OMe)}_3}$$

92%

JOC 42, 3180 (1977)

$$\xrightarrow[\underline{t}\text{-AmONa, THF}]{\text{NaH/Ni(OAc)}_2}$$

93%

JOC **44**, 2203 (1979)

$$\xrightarrow[\text{TiCl}_4]{\text{Me}_3\text{SiH}}$$

74%

Chem Pharm **25**, 1468 (1977)

$$\xrightarrow[\text{2) H}_3\text{O}^{\oplus}]{\text{1) Ph}_2\text{SiH}_2}$$

85%

Bull Akad USSR Chem **26**, 995 (1977)

$$\xrightarrow[\text{Pd/C}]{\text{HCO}_2^{\ominus}\ \text{HNBu}_3^{\oplus}}$$

69%

JOC **43**, 3985 (1978)

Ph⟍＝⟍—C(=O)—CH₃ → 1) C₈K, THF 2) H₃O⊕ → Ph⟍—CH₂—C(=O)—CH₃   85%

Synthesis, 30 (1979)

Ph⟍＝⟍—COOH → 1) C₈K, THF 2) H₃O⊕ → Ph⟍—⟍—COOH   86%

Synthesis, 30 (1979)

cyclohex-2-enone → NaHFe₂(CO)₈ → cyclohexanone   100%

JACS 100, 1119 (1978)

(CH₃)₂C＝CH—C(=O)—CH₃ → H₂, Rh-Al₂O₃ 140° C → (CH₃)₂CH—CH₂—C(=O)—CH₃   100%

Comptes Rendus (C) 284, 577 (1977)

→ H₂, K₃[Co(CN)₅H] NaOH, phase transfer →   93%

Tetr Lett, 115 (1979)

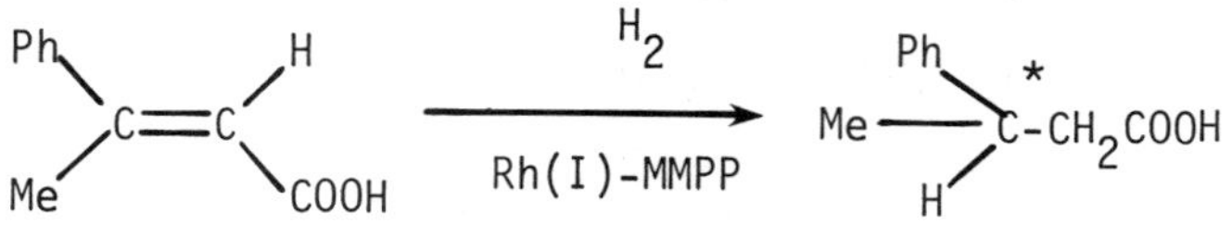

$$>92\%$$

$$\sim70\%ee$$

(MMPP = $\underline{l}$-menthylmethylphenylphosphine)

Tetr Lett, 2487 (1977)

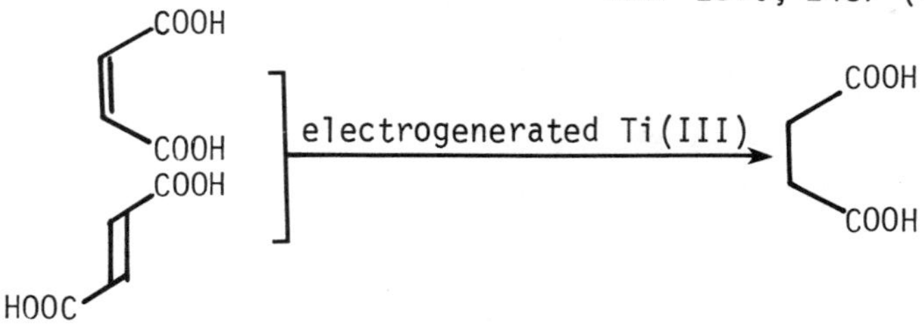

59%

87%

Can J Chem 56, 2269 (1978)

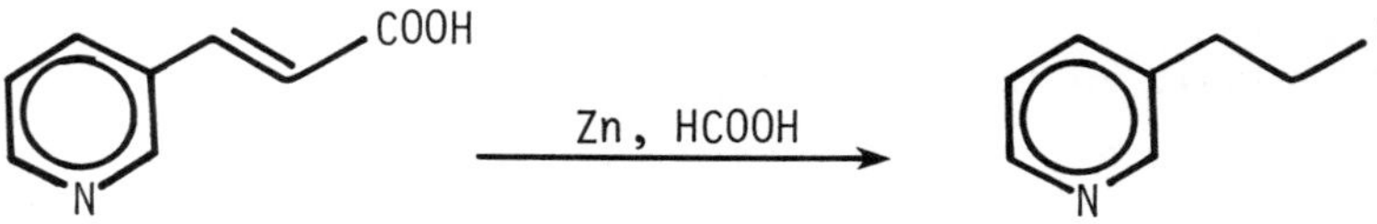

48%

Coll Czech Chem Comm 43, 1628
(1978)

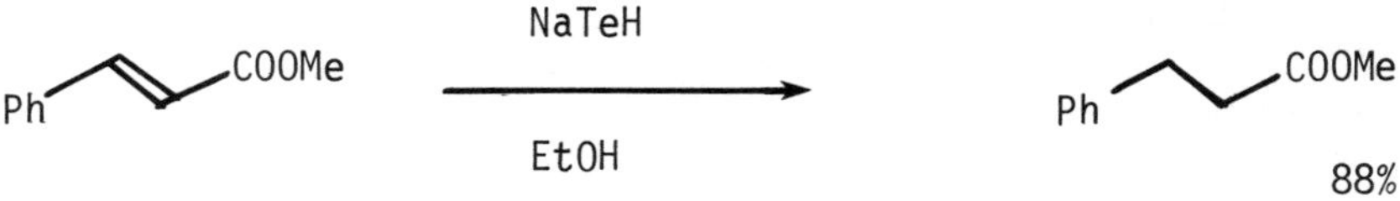

88%

Synthesis, 545 (1978)

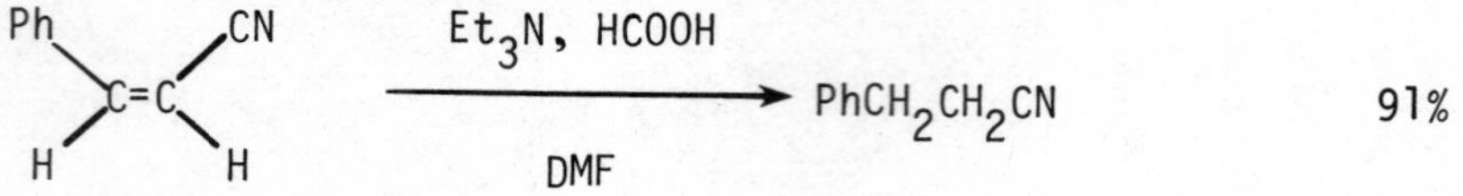

Chem Pharm 25, 2396 (1977)

74E: Conjugate alkylations

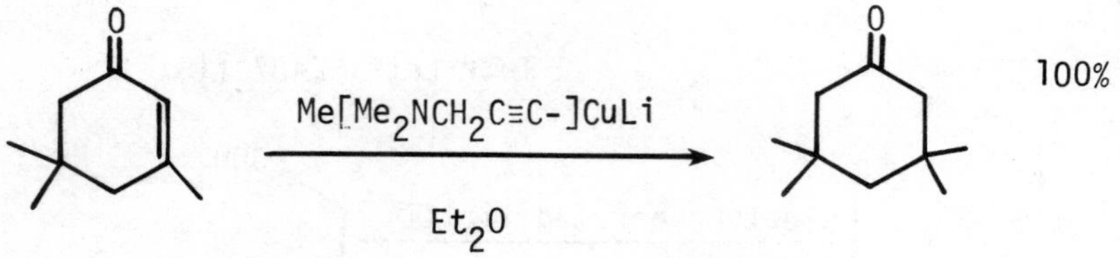

100%

Me[Me₂NCH₂C≡C-]CuLi

Et₂O

JOC 44, 1006 (1979)

RMgX   CuCl, Et₂O
  +   ─────────────→   RCH₂CH₂COOEt   41-80%
CH₂=CHCO₂Et   -40°

R = alkyl, Ph, benzyl, c-Hx

JOC 42, 3209 (1977)

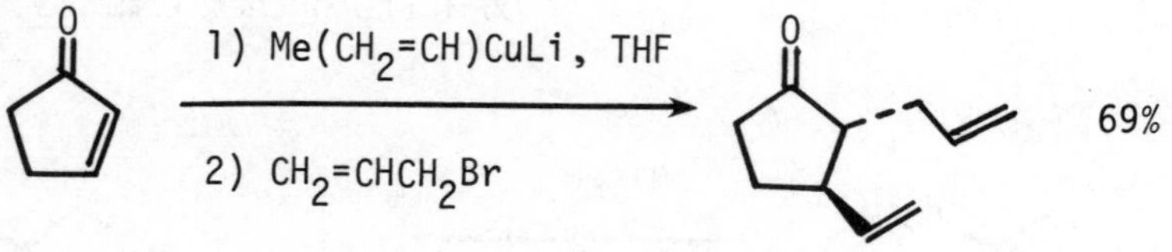

1) Me(CH₂=CH)CuLi, THF

2) CH₂=CHCH₂Br

69%

Tetr Lett, 3215 (1977)

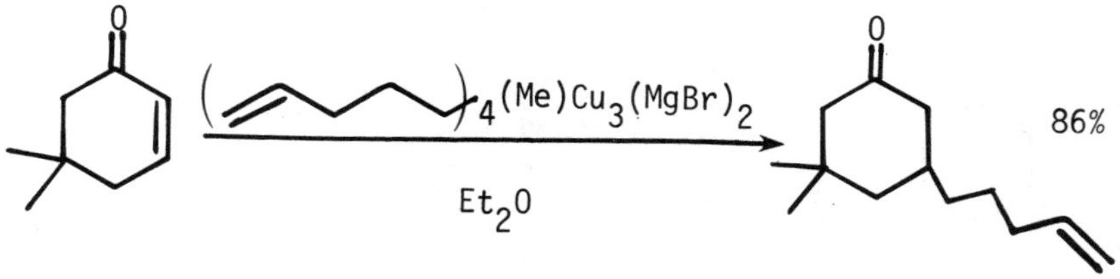

Nouveau J Chem 2, 271 (1978)

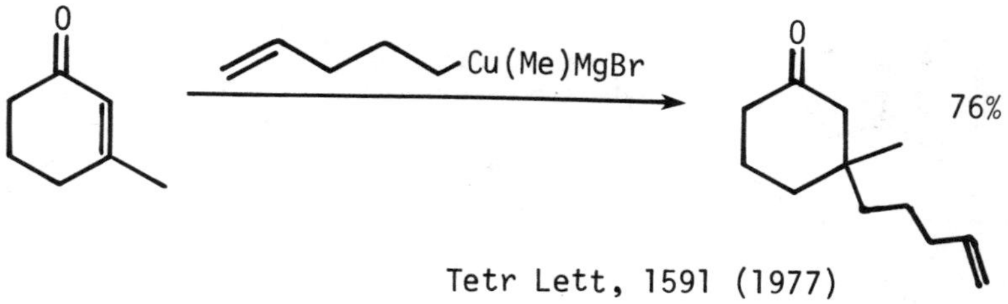

Tetr Lett, 1591 (1977)

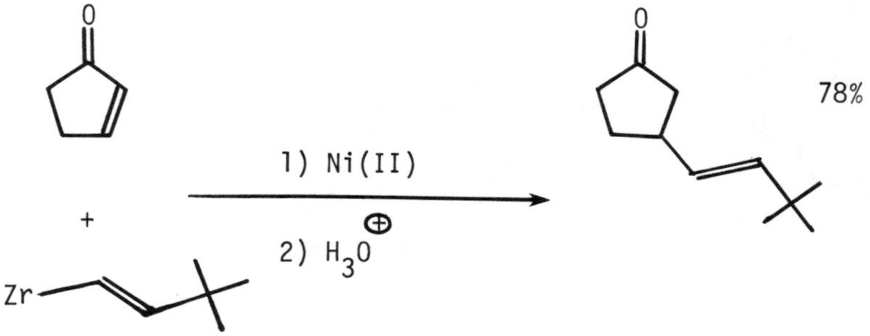

JACS 99, 8045 (1977)

electrolysis

s-Bu₃B $\xrightarrow[\text{CH}_3\text{CN}]{\text{Bu}_4\text{NI}}$ s-BuCH₂CH₂CO₂Et

+

CH₂=CHCO₂Et

59%

BCS Japan 51, 339 (1978)

$$[\underline{t}\text{-Bu-Cu-C}\equiv\text{C-CMe}_2\text{OMe}]\text{Li}$$

95%

JOC 43, 3418 (1978)

$$\text{Bu}_3\text{ZnLi-2LiCl}$$

$$-78° \text{ to } 0°$$

92%

Chem Lett, 679 (1977)

1) <u>t</u>-BuOK

2) <u>t</u>-BuLi

3) CH_3I

92%

JACS 100, 292 (1978)

74F: Cyclopropanations

$$\overset{\oplus}{\underset{}{\text{CpFe(CO)}_2\text{CH}_2\overset{\oplus}{\text{SMe}_2}}} \quad \overset{\ominus}{\text{BF}_4}$$

96%

JACS 101, 6473 (1979)

$$Ph\text{-}CH\text{=}CH_2 \xrightarrow[\text{I}_2, \text{ benzene}]{\text{CH}_2\text{I}_2, \text{ Cu}} Ph\text{---}\triangle \quad 90\%$$

JACS 101, 2139 (1979)

$$\xrightarrow{\text{H}_2\text{C=CH-CCl}_2\text{Li}} \quad 39\%$$

Synthesis, 425 (1979)

$$\xrightarrow[\text{Cu}]{\text{FCHI}_2} \quad\quad\text{F} \quad 80\%$$

Tetrahedron 35, 1919 (1979)

$$\xrightarrow[\text{H}_2\text{O, BzEt}_3\text{NCl}]{\text{CHCl}_3/\text{NaOH}} \quad\quad\begin{array}{c}\text{Cl}\\\text{Cl}\end{array}$$

77%

Synthesis, 682 (1977)

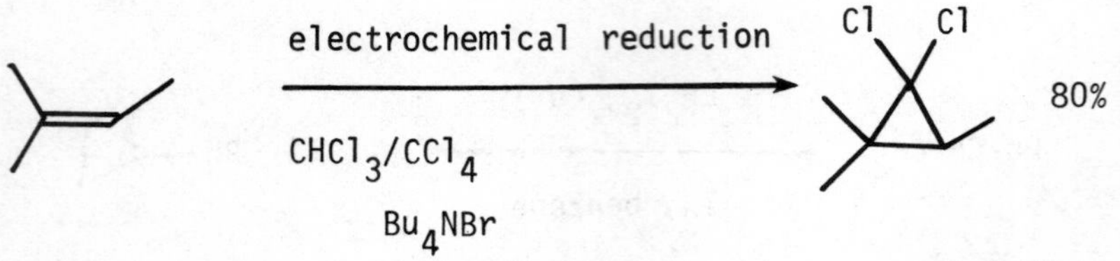

electrochemical reduction

$CHCl_3/CCl_4$

Bu_4NBr

Cl Cl

80%

Liebigs Ann Chem, 1416 (1978)

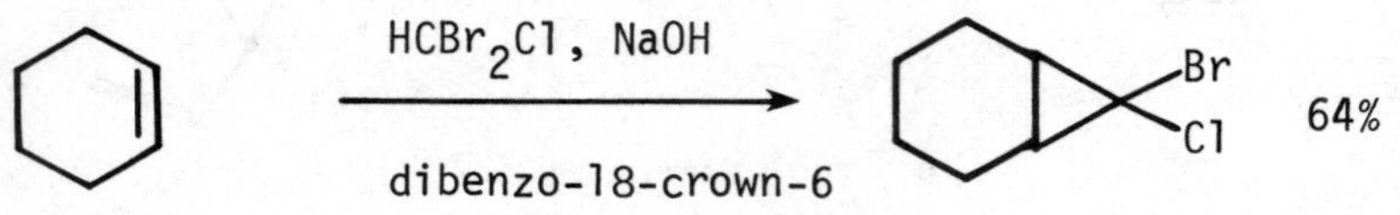

$HCBr_2Cl$, NaOH

dibenzo-18-crown-6

Br
Cl

64%

Synthesis, 783 (1977)

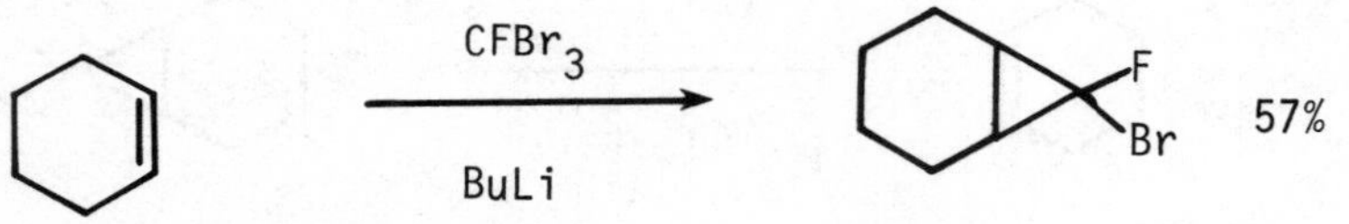

$CFBr_3$

BuLi

F
Br

57%

JOC 42, 828 (1977)

Section 75   <u>Alkyls and Methylenes from Miscellaneous Compounds</u>

No additional examples

CHAPTER 6
PREPARATION OF AMIDES

Section 76   Amides from Acetylenes

No additional examples

Section 77   Amides from Carboxylic Acids

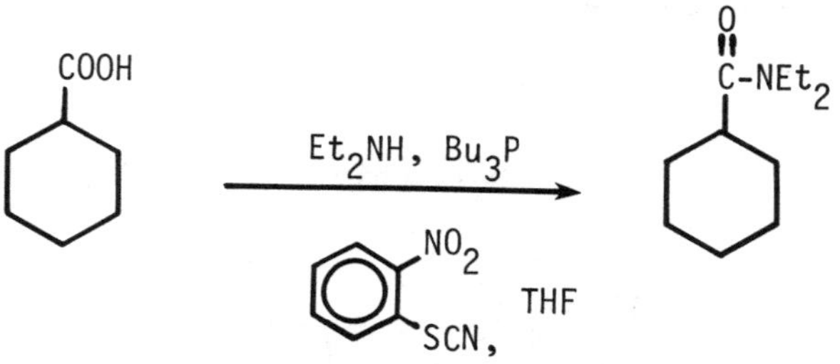

96%

A very general reaction for formation of amides.

JOC **44**, 2945 (1979)

$$R-COOH \xrightarrow[\text{DME}]{\text{DMPADC}} R-\overset{\overset{\displaystyle O}{\|}}{C}-NMe_2 \qquad \sim 60\text{-}90\%$$

DMPADC = $\underline{N}$, $\underline{N}$-dimethylphosphoramidic dichloride

Synth Comm $\underline{9}$, 31 (1979)

$$CH_3(CH_2)_7COOH \xrightarrow[\text{2) }]{\text{1) Catecholborane}} CH_3(CH_2)_7\overset{\overset{\displaystyle O}{\|}}{C}-N\big\langle \qquad 85\%$$

JOC $\underline{43}$, 4393 (1978)

$$R-COOH \xrightarrow[\text{2) } R_2'NH]{\text{1) } Et_2N-C\equiv C-\overset{\overset{\displaystyle O}{\|}}{C}-Me} R-\overset{\overset{\displaystyle O}{\|}}{C}-NR_2' \qquad \sim 90\%$$

Helv Chim Acta $\underline{61}$, 2428 and 2437 (1978)

$$H_2N-(CH_2)_3-COOH \xrightarrow[\text{2) MeOH}]{\text{1) } Me_3Si_2NH} \qquad 87\%$$

Synthesis, 614 (1978)

$$Ph-\overset{\overset{\displaystyle O}{\|}}{C}-NH-CH_2-\overset{\overset{\displaystyle O}{\|}}{C}-O-N= \xrightarrow{HOOC-CH_2CH_2NH_2} Ph-\overset{\overset{\displaystyle O}{\|}}{C}-NH-CH_2-\overset{\overset{\displaystyle O}{\|}}{C}-NH$$

$$HOOC-CH_2CH_2$$

Synthesis, 726 (1979)

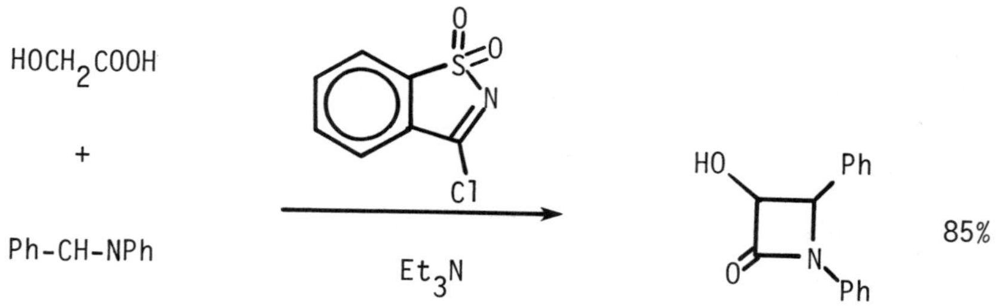

HOCH$_2$COOH

+

Ph-CH-NPh

Et$_3$N

85%

Synthesis, 407 (1977)

Related methods: Amides from Amines (Section 82)

Section 78   Amides from Alcohols

No additional examples

Section 79   Amides from Aldehydes

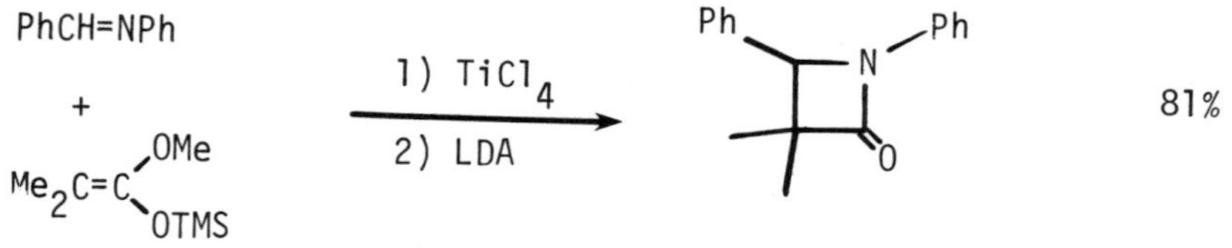

PhCH=NPh

+

Me$_2$C=C⟨OMe, OTMS

1) TiCl$_4$

2) LDA

81%

Tetr Lett, 3643 (1977)

Section 80   Amides from Alkyls, Methylenes and Aryls

No additional examples

Section 81   <u>Amides from Amides</u>

$$Ph-\overset{\overset{\textstyle O}{\|}}{C}-NH_2 \quad \xrightarrow[\text{2) MeI}]{\text{1) KOH, DMSO}} \quad Ph-\overset{\overset{\textstyle O}{\|}}{C}-NMe_2 \qquad 99\%$$

Tetrahedron <u>35</u>, 2169 (1979)

$$\xrightarrow[\text{Bu}_4\text{NBr}]{C_6H_{13}Br,\ KOH} \qquad 90\%$$

Tetr Lett, 615 (1978)

$$\xrightarrow[\text{2) BzBr}]{\text{1) NaOH, TEBA}} \qquad 66\%$$

Z Chem <u>17</u>, 260 (1977)

$$Et-\overset{\overset{\textstyle O}{\|}}{C}-NHBz \quad \xrightarrow[\underset{\oplus\ \ \ominus}{Bu_4NHSO_4,\ benzene}]{BuBr,\ NaOH/H_2O} \quad Et-\overset{\overset{\textstyle O}{\|}}{C}-\underset{\underset{\textstyle Bu}{|}}{N}Bz \qquad 82\%$$

Synthesis, 527 (1979)

$$\underline{t}\text{-Bu-}\overset{\overset{\displaystyle O}{\|}}{\underset{\underset{\displaystyle NNHTs}{|}}{C}} \quad \xrightarrow[\text{2) BzNH}_2]{\text{1) Pb(OAc)}_4} \quad \underline{t}\text{-Bu-}\overset{\overset{\displaystyle O}{\|}}{C}\text{-NHBz} \qquad 98\%$$

Angew Int Ed <u>16</u>, 728 (1977)

$$\xrightarrow{\text{TeF}_5\text{OC}_4\text{H}_9}$$

N-C$_4$H$_9$ >80%

JCS Perkin I, 2005 (1979)

$$C_5H_{11}\text{-}\overset{\overset{\displaystyle O}{\|}}{C}\text{-NHCH}_2\text{OH} \quad \xrightarrow[\text{TFA}]{\text{NaCNBH}_3} \quad C_5H_{11}\text{-}\overset{\overset{\displaystyle O}{\|}}{C}\text{-NHMe} \qquad 86\%$$

Synth Comm <u>7</u>, 549 (1977)

$$\xrightarrow[\text{2) H}_2\text{O}]{\text{1) NaH}}$$

75%

Synthesis, 31 and 33 (1977)

Conjugate reductions of unsaturated amides are listed in Section 74 (Alkyls from Olefins).

Section 82   Amides from Amines

$$Ph-NH_2 \xrightarrow{\overset{\overset{\displaystyle O}{\|}}{Me-C-O-(salicylic\ acid\ polymer)}} Ph-NH-Ac \qquad 90\%$$

Synth Comm <u>7</u>, 57 (1977)

97%

Synth Comm <u>7</u>, 393 (1977)

$$HO-CH_2CH_2CH_2NH_2 \xrightarrow[DMF]{C_6F_5OAc} HO-CH_2CH_2CH_2\underset{\underset{O}{\overset{|}{C}CH_3}}{N}H \qquad 91\%$$

JOC <u>44</u>, 654 (1979)

$$PhCH_2NH_2 \xrightarrow[2)\ O_2]{1)\ \text{(quinone)}} Ph\overset{\overset{\displaystyle O}{\|}}{C}NH_2 \qquad 50\%$$

JCS Chem Comm, 970 (1979)

Related methods: Amides from Carboxylic Acids (Section 77)
                   Protection of Amines (Section 105A)

Section 83   Amides from Esters

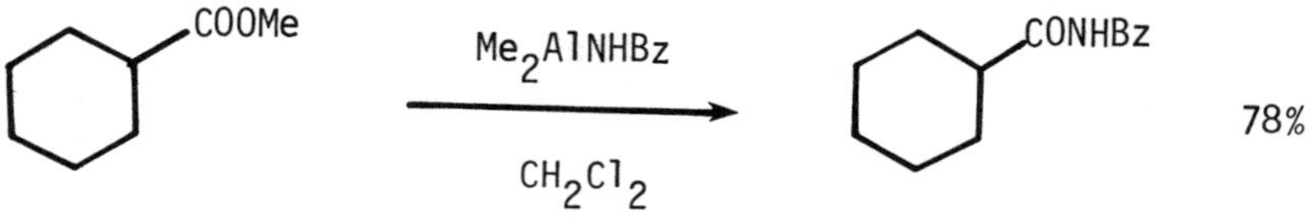

Tetr Lett, 4171 (1977)

Section 84   Amides from Ethers and Epoxides

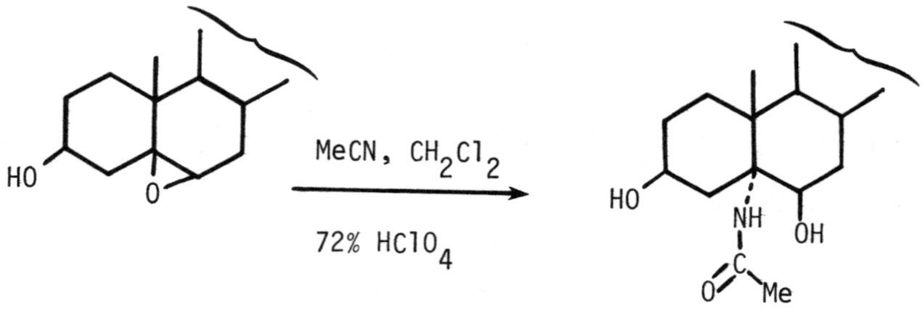

Synthesis, 35 (1977)

Section 85 Amides from Halides

I
NOPF6
CH3CN
NH-C-CH3
O
91%

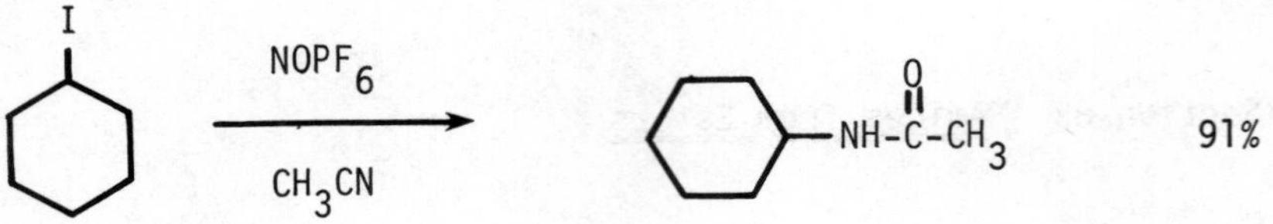

Synthesis, 274 (1979)

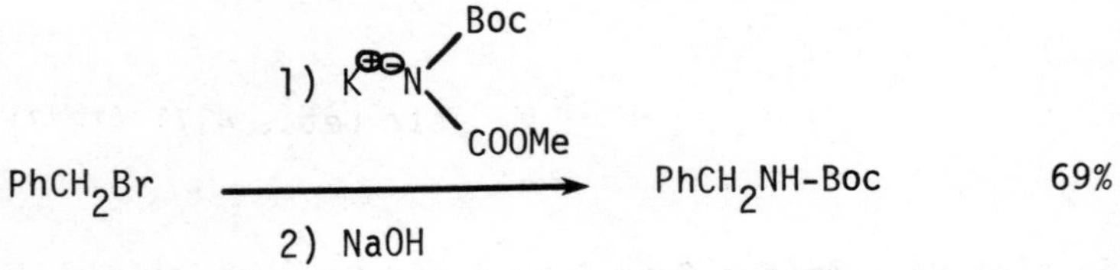

1) K N
Boc
COOMe
PhCH2Br
2) NaOH
PhCH2NH-Boc
69%

JCS Chem Comm, 758 (1977)

Section 86 Amides from Hydrides

1- adamantyl-H $\xrightarrow[\text{MeCN}]{Br_2,\ H_2SO_4}$ 1-adamantyl-NHAc 92%

Synthesis, 632 (1977)

Section 87   <u>Amides from Ketones</u>

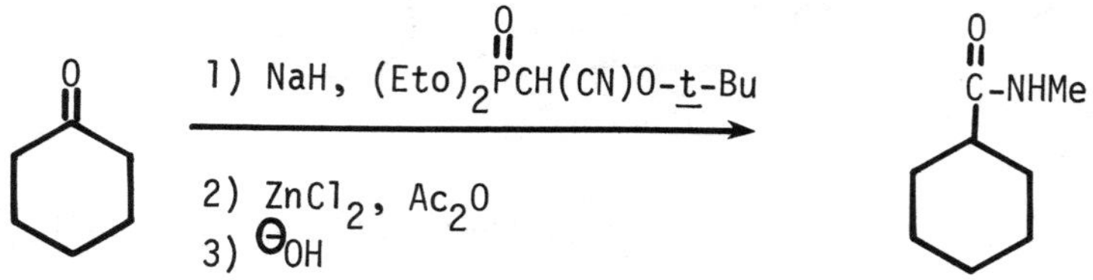

1) NaH, (Eto)$_2$PCH(CN)O-$\underline{t}$-Bu

2) ZnCl$_2$, Ac$_2$O
3) $\ominus$OH

94%

JACS <u>99</u>, 182 (1977)

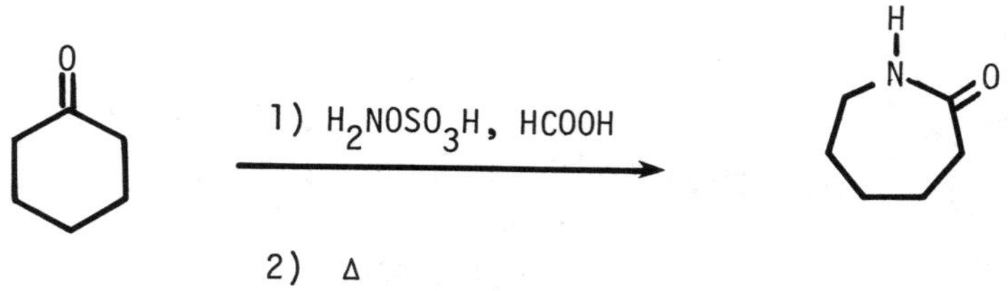

1) H$_2$NOSO$_3$H, HCOOH

2) Δ

82%

Synthesis, 537 (1979)

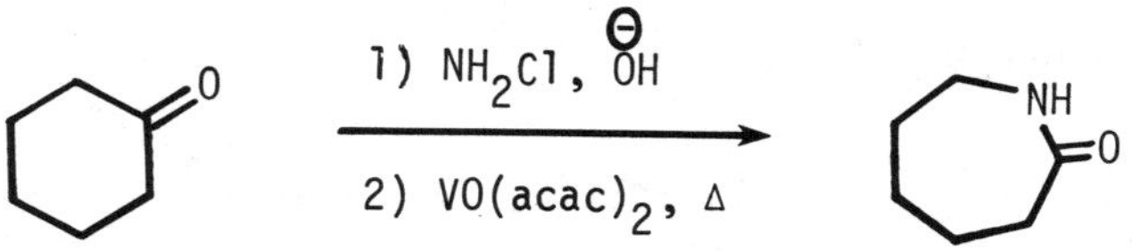

1) NH$_2$Cl, $\overset{\ominus}{\text{OH}}$

2) VO(acac)$_2$, Δ

88%

J Prakt Chem <u>319</u>, 274 (1977)

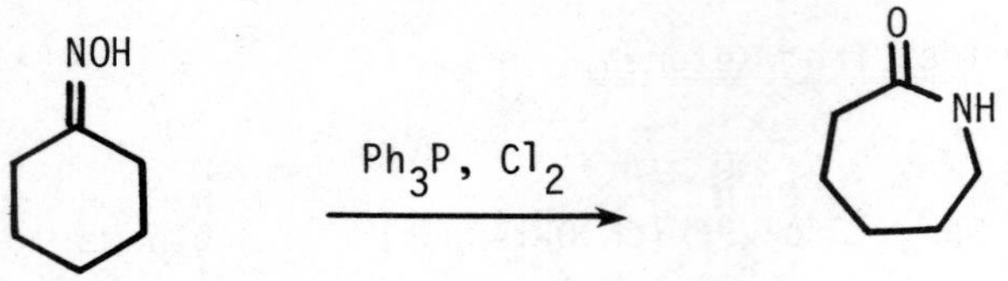

86%

Bull Chem Soc Japan <u>52</u>, 3381 (1979)

Section 88   <u>Amides from Nitriles</u>

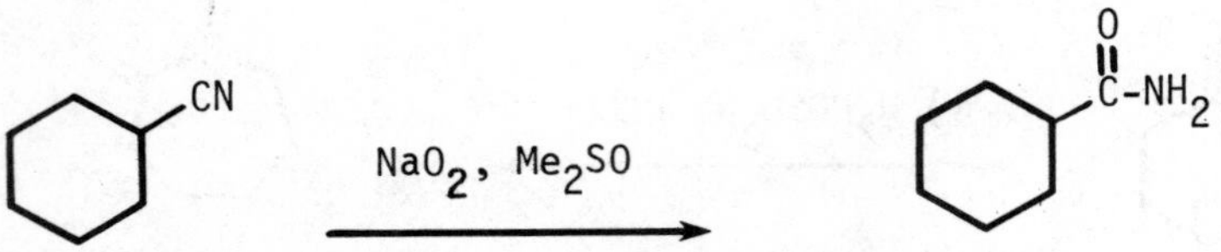

85%

JOC <u>44</u>, 4727 (1979)

Ph-CN $\xrightarrow[\text{2) } \underline{n}\text{-C}_6\text{H}_{13}\text{-I}]{\text{1) KOH, } \underline{t}\text{-BuOH}}$ Ph-C-NH-C$_6$H$_{13}$ 94%

Synthesis, 303 (1978)

Section 89   <u>Amides from Olefins</u>

No additional examples

Section 90   Amides from Miscellaneous Compounds

Synthesis, 118 (1977)   97%

JOC 42, 3755 (1977)   63%

Synth Comm 9, 281 (1979)   100%

Section 90A   Protection of Amides

Review: "Advances in the Chemistry of the Acetals of Acid Amides and Lactams"

Russ Chem Rev 46, 361 (1977)

CHAPTER 7
PREPARATION OF AMINES

Section 91   Amines from Acetylenes

No additional examples

Section 92   Amines from Carboxylic Acids and Acid Halides

No additional examples

Section 93   Amines from Alcohols

$$C_8H_{17}OH \xrightarrow[\text{CuO, } Cr_2O_3\text{, } Na_2O]{\text{HNMe}_2\text{, } 230^0} C_8H_{17}NMe_2 \qquad 97\%$$

Tetr Lett, 1937 (1977)

$$Ph(CH_2)_3OH \xrightarrow[\text{CuO, } 240^0]{\text{HNMe}_2, \text{ H}_2} Ph(CH_2)_3NMe_2$$ 90%

Synth Comm **8**, 27 (1978)

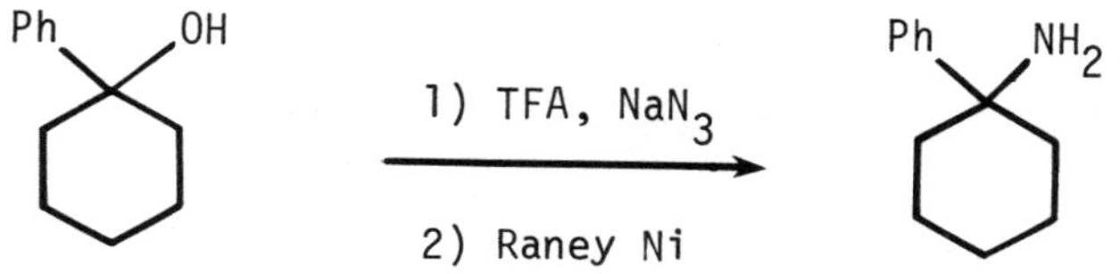

1) TFA, NaN$_3$

2) Raney Ni

79%

Synthesis, 24 (1978)

1. Pyridinium salt, Et$_3$N

(R)-2-octanol $\xrightarrow{\hspace{4cm}}$ (S)-2-aminooctane

2) LiN$_3$, HMPA
3) LiAlH$_4$ 77%

Pyridinium salt =

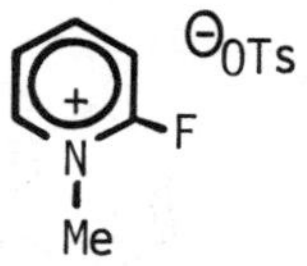

Chem Lett, 635 (1977)

Section 94   Amines from Aldehydes

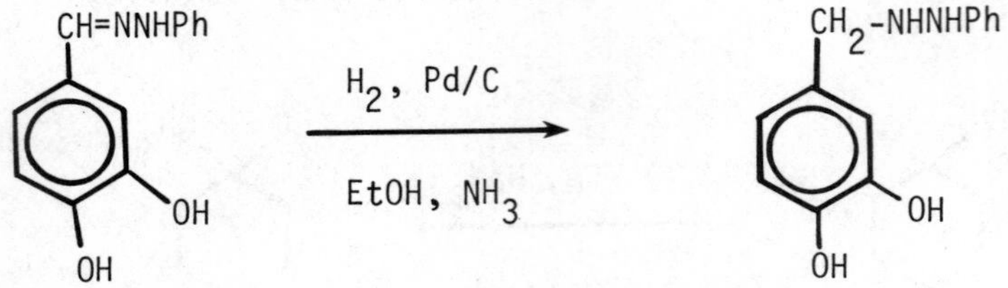

Synth Comm 7, 71 (1977)

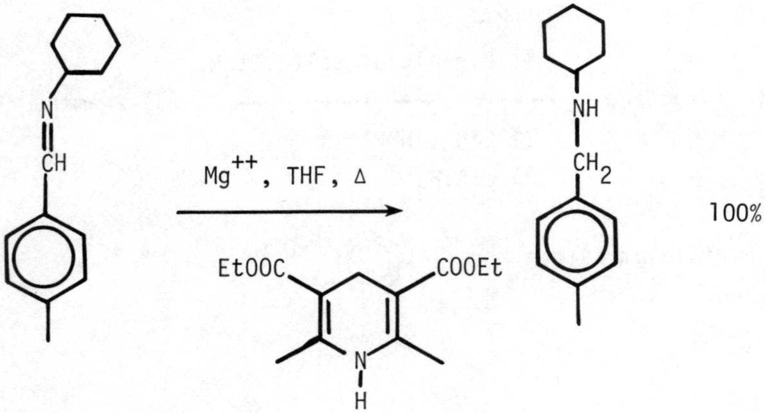

100%

Tetr Lett, 913 (1977)

HC=NPh

OMe

1) C_8K, THF

2) $H_3O^{\oplus}$

CH_2NHPh

OMe

90%

Synthesis, 30 (1979)

PhCH=NOMe

$\xrightarrow[\text{THF}]{\text{NaBH}_3(\text{OCOCF}_3)}$

$PhCH_2-NH_2$ 90%

Chem Pharm 26, 2897 (1978)

Ph-CH=NOH

$\xrightarrow[\underline{i}\text{-Pr-COOH}]{\text{NaBH}_4}$

$Ph-CH_2-\overset{\overset{\displaystyle OH}{|}}{N}-CH_2-\underline{i}-Pr$ 65%

Synthesis, 856 (1977)

$Me_2\overset{\underset{\displaystyle Cl}{|}}{C}$

$\xrightarrow[\text{EtOH}]{\text{NaBH}_4}$

98%

Rec Trav Chim 96, 242 (1977)

Related methods: Amines from Ketones (Section 102)

Section 95   Amines from Alkyls, Methylenes and Aryls

No Examples

Section 96   Amines from Amides

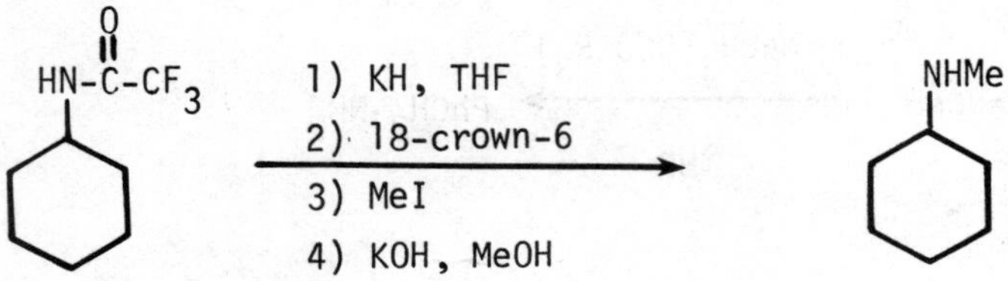

Tetr Lett, 4987 (1978)

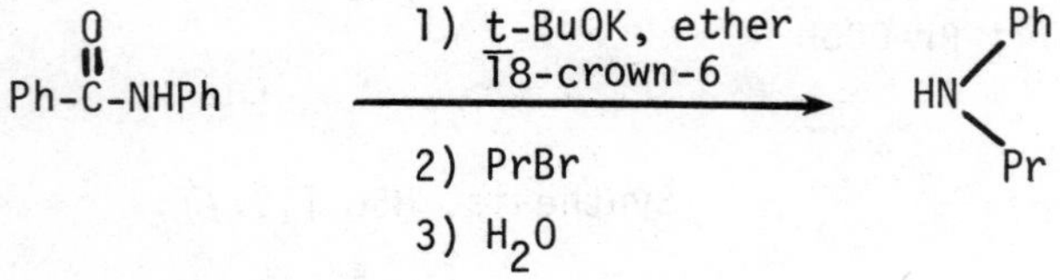

Synth Comm 9, 757 (1979)

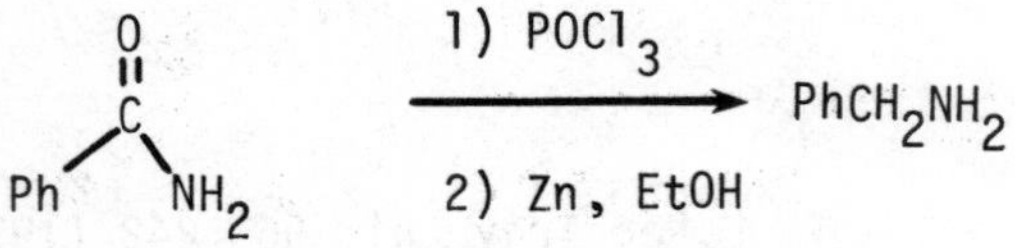

Experientia 33, 101 (1977)

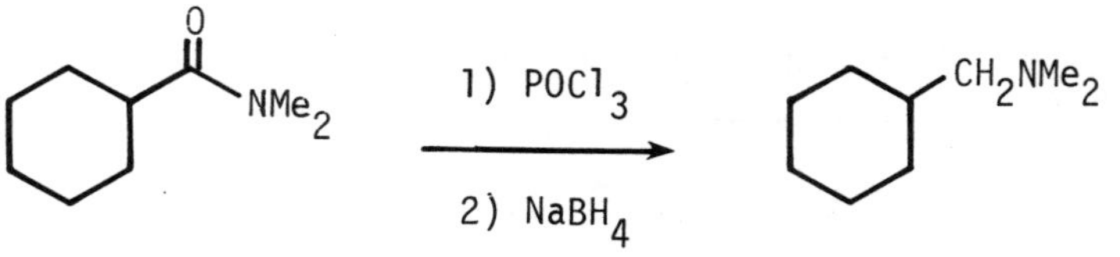

JOC 42, 2082 (1977)

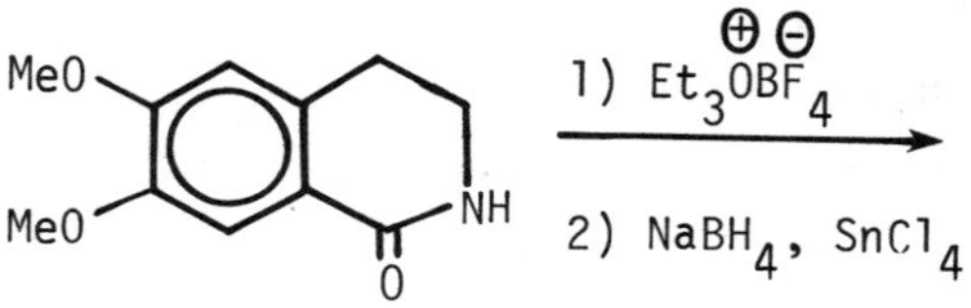

80%

Synthesis, 652 (1977)

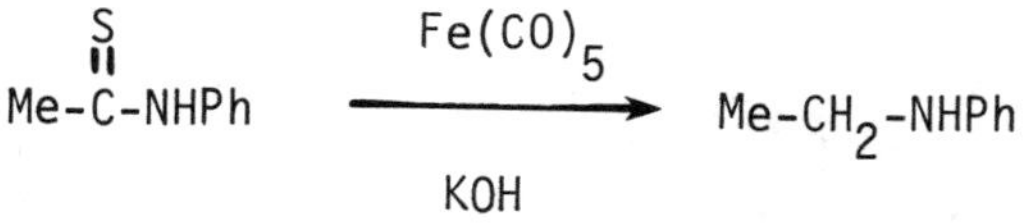

51%

JOC 42, 3522 (1977)

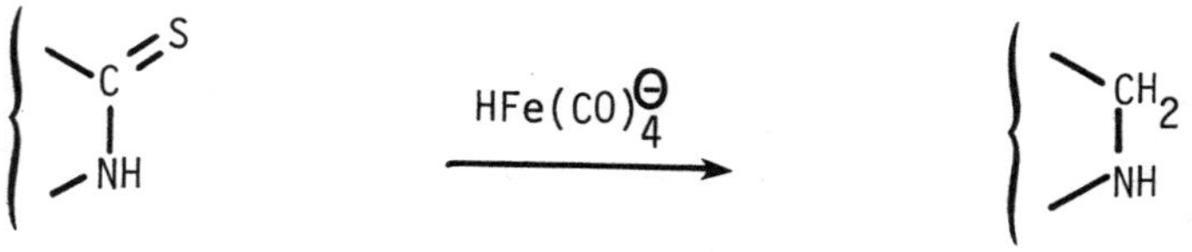

JOC 42, 3522 (1977)

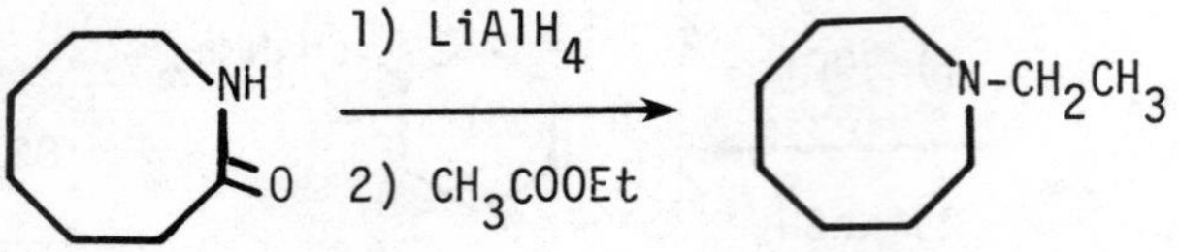

72%

Tetr Lett, 3395 (1979)

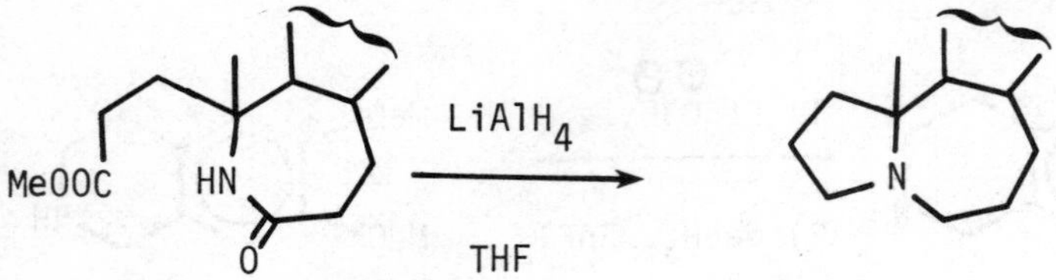

91%

Tetr Lett, 1077 (1978)

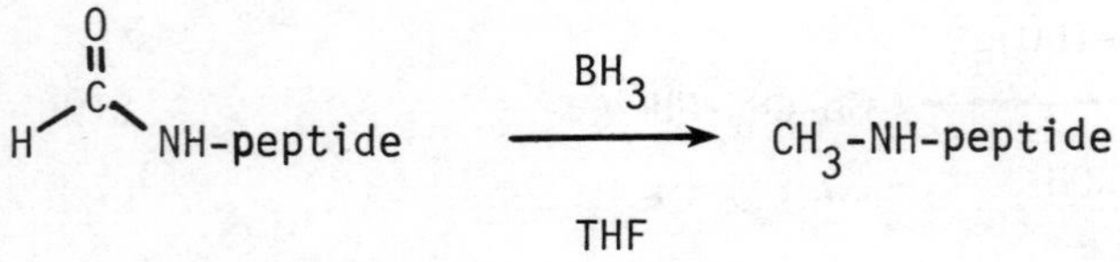

38-53%

JOC $\underline{42}$, 4148 (1977)

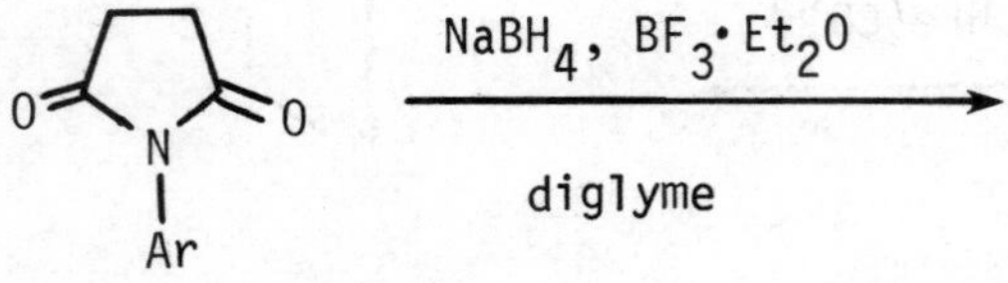

~80%

Liebigs Ann Chem, 461 (1979)

HC(OMe)$_2$ $\xrightarrow[\text{2) H}_2\text{O, NH}_4\text{Cl}]{\text{1) 2 EtMgBr}}$ HCEt$_2$
| |
NBu$_2$ NBu$_2$

Synthesis, 757 (1978)

  Me   1) BuBr, NaOH   Me
  | Bu$_4$NHSO$_4^{\ominus}$ |
PhCH-NHCHO $\xrightarrow{\hspace{2cm}}$ PhCH-NH$_2$Bu   91%
                  2) H$_2$SO$_4$ $\overset{\oplus}{}$ Cl$^{\ominus}$
                  3) NaOH
                  4) HCl gas

Synthesis, 549 (1979)

$$\text{cyclohexyl-C(=O)-NH}_2 \xrightarrow[\text{H}_2\text{O}]{\text{C}_6\text{H}_5\text{I(OCOCF}_3)_2} \text{cyclohexyl-NH}_3^{\oplus}$$

96%

JOC <u>44</u>, 1746 (1979)

Related methods: Protection of Amines (Section 105A)

Section 97   Amines from Amines

$$\text{(piperidine)} \xrightarrow[\text{($\underline{t}$-BuO)}_3\text{Al}]{\text{$\underline{i}$-PrOH, Ra-Ni}} \text{(N-isopropyl piperidine)} \qquad 94\%$$

Synthesis, 722 (1977)

$$2C_6H_{13}\text{-}NH_2 \xrightarrow[\text{xylene, reflux}]{\text{Raney-Ni}} (C_6H_{13})_2NH \qquad 75\%$$

Synthesis, 70 (1979)

$$\text{steroid-}NH_2 \xrightarrow[\text{2) NaCNBH}_3]{\text{1) CH}_2\text{O, MeOH}} \text{steroid-}NMe_2 \qquad {\sim}90\%$$

Tetr Lett, 3469 (1977)

$$\text{Ph-}NH_2 \xrightarrow[\substack{\text{2) BuBr, NaOH} \\ \text{Bu}_4\overset{\oplus}{N}H\overset{\ominus}{S}O_4 \\ \text{3) HCl}}]{\substack{\text{1) (EtO)}_2\overset{\text{O}}{\overset{\|}{P}}H}} \text{Ph-NH-Bu} \qquad 83\%$$

Angew Int Ed 16, 107 (1977)

J Prakt Chem 321, 680 (1979)

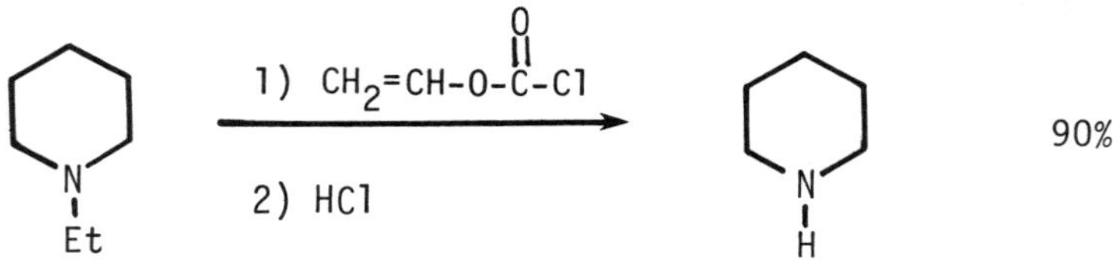

Tetr Lett, 1567 (1977)

Section 98   Amines from Esters

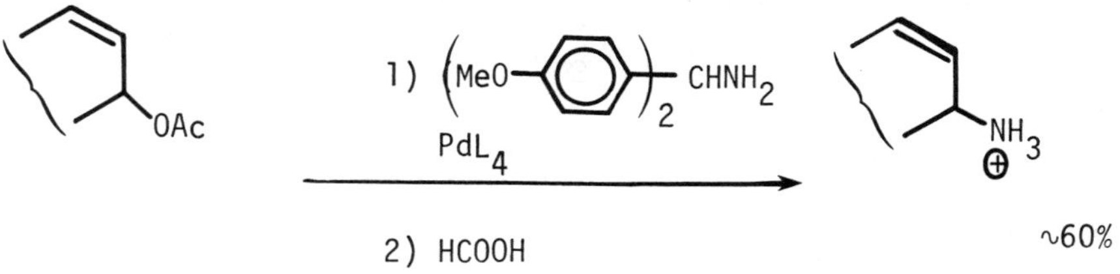

JOC 44, 3451 (1979)

Section 99 Amines from Ethers

41%

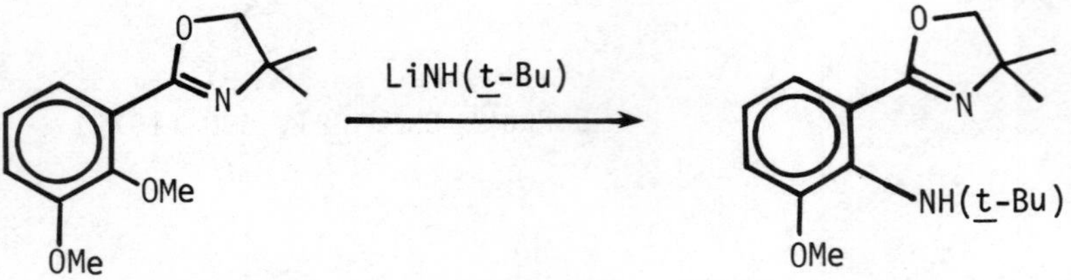

JOC 42, 2653 (1977)

Section 100 Amines from Halides

2 BuBr —(H_2N-CN, NaOH, H_2O, Aliquat 336)→ Bu_2N-CN (80%) —(hydrolysis)→ Bu_2NH

Synthesis, 882 (1978)

~50%

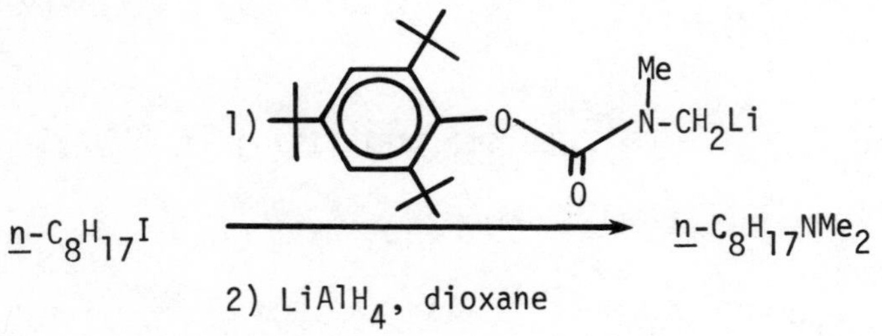

Angew Int Ed 17, 274 (1978)

$$
\text{BuLi} \xrightarrow[\hspace{3cm}]{\overset{\oplus}{\text{Me}_2\text{N}=\text{CH}_2} \quad \overset{\ominus}{\text{OCOCF}_3}} \text{BuCH}_2\text{NMe}_2 \qquad 72\%
$$

Synth Comm <u>6</u>, 539 (1976)

1) PhCH$_2$NHOH

2) [pyridinium with N-Me, F substituent] $\overset{\ominus}{\text{OTs}}$

$$
\text{PhOCH}_2\text{CH}_2\text{Br} \xrightarrow{\hspace{3cm}} \text{PhOCH}_2\text{CH}_2\text{NH}_2 \qquad 60\%
$$

3) HCl

Chem Lett, 1057 (1978)

$$
\overset{\text{O}}{\underset{}{\overset{\|}{\text{Ph}_2\text{P}}}}-\text{N}=\text{CH}-\text{OEt} \xrightarrow[\substack{2)\ \text{aq. NH}_4\text{Cl} \\ 3)\ \text{HCl/THF}}]{1)\ 3\ \text{BuMgBr}} \text{Bu}_2\text{CHNH}_3^{\oplus} \qquad 80\%
$$

Synthesis, 691 (1979)

$$
\overset{\text{O}}{\overset{\|}{\text{Ph}_2\text{P}}}-\text{NH}_2 \xrightarrow[\text{phase-Trans.}]{\text{EtBr, NaOH}} \overset{\text{O}}{\overset{\|}{\text{Ph}_2\text{PNHEt}}} \xrightarrow[\text{phase-trans.}]{\underline{n}\text{-PrBr, NaOH}} \overset{\text{O}}{\overset{\|}{\text{Ph}_2\text{PNEtPr}}}
$$

Ph$_2$PNHEt $\Big\downarrow$ HCl / THF → Et-NH$_2$ 76%

Ph$_2$PNEtPr $\Big\downarrow$ HCl / THF → Et-NH-Pr   88%

Angew Int Ed <u>16</u>, 702 (1977)

Section 101   <u>Amines from Hydrides</u>

No additional examples

Section 102   <u>Amines from Ketones</u>

$$\underset{Et}{\overset{Ph}{\diagdown}}C=NNHTs \xrightarrow{BH_3 \cdot pyridine} \underset{Et}{\overset{Ph}{\diagdown}}CH-NHNHTs \qquad 95\%$$

Synth Comm <u>9</u>, 49 (1979)

$$\begin{array}{c} Ph-CH_2NH_2 \\ + \\ CH_3-\underset{\underset{O}{\parallel}}{C}-CH_3 \end{array} \quad \xrightarrow[\text{2)}\ \ 50^{\circ}]{\text{1) }NaBH_4,\ AcOH} \quad Ph-CH_2-\overset{Et}{\underset{}{N}}-CHMe_2 \qquad 82\%$$

Synthesis, 766 (1978)

$$\underset{OH}{\overset{NNHTs}{\underset{}{Ph-CH-C}}}\diagdown Ph \xrightarrow[H^{\oplus}/THF]{NaBH_3CN} \underset{OH}{\overset{NHNHTs}{\underset{}{Ph-CH-CHPh}}} \qquad 85\%$$

Synthesis, 789 (1979)

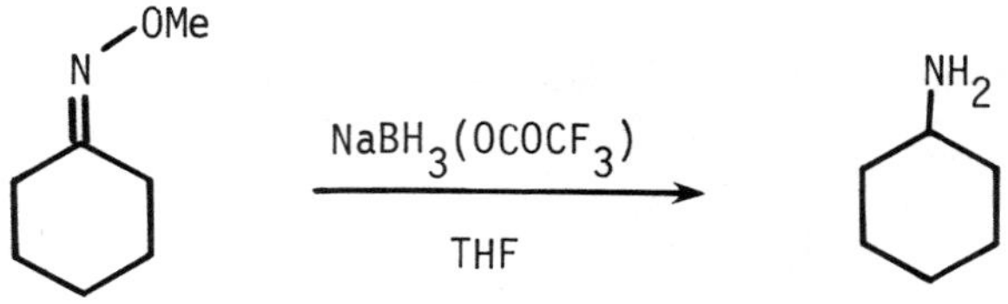

Chem Pharm <u>26</u>, 2897 (1978)

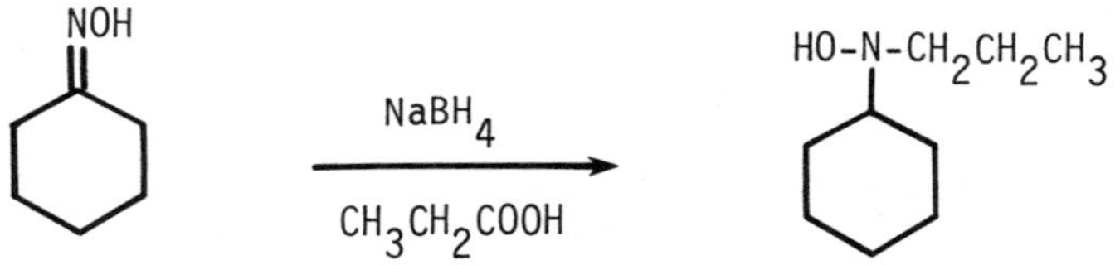

Synthesis, 856 (1977)

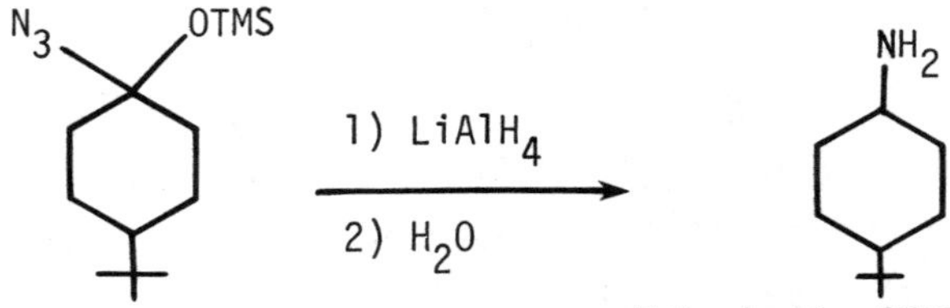

Tetr Lett, 2737 (1977)

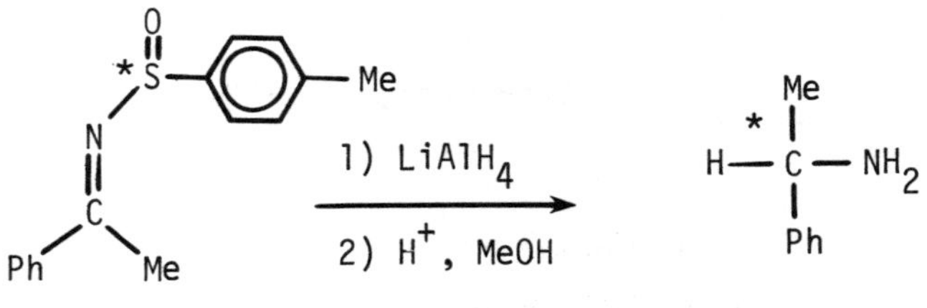

78%ee

JCS Chem Comm, 723 (1977)

Related methods: Amines from Aldehydes (Section 94)

Section 103   Amines from Nitriles

$$Ph-CN \xrightarrow{BH_3OH^{\ominus}} Ph-CH_2NH_2 \qquad\qquad 89\%$$

JOC 42, 3963 (1977)

$$Me(CH_2)_4CN \xrightarrow[THF]{H_2, [RhH(\underline{i}-Pr_3P)_3]} Me(CH_2)_4CH_2NH_2 \qquad 100\%$$

JCS Chem Comm, 870 (1979)

1) MeMgI

2) ⟋⟍MgBr

3) H₂O

75%

Tetr Lett, 23 (1977)

1) BuLi

2) BzBr

3) H₃O⊕

61%

Liebigs Ann Chem, 40 (1977)

Section 104   Amines from Olefins

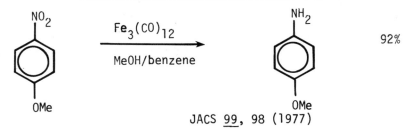

$$C_8H_{17}-CH=CH_2 \xrightarrow[\begin{array}{l}3)\ MCPBA \\ 4)\ KBH_4\end{array}]{\begin{array}{l}1)\ PdCl_2(PhCN)_2 \\ 2)\ Me_2NH\end{array}} C_8H_{17}-\underset{\underset{CH_2NMe_2}{|}}{CH-NMe_2} \qquad 81\%$$

Tetr Lett, 163 (1978)

51%

J Prakt Chem 320, 413 (1978)

$$C_8H_{17}-CH=CH_2 \xrightarrow[2)\ Br_2]{1)\ MeNH_2,\ PdCl_2(PhCN)_2} \qquad 43\%$$

JCS Chem Comm, 413 (1977)

Section 105   Amines from Miscellaneous Compounds

$$\xrightarrow[MeOH/benzene]{Fe_3(CO)_{12}} \qquad 92\%$$

JACS 99, 98 (1977)

$$Ph\text{-}NO_2 \xrightarrow[\text{glyme, } Et_3N]{H_2O,\ CO,\ Fe(CO)_5} Ph\text{-}NH_2 \qquad 100\%$$

JACS 100, 3969 (1978)

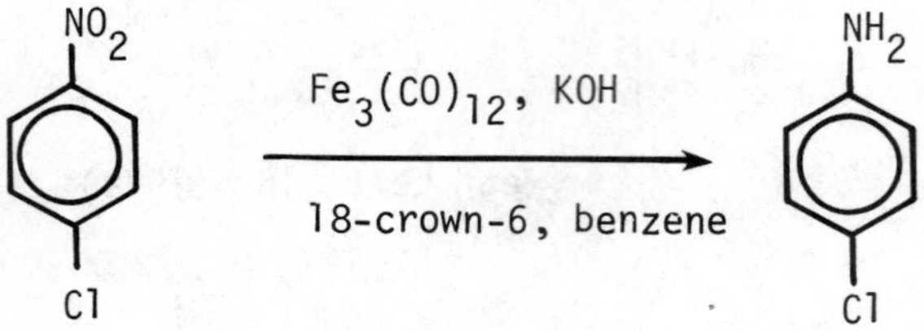
NO2
Cl
Fe3(CO)12, KOH
18-crown-6, benzene
NH2
Cl

79%

Angew Int Ed 16, 41 (1977)

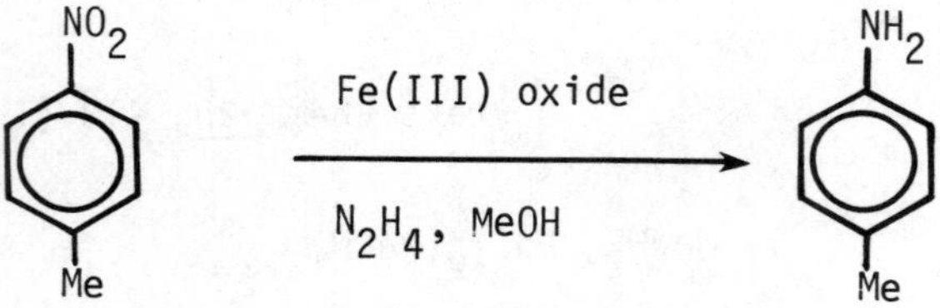
NO2
Me
Fe(III) oxide
N2H4, MeOH
NH2
Me

92%

Synthesis, 834 (1978)

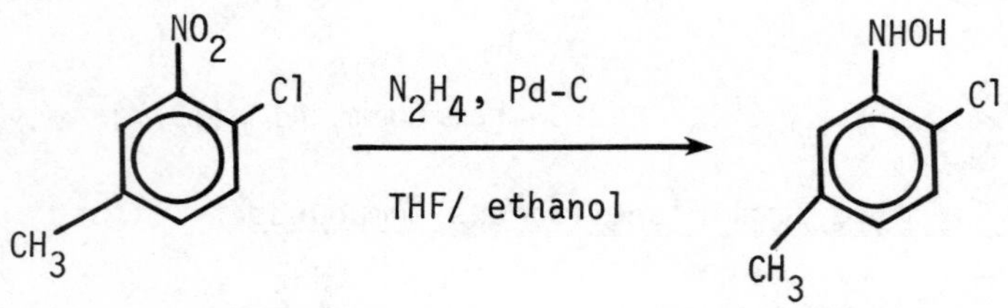
NO2
Cl
CH3
N2H4, Pd-C
THF/ ethanol
NHOH
Cl
CH3

100%

Synthesis, 850 (1977)

$$Ph-NO_2 \xrightarrow[\text{Pt/C}]{Et_3N \cdot HCOOH} Ph-NH_2 \qquad 100\%$$

JOC <u>42</u>, 3491 (1977)

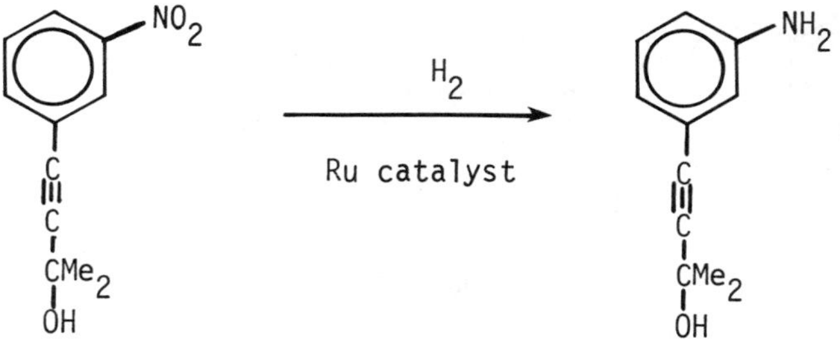

JOC <u>44</u>, 1233 (1979)

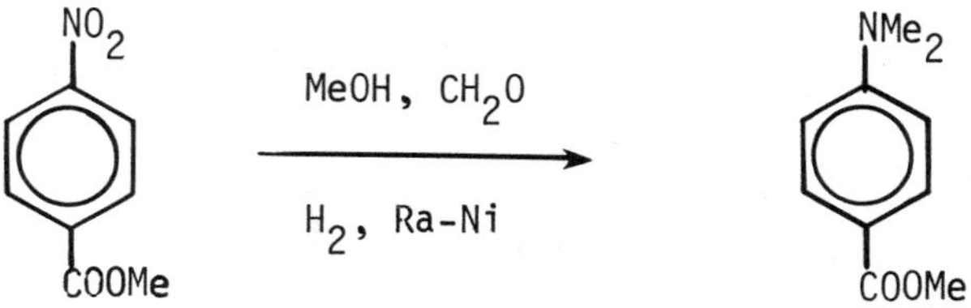

Indian J Chem <u>14B</u>, 904 (1976)

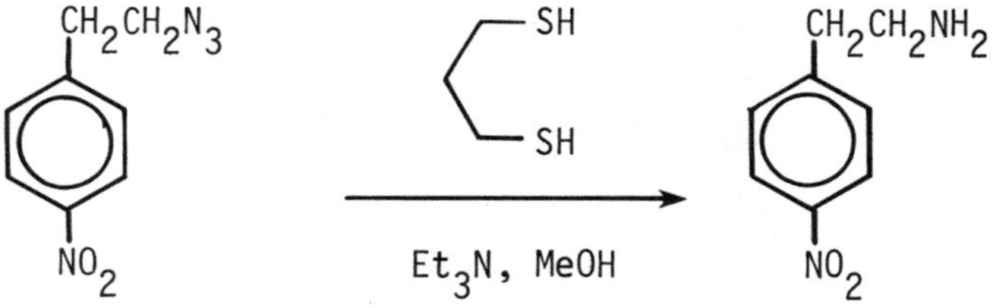

Tetr Lett, 3633 (1978)

$$Ph-NO_2 \xrightarrow[CH_3OH]{CrCl_2} Ph-NH_2 \qquad 97\%$$

Synthesis, 792 (1977)

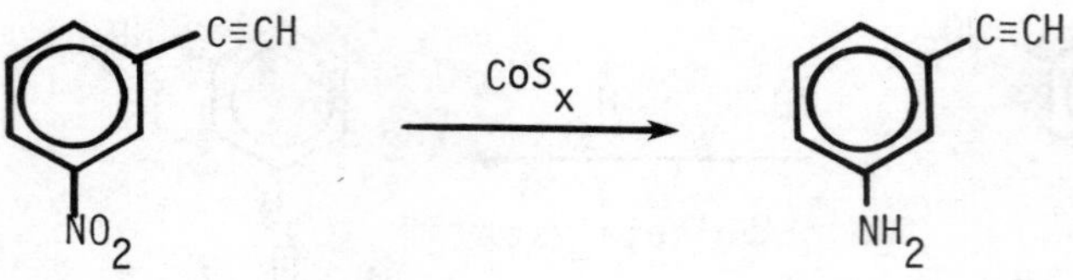

JOC **44**, 3671 (1979)

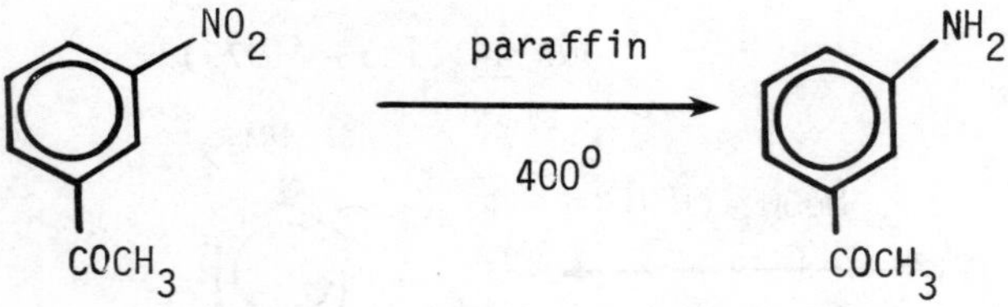

Synthesis, 23 (1978)

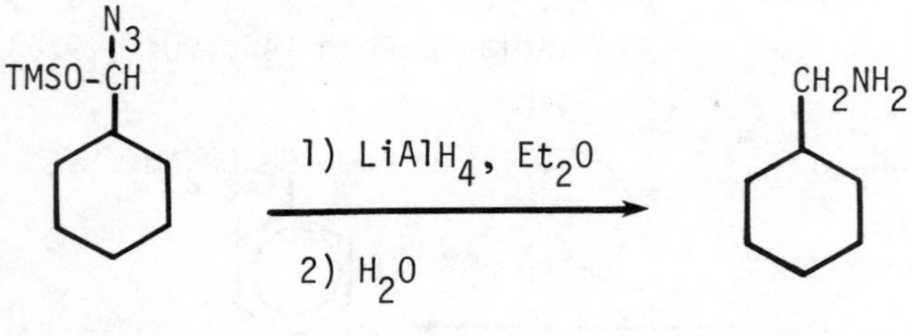

Tetr Lett, 2737 (1977)

Review: "o-Mesitylenesulfonylhydroxylamine and Related Compounds--Powerful Aminating Reagents"

Synthesis, 1 (1977)

Review: "General Methods of Alkaloid Synthesis"

Accounts Chem Res <u>10</u>, 193 (1977)

Section 105A   <u>Protection of Amines</u>

Related methods: Amides from Amines (Section 82); Amines from Amides (Section 96)

$$H_2N-\underset{\underset{CH_2CHMe_2}{|}}{CH}-COO-\underline{t}-Bu \quad \xrightarrow[DCC]{HCOOH} \quad H-\underset{O}{\overset{O}{\overset{||}{C}}}-NH-\underset{\underset{CH_2CHMe_2}{|}}{CH}-COO-\underline{t}-Bu \qquad 87\%$$

JOC <u>42</u>, 2019 (1977)

$$H_2N-\underset{\underset{R}{|}}{CH}-COOR' \quad \xrightarrow[O\diagdown NMe/CH_2Cl_2]{[H-\overset{O}{\overset{||}{C}}-O-\overset{O}{\overset{||}{C}}-H]\ in\ situ} \quad H-\overset{O}{\overset{||}{C}}-\underset{\underset{H}{|}}{N}-\underset{\underset{R}{|}}{CH}-COOR'$$

Synthesis, 709 (1979)

$$H_2N(CH_2)_8NH_2 \quad \xrightarrow[\substack{2)\ PhCOCl \\ 3)\ TFA}]{1)\ \text{(P)}-\!\!\text{⬡}\!\!-CH_2O\overset{O}{\overset{||}{C}}-\text{⬡}-NO_2} \quad H_2N(CH_2)_8NH\overset{O}{\overset{||}{C}}Ph \qquad 81\%$$

Israel J Chem <u>17</u>, 248 (1979)

$$\underset{\underset{\text{H}_2\text{N-CH-COOH}}{|}}{\overset{R}{}} \quad \xrightarrow[\text{Et}_3\text{N, MeOH}]{\overset{\overset{O}{\|}}{\text{EtO-C-CF}_3}} \quad \underset{\underset{\text{F}_3\text{C-C-NH-CH-COOH}}{}}{\overset{O \quad R}{\overset{\|}{} \quad |}} \qquad \sim 80\text{-}90\%$$

JOC <u>44</u>, 2805 (1979)

Use of the p-hydroxybenzyloxycarbonyl protecting group in peptide synthesis. Removed by H_2O_2/NH_3.

Tetrahedron <u>34</u>, 3105 (1978)

$$\text{R-NH}_2 \quad \underset{\xleftarrow{\text{Zn/HOAc}}}{\xrightarrow{\overset{\overset{O}{\|}}{\text{Cl}_3\text{C-C(Me}_2)\text{-O-C-Cl}}}} \quad \text{R-NH-}\overset{\overset{O}{\|}}{\text{C}}\text{-O-C(Me}_2)\text{CCl}_3$$

Angew Int Ed <u>17</u>, 361 (1978)

Use of $\underline{t}$-BuO-$\overset{\overset{O}{\|}}{\text{C}}$-N=C$\overset{\text{Ph}}{\underset{\text{COOEt}}{}}$ as a $\underline{t}$-butyloxycarbonylating reagent.

BCS Japan <u>50</u>, 718 (1977)

$$\text{Z-peptide} \quad \xrightarrow[\text{Pd/C}]{} \quad \text{H}_2\text{N-peptide}$$

JCS Perkin I, 490 (1977)

Boc- Gly-OBzl $\xrightarrow[\text{2) } H_2O]{\text{1) } Me_3SiSiMe_3/I_2}$ glycine   100%

Angew Int Ed 18, 612 (1979)

Boc-Trp-OH $\xrightarrow[\text{dioxane/anisole}]{\text{TsOH}}$ H_2N-Trp-OH   100%

Chem Pharm Bull 26, 2198 (1978)

Boc and Z(OMe) protecting groups are removed by ethanesulfonic acid in acetic acid or methylene chloride. Z, benzyl ester, S-p-methoxybenzyl, and N^G_p methoxybenzenesulfonyl groups are unaffected.

Chem Pharm Bull 25, 740 (1977)

1, 4-cyclohexadiene/Pd-C removes N-benzyloxycarbonyl protecting groups.

JOC 43, 4194 (1978)

JOC $\underline{42}$, 399 (1977)

JCS Chem Comm, 450 (1978)

$$O_2N-\langle\bigcirc\rangle-O-\overset{\overset{\displaystyle O}{\|}}{C}-O-CH_2-\langle\bigcirc\rangle N$$

$$R-NH_2 \qquad\qquad R-NH-\overset{\overset{\displaystyle O}{\|}}{C}-OCH_2-\langle\bigcirc\rangle N$$

H$_2$/cat.

or Zn/HCl

JOC **42**, 3286 (1977)

$$Me_3SiCH_2CH_2O\overset{\overset{\displaystyle O}{\|}}{C}-Cl$$

$$R-NH_2 \qquad\qquad R-NH\overset{\overset{\displaystyle O}{\|}}{C}OCH_2CH_2TMS$$

Et$_4$NF, MeCN

JCS Chem Comm, 358 (1978)

Peoc group: $R_3\overset{\oplus}{P}CH_2CH_2-O-\overset{\overset{\displaystyle O}{\|}}{C}-$

Enhances the water-solubility of protected amino acids and peptides; extremely resistant to acids, including TFA. Removed by weak bases.

Angew Int Ed **17**, 67 (1978)

Use of the 1-methylcyclobutyloxycarbonyl protecting group in peptide synthesis. More stable than t-Boc toward HOAc.

$$-\overset{\displaystyle}{\underset{\displaystyle H}{N}}-\overset{\overset{\displaystyle O}{\|}}{C}-O-\overset{\displaystyle Me}{\bigangle}\square$$

JOC **42**, 143 (1977)

$\underline{Adpoc}$ = $\overset{\overset{\text{O}}{\|}}{-\text{C}}$-O-C(Me$_2$)-1-adamantyl

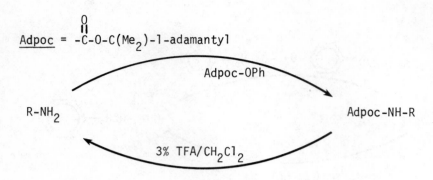

R-NH$_2$ Adpoc-NH-R

Stable toward hydrogenolysis.

Angew Int Ed <u>17</u>, 944 (1978)

$\underset{\sim}{I}$ is more stable to TFA than is the $\underline{t}$-Boc group.

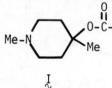

$\underset{\sim}{I}$

Cleaved by HBr/HOAc

JCS Perkin I, 1459 (1979)

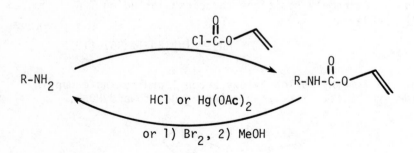

R-NH$_2$ R-NH-C-O

Tetr Lett, 1563 (1977)

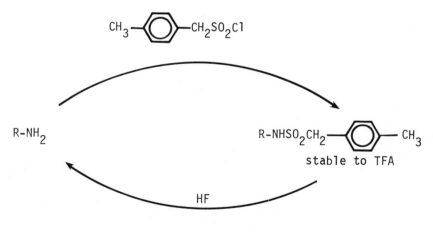

JCS Perkin I, 627 (1977)

JCS Chem Comm, 220 (1978)

$$R_2NSO_2Bz \xrightarrow[\underline{i}\text{-PrOH}]{h\nu} R_2NH$$

Tetr Lett, 1029 (1978)

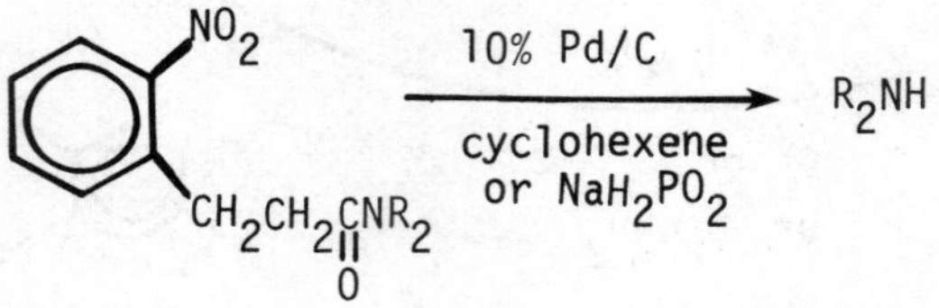

Tetr Lett, 555 (1979)

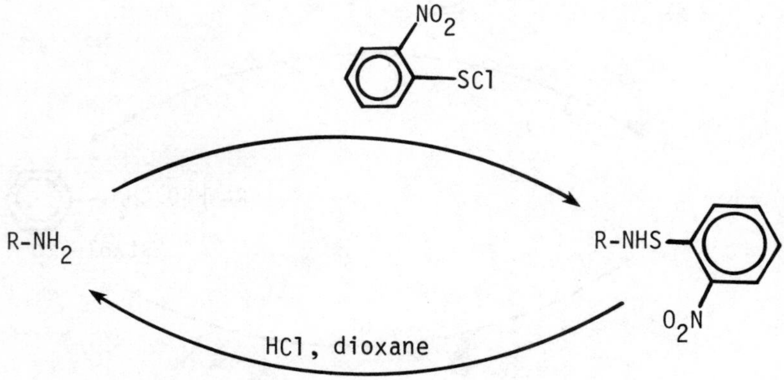

Chem Pharm Bull <u>26</u>, 296 (1978)

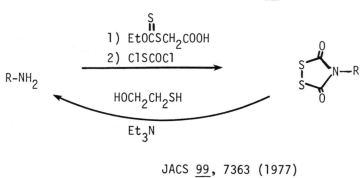

Chem Lett, 7 (1978)

R = peptide

AZOC, BPOC, BOC, t-butyl ester, MSOC, and Tos are unaffected by the reaction conditions.

Helv Chim Acta 61, 1086 (1978)

JACS 99, 7363 (1977)

Tetr Lett, 325 (1979)

Chem Pharm Bull 26, 660 (1978)

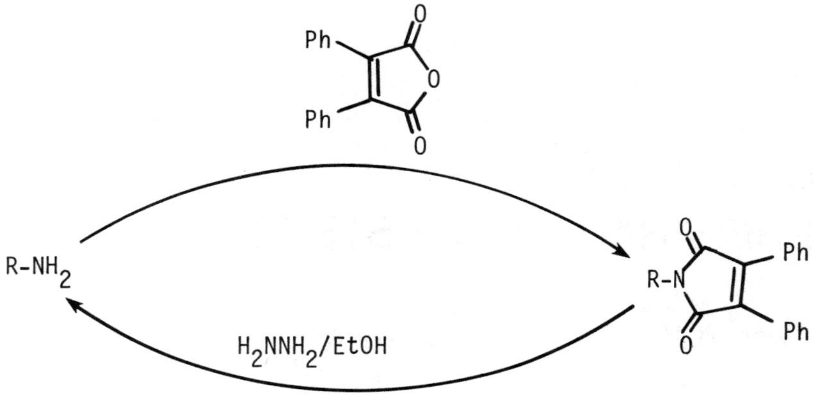

Yellow, fluorescent derivatives.

JOC 42, 2819 (1977)

Review: "Protecting Groups in Peptide Synthesis"

Chem & Ind, 617 (1979)

CHAPTER 8
PREPARATION OF ESTERS

Section 106   Esters from Acetylenes

No additional examples

Section 107   Esters from Carboxylic Acids and Acid Halides

The following types of reactions are found in this section:
1. Esters from carboxylic acids (and acid halides) and alcohols.
2. Lactones from hydroxy acids.
3. Esters from carboxylic acids and halides, sulfonates, and
miscellaneous compounds.

$$CH_3CH_2I$$

anion-exchange resin   94%

JOC 44, 2425 (1979)

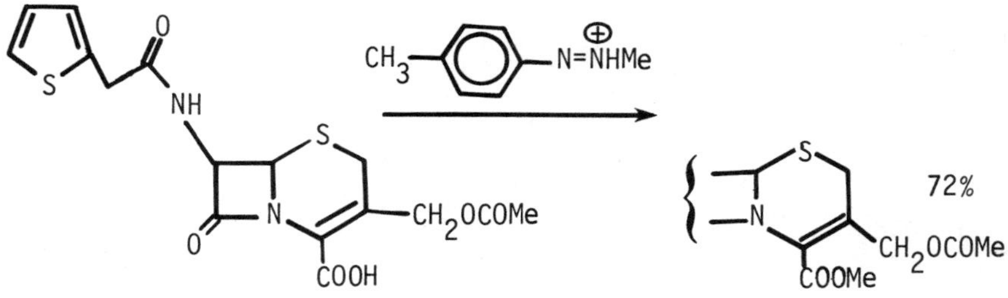

Tetr Lett, 5219 (1978)

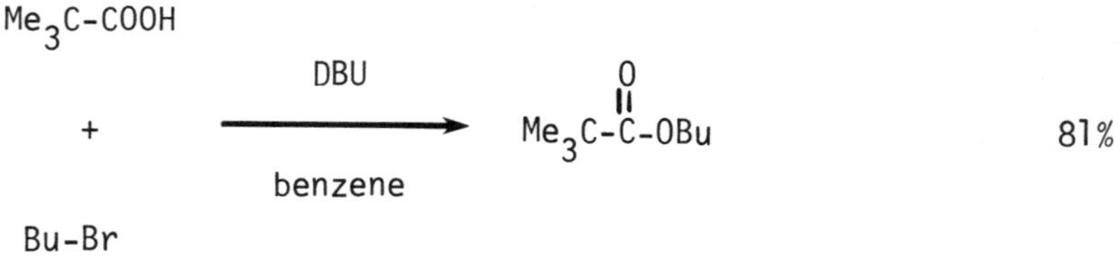

BCS Japan 51, 2401 (1978)

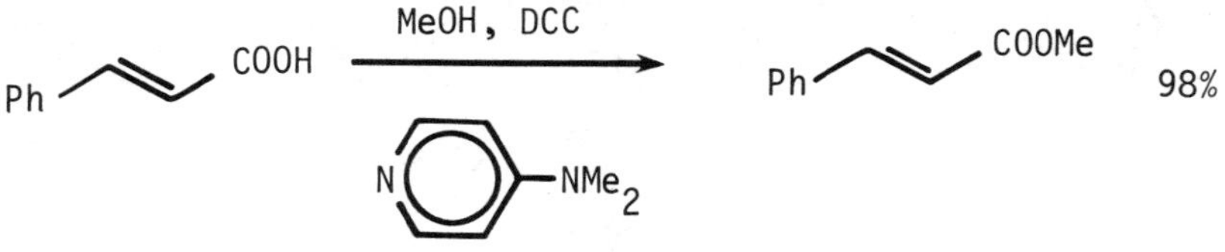

Angew Int Ed 17, 522 (1978)

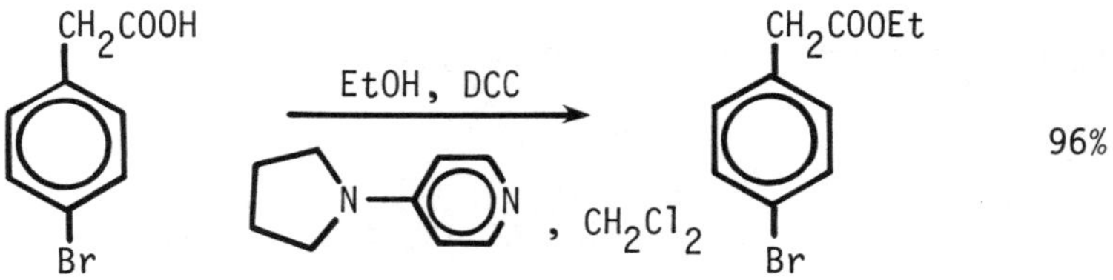

Tetr Lett, 4475 (1978)

Ph-CH$_2$CH$_2$COOH

+

i-PrOH

$\xrightarrow[\text{pyridine}]{\text{DCC, TsOH}}$

$$\text{Ph-CH}_2\text{CH}_2\overset{\overset{\text{O}}{\|}}{\text{C}}\text{-O-}\underline{i}\text{-Pr}$$ 99%

Acta Chem Scand B, 33, 410 (1979)

lauric acid

+

cholesterol

$\xrightarrow[\text{Me}_2\text{N}\text{—}\langle\text{N}\rangle]{\text{DCC}}$

cholesteryl laurate

67%

Synth Comm 9, 539 (1979)

85%

Synthesis, 429 (1979)

Ph-COOH $\xrightarrow{\text{P(OMe)}_5}$ $\text{Ph-}\overset{\overset{\text{O}}{\|}}{\text{C}}\text{-OMe}$ 90%

JOC 43, 4672 (1978)

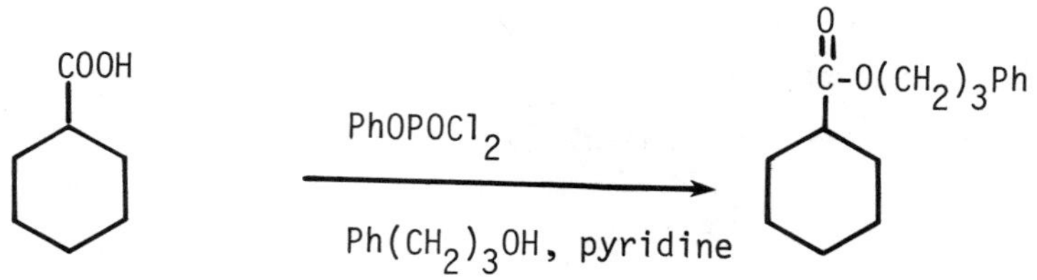

Tetr Lett, 4461 (1978)

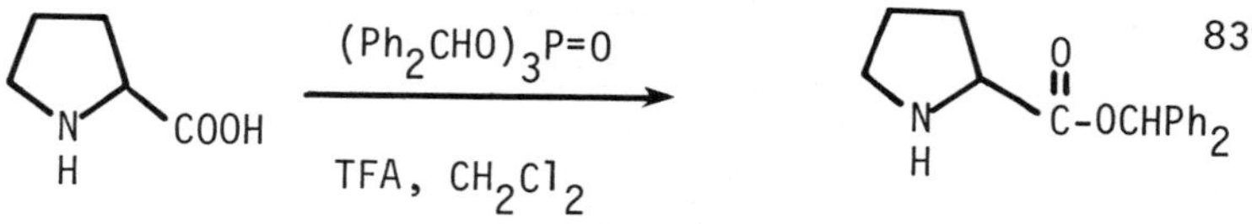

Tetr Lett, 4697 (1978)

Z-Gly-Tyr-OH $\xrightarrow[\substack{(PhO)_3PBu \\ | \\ Br}]{EtOH, \text{ pyridine}}$ Z-Gly-Tyr-OEt   95%

Synthesis, 355 (1979)

R-COOH $\xrightarrow{Me_3OBF_4}$ R-$\overset{\overset{\text{O}}{\|}}{C}$-OMe   ~90%

R = alkyl, aryl, vinyl

JOC <u>44</u>, 1149 (1979)

$$\text{R-COOH} \xrightarrow{\text{Me}_3\text{SOH}} \text{R-COOMe} \qquad 100\%$$

JOC <u>44</u>, 638 (1979)

$$\underset{\text{R-C-OH}}{\overset{O}{\|}} \xrightarrow[\text{2. EtOH}]{\text{1. Ph}_3\text{P,}} \underset{\text{R-C-OEt}}{\overset{O}{\|}} \quad \sim 70\text{-}80\%$$

R = cephalosporinic acid.

Chem Lett, 979 (1978)

99%

BzEt$_3$NCl

Chem Lett, 441 and 763 (1977)

Chem Lett, 885 (1978)

PhCH$_2$CH$_2$COOH $\xrightarrow[\text{2) EtOH, Bu}_3\text{N}]{\text{1)}}$ PhCH$_2$CH$_2$C-OEt   73%

BCS Japan <u>50</u>, 1863 (1977)

Me$_2$CHCOOH $\xrightarrow[\text{2) \underline{t}-BuOH}]{\text{1)}}$ Me$_2$CH-C-O-<u>t</u>-Bu   91%

Chem Lett, 145 (1979)

H$_3$N-CH-COO$^{\ominus}$ $\xrightarrow[\text{reflux}]{\text{TsOH, EtOH}}$ TsONH$_3$-CH-COOEt

Bull Chem Soc Japan <u>52</u>, 1879 (1979)

Boc-Pro-Pro-OH $\xrightarrow[\text{2) BzBr}]{\text{1) Cs}_2\text{CO}_3}$ Boc-Pro-Pro-OBz   96%

JOC <u>42</u>, 1286 (1977)

$$Me_3C\text{-}COOH \xrightarrow[\text{KF, DMF}]{\overset{\displaystyle \overset{O}{\parallel}}{Br\text{-}CH_2\text{-}C\text{-}Ph}} Me_3C\text{-}\overset{O}{\overset{\parallel}{C}}\text{-}OCH_2\overset{O}{\overset{\parallel}{C}}Ph \qquad 95\%$$

Tetr Lett, 599 (1977)

$$Boc\text{-}Trp\text{-}OH \xrightarrow[\text{CH}_2\text{Cl}_2, \text{ PhOH}]{\text{BOP, Et}_3\text{N}} Boc\text{-}Trp\text{-}OPh \qquad 96\%$$

BOP =

Synthesis, 413 (1977)

Further examples of the reaction RCOOH + ROH → RCOOR are included in Section 108 (Esters from Alcohols and Phenols) and Section 10A (Protection of Carboxylic Acids).

Section 108   Esters from Alcohols and Phenols

Tetrahedron 34, 2069 (1978)

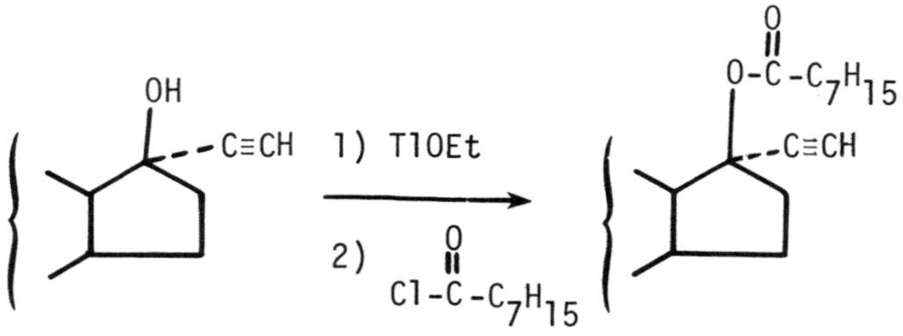

70%

Synth Comm 7, 383 (1977)

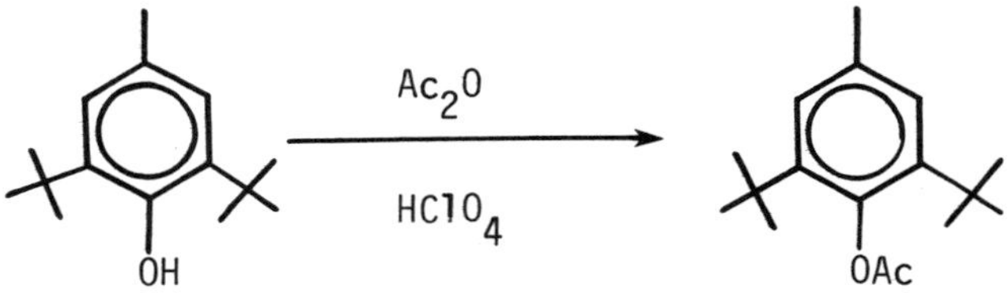

85%

JOC (USSR) 13, 608 (1977)

$H_3\overset{\oplus}{N}(CH_2)_4OH \; Cl^{\ominus} \xrightarrow[\text{MeCN/benzene}]{\text{PhCOCl}} H_3\overset{\oplus}{N}(CH_2)_4O\text{-}\overset{O}{\overset{\|}{C}}\text{-Ph} \; Cl^{\ominus}$ 58%

Synth Comm 8, 327 (1978)

$BuOH \xrightarrow{\text{Ph-}\overset{O}{\overset{\|}{C}}\text{-O-(salicylic acid polymer)}} \text{Ph-}\overset{O}{\overset{\|}{C}}\text{-OBu}$ 77%

Synth Comm 7, 57 (1977)

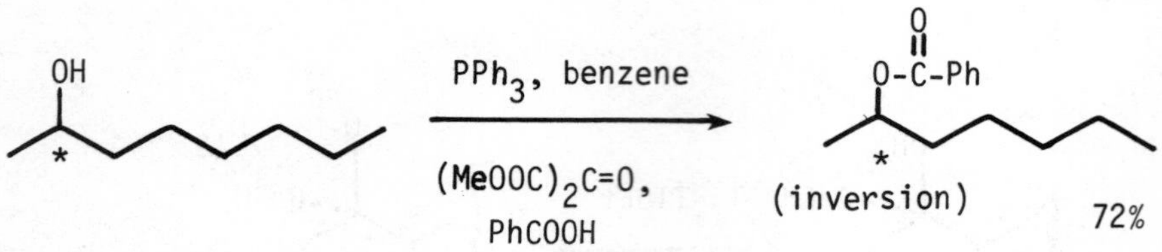

PPh$_3$, benzene

(MeOOC)$_2$C=O,
PhCOOH

(inversion)

72%

Tetr Lett, 3179 (1977)

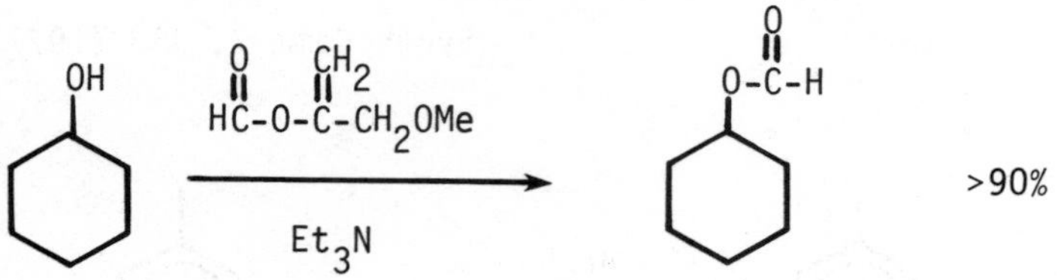

>90%

Rec Trav Chim 98, 324 (1979)

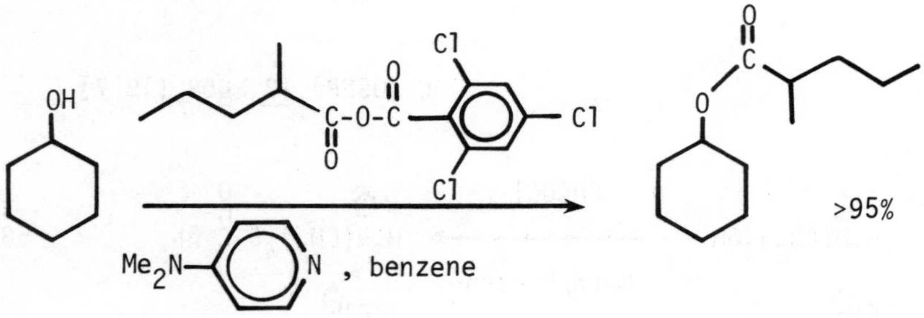

Me$_2$N— N , benzene

>95%

Bull Chem Soc Japan 52, 1989 (1979)

AcIm = N-acetylimidazole

JACS 99, 3531 (1977)

$Ph-\overset{O}{\underset{||}{C}}-CN$ acylates steroid alcohols selectively in the order
21>17β>3β>6α, 3α>>20α>20β>7β>>7α>6β>22β>22α

Coll Czech Chem Commun 44, 2443 (1979)

Review: "4-Dialkylaminopyridines as Highly Active Acylation
Catalysts"

Angew Int Ed 17, 569 (1978)

$$C_5H_{11}CH_2OH \xrightarrow[\text{H}_2\text{O, } \underline{t}\text{-BuOH}]{\text{KI, electrolysis}} C_5H_{11}\overset{O}{\underset{||}{C}}\text{-O-}\underline{t}\text{-Bu} \qquad 83\%$$

Tetr Lett, 165 (1979)

Further examples of the reaction ROH → R'COOR are included in
Section 107 (Esters from Carboxylic Acids and Acid Halides) and
Section 45A (Protection of Alcohols and Phenols).

Section 109   Esters from Aldehydes

$$\text{cyclohexyl-CHO} \xrightarrow[\text{MeOH/THF}]{O_3, \text{ MeLi}} \text{cyclohexyl-C(=O)-OMe}$$

63%

Tetr Lett, 1627 (1978)

$$CH_3(CH_2)_8CHO \xrightarrow{\text{NBS, } h\nu} CH_3(CH_2)_8\overset{O}{\overset{\|}{C}}\text{-OEt}$$

83%

+

EtO-TMS

JOC 43, 371 (1978)

$$MeO-\text{C}_6\text{H}_4-CHO \xrightarrow[\text{2) EtOH, HCl}]{\text{1) } \overset{O}{\overset{\|}{CH_2SCH_3}}\text{, base}} $$

~85%

Bull Chem Soc Japan 52, 2013 (1979)

Me$_2$C=C(OMe)$_2$

+

Ph-CHO

$$\xrightarrow[\text{2) H}_3\text{O}^{\oplus}]{\text{1) ZnCl}_2}$$

Ph-CH-CMe$_2$
  | |
  HO   COOMe

80%

Synthesis, 400 (1978)

$$\xrightarrow{\text{CrO}_3,\ \text{pyr}\cdot\text{HCl}}$$

90%

Tetr Lett, 3483 (1977)

Pr-CH (dioxolane)

$$\xrightarrow[\text{MeOH, KOH}]{\text{anodic oxidation}}$$

Pr   OMe

72%

Synthesis, 283 (1978)

$\underline{\text{i}}$-Bu-CHO

$$\xrightarrow[\text{ZnCl}_2]{\overset{\overset{\text{O}}{\|}}{\text{CH}_3\text{-C-Br}}}$$

$\underline{\text{i}}$-Bu-CH-OAc
         |
         Br

99%

Synthesis, 593 (1978)

1) $Et_3SiOP(NMe_2)_2$
2) BuLi

3)

JACS 101, 371 (1979)

65%

Related methods: Esters from Ketones (Section 117)

Section 110   Esters from Alkyls, Methylenes and Aryls

No examples of the reaction RR → RCOOR' or R'COOR (R,R'=alkyl, aryl, etc.) occur in the literature. For the reaction RH → RCOOR' or R'COOR see Section 116 (Esters from Hydrides).

Section 111   Esters from Amides

1) MeI, MeOH

2) aq.K_2CO_3

100%

JACS 101, 1316 (1979)

Section 112   Esters from Amines

Ph-CH$_2$CH$_2$NH$_2$

+ $\xrightarrow{\text{i-AmONO}}$ CH$_3$-$\overset{\overset{\text{O}}{\|}}{\text{C}}$-OCH$_2CH_2$Ph   84%

CH$_3$COOH

Synth Comm <u>8</u>, 33 (1978)

BuCH$_2$NH$_2$ $\xrightarrow[\text{2) PhCOONa}]{\text{1)}}$ Bu-O-$\overset{\overset{\text{O}}{\|}}{\text{C}}$-Ph   85%

JCS Chem Comm, 701 (1977)

Section 113   Esters from Esters

Conjugate reductions and conjugate alkylations of unsaturated
esters are found in Section 74 (Alkyls from Olefins).

Ph-CH$_2$-COOEt $\xrightarrow[\text{2) MeI}]{\text{1) C}_8\text{K}}$ Ph-CH-COOEt   70%
                                              |
                                              Me

Tetr Lett, 653 (1977)

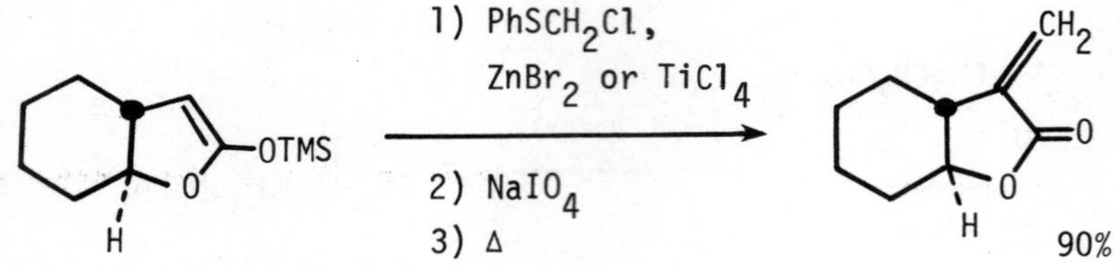

1) PhSCH$_2$Cl,
   ZnBr$_2$ or TiCl$_4$

2) NaIO$_4$
3) Δ

90%

Tetr Lett, 993 and 995 (1979)

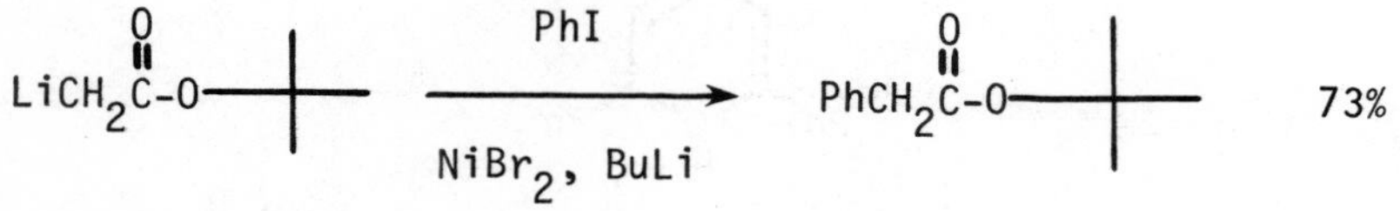

$$\underset{\text{LiCH}_2\overset{\displaystyle O}{\overset{\|}{C}}-O}{} \quad \xrightarrow[\text{NiBr}_2,\ \text{BuLi}]{\text{PhI}} \quad \underset{\text{PhCH}_2\overset{\displaystyle O}{\overset{\|}{C}}-O}{} \qquad 73\%$$

JACS 99, 4833 (1977)

boric acid

180°

92%

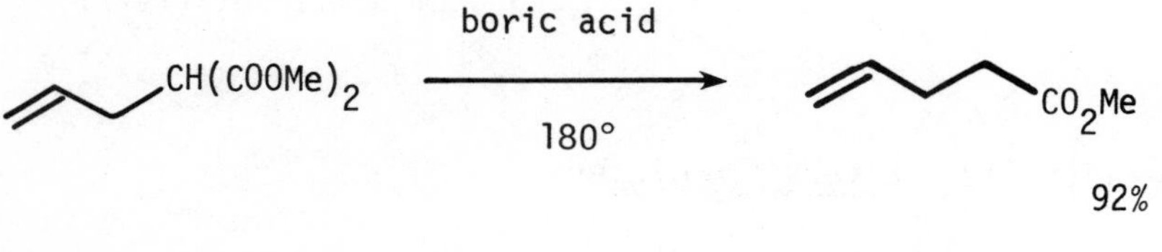

Synth Comm 9, 609 (1979)

Angew Chem Int Ed <u>18</u>, 793 (1979)

Section 114   Esters from Ethers

$$R-OBz \xrightarrow{Ph-I(OCOCF_3)_2} R-O-\overset{\overset{\displaystyle O}{\|}}{C}-CF_3 \qquad 42-100\%$$

R = <u>n</u>-alkyl, Ph, Bz

JCS Chem Comm, 615 (1979)

$$\underline{n}\text{-Bu-O-CH}_2\text{Ph} \xrightarrow[\text{CH}_2\text{Cl}_2]{\overset{\oplus}{\text{BzEt}_3\text{NMnO}_4}\overset{\ominus}{}} \text{Bu-O-}\overset{\overset{\displaystyle O}{\|}}{\text{C}}\text{-Ph} \qquad 90\%$$

Angew Int Ed <u>18</u>, 68 and 69 (1979)

Section 115   Esters from Halides and Sulfonates

$$\text{I-(CH}_2)_{15}\text{-COOH} \xrightarrow[\text{DMF}]{\text{Cs}_2\text{CO}_3} \qquad 85\%$$

JCS Chem Comm, 286 (1979)

$$\text{Bu-MgBr} \xrightarrow[\text{2) BzOH, I}_2]{\text{1) Fe(CO)}_5} \underset{\underset{\text{O}}{\overset{\text{O}}{\|}}}{\text{Bu-C-OBz}} \qquad 73\%$$

Tetr Lett, 1477 (1978)

Me$_2$C-COOMe

70%

Tetr Lett, 1455 (1978)

Related methods: Carboxylic Acids from Halides (Section 25)

Section 116   <u>Esters from Hydrides</u>

This section contains examples of the reaction RH → RCOOR' or R'COOR (R=alkyl, aryl, etc.).

$$\xrightarrow[\text{TFA, TFAA}]{\text{AgO}}$$

59%

Synth Comm <u>6</u>, 543 (1976)

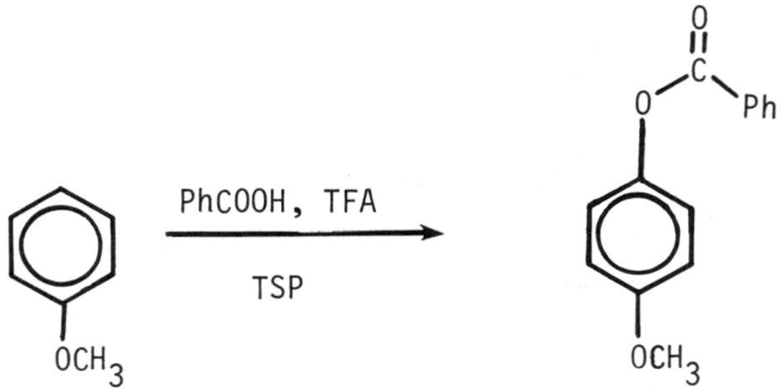

(TSP = 2-trifluoromethanesulfonyloxypyridine)

Chem Lett, 1099 (1977)

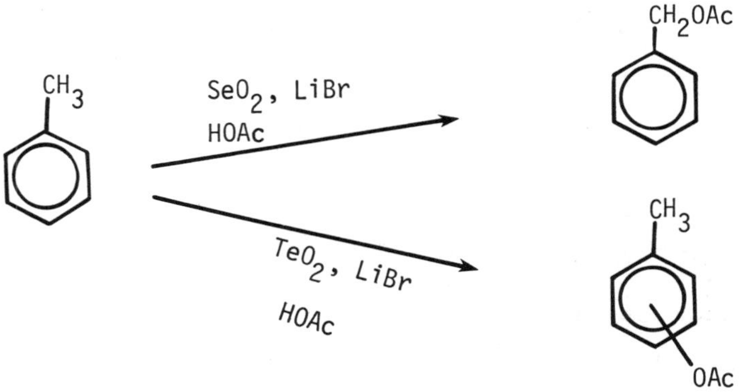

Tetr Lett, 3279 (1978)

Also via: Carboxylic acids, Section 26; Alcohols, Section 41

Section 117   Esters from Ketones

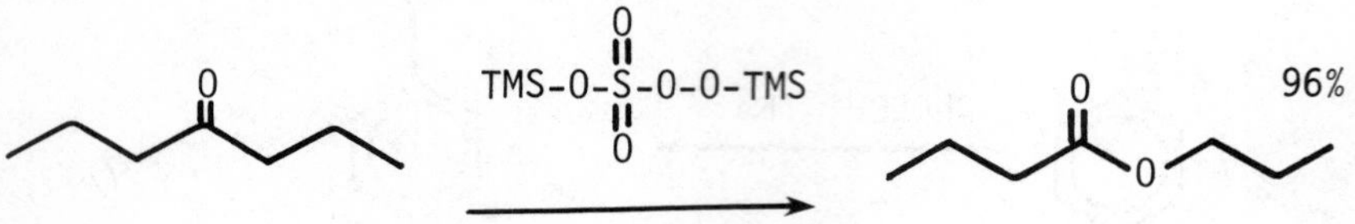

$$TMS-O-S-O-O-TMS$$

96%

JOC 44, 4969 (1979)

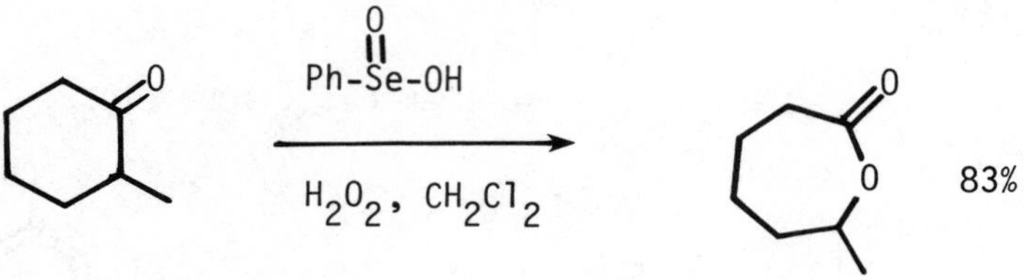

$$Ph-Se-OH$$

$$H_2O_2, CH_2Cl_2$$

83%

JCS Chem Comm, 870 (1977)

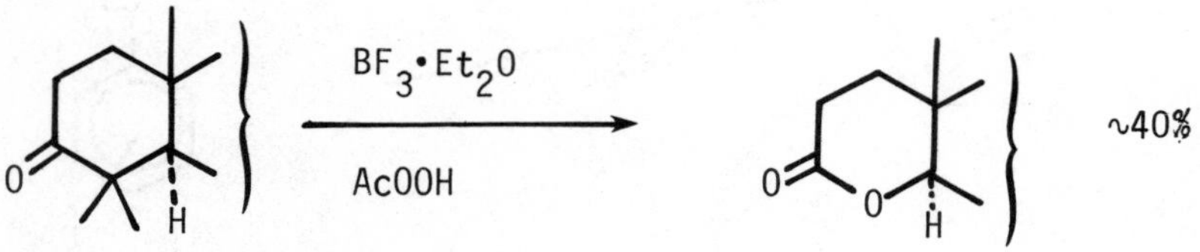

$$BF_3 \cdot Et_2O$$

AcOOH

~40%

Acta Chem Scand (B) 32, 467 (1978)

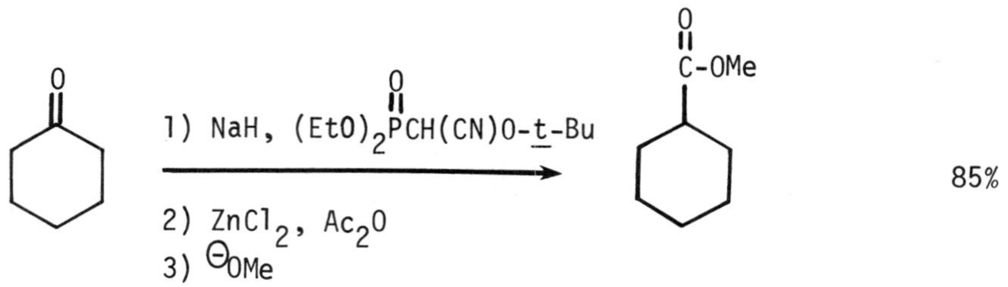

1) NaH, $(EtO)_2\overset{O}{\overset{\|}{P}}CH(CN)O-\underline{t}-Bu$

2) $ZnCl_2$, Ac_2O
3) $^{\ominus}OMe$

85%

JACS <u>99</u>, 182 (1977)

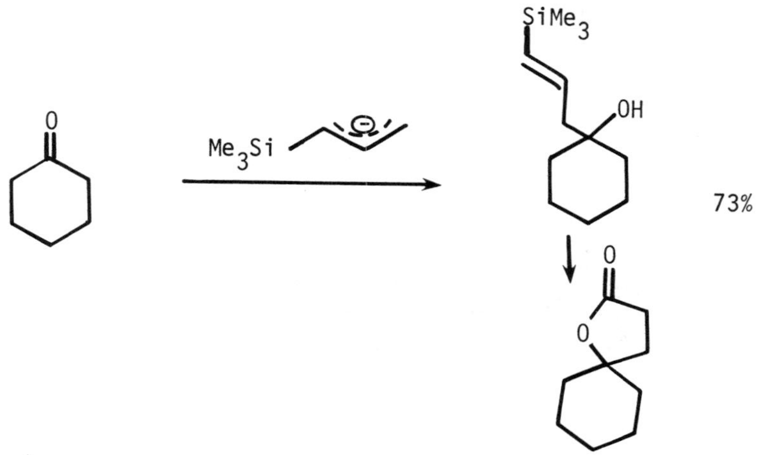

Me_3Si

73%

JCS Chem Comm, 772 (1977)

Also via Carboxylic acids, Section 27

Section 118   <u>Esters from Nitriles</u>

No additional examples

Section 119   Esters from Olefins

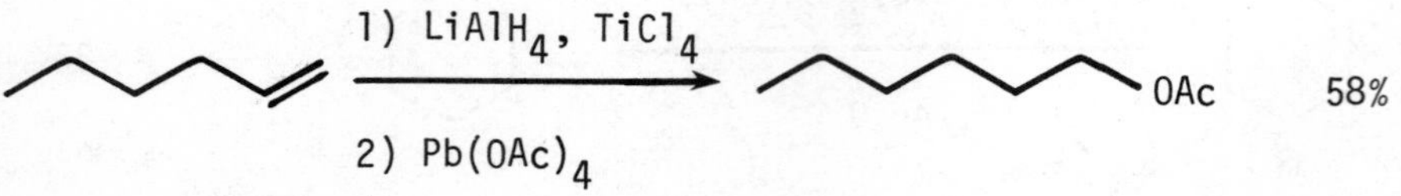

Tetr Lett, 1405 (1979)

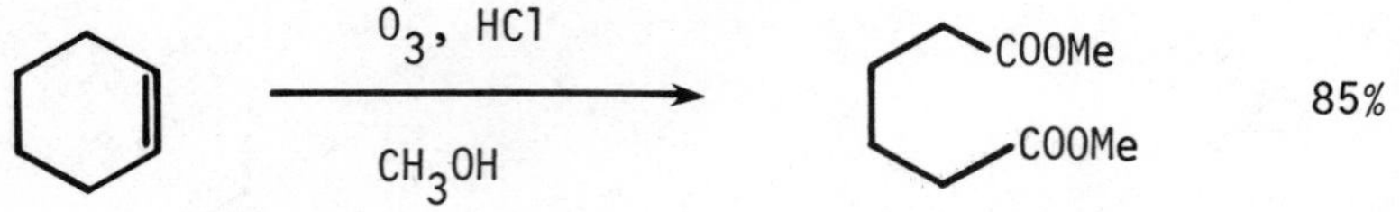

Angew Int Ed 17, 939 (1978)

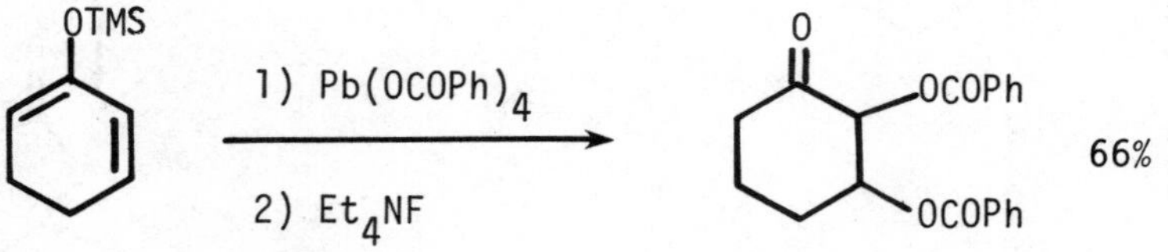

JOC 42, 1051 (1977)

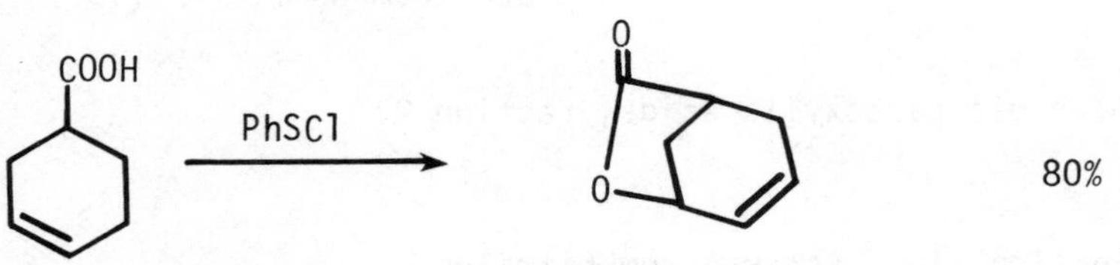

JACS 101, 3884 (1979)

Also via Alcohols, Section 44

Section 120   Esters from Miscellaneous Compounds

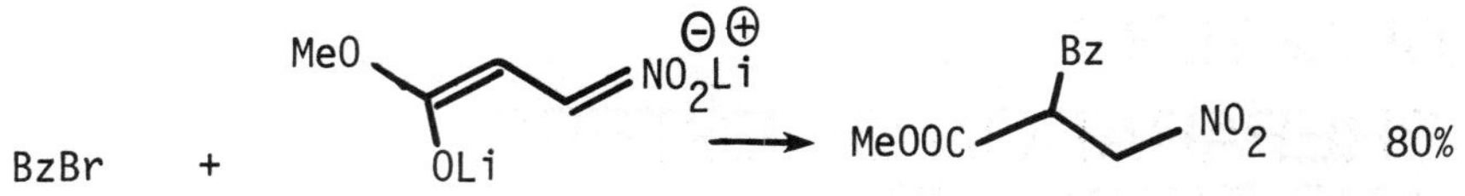

BzBr   +   80%

Tetr Lett, 1161 (1977)

Section 120A   Protection of Esters

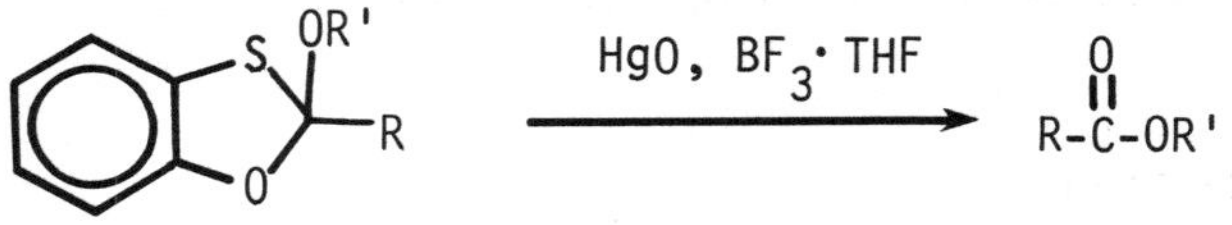

Synthesis, 223 (1979)

CHAPTER 9
PREPARATION OF ~~ESTERS~~ *ETHERS*
AND EPOXIDES

Section 121   Ethers and Epoxides from Acetylenes

No examples

Section 122   Ethers and Epoxides from Carboxylic Acids

No additional examples

Section 123   Ethers and Epoxides from Alcohols and Phenols

$$R-OH \xrightarrow[\text{silica gel}]{CH_2N_2} R-OCH_3 \qquad 25\text{-}98\%$$

Tetr Lett, 4405 (1979)

$$Ph_2CHOH \xrightarrow[\text{2) MeI}]{\text{1) KOH, DMSO}} Ph_2CHOMe \qquad 97\%$$

Tetrahedron **35**, 2169 (1979)

$$H\text{-}(OCH_2CH_2)_4\text{-}OH \xrightarrow[\text{NaOH, } H_2O]{CH_3Cl, \text{ benzene}} H_3C\text{-}(OCH_2CH_2)_4\text{-}OCH_3 \qquad 95\%$$

Synthesis, 123 (1979)

$$\underline{n}\text{-}C_6H_{13}\text{-}OH$$

$$+ \xrightarrow{Ni(acac)_2} \underline{n}\text{-}C_6H_{13}\text{-}O\text{-}Bz \qquad 81\%$$

$$Bz\text{-}Cl$$

Synthesis, 803 (1977)

$$\text{(allyl)}OH \xrightarrow[\text{KOH/DMSO}]{Et_2SO_4} \text{(allyl)}OEt \qquad 70\%$$

Synthesis, **428** (1979)

$$\overset{\oplus\ominus}{Et_3OBF_4}$$

CH_2OH $\xrightarrow{\quad}$ CH_2OEt   89%

JOC **42**, 1801 (1977)

$$\text{Bu-OH} \xrightarrow[\text{TFA, CH}_2\text{Cl}_2]{\overset{\overset{\text{O}}{\|}}{\text{P(OCHPh}_2)_3}} \text{Bu-OCHPh}_2 \qquad 74\%$$

Tetr Lett, 3943 (1978)

$$\text{R-OH} \xrightarrow[\text{Et}_3\text{N, DMF}]{\text{Me}_2\text{N}\!-\!\!\langle\bigcirc\rangle\!\!-\!\text{N, Ph}_3\text{CCl}} \text{R-O-CPh}_3$$

Tetr Lett, 95 and 99 (1979)

$$\text{Ph-OH} \xrightarrow{\text{P(OMe)}_5} \text{Ph-OMe} \qquad 90\%$$

JOC $\underline{43}$, 4672 (1978)

$$\text{Ar-OH} \xrightarrow{\overset{\oplus\ \ominus}{\text{Me}_3\text{SOH}}} \text{Ar-OMe} \qquad 100\%$$

JOC $\underline{44}$, 638 (1979)

$$\text{Ph-OH} \xleftarrow{\overset{\oplus\ \ominus}{\text{Me}_3\text{SeOH}}} \text{Ph-OMe} \qquad 100\%$$

Tetr Lett, 1787 (1979)

Ph-OH $\xrightarrow[\text{KF/Celite}]{\text{Bu-I}}$ Ph-O-Bu   83%

Chem Lett, 45 (1979)

PhOH + t-BuBr $\xrightarrow[30°]{\text{pyridine}}$ Ph-O-t-Bu   90%

Chem Lett, 57 (1978)

Ph-OH
+ $\xrightarrow[\text{DMF}]{\text{Et}_4\text{NF}}$ Ph-O-Et   75%
EtI

Can J Chem 57, 1887 (1979)

Related methods: Protection of Alcohols (Section 45A).

95%

100%

Tetr Lett, 5153 (1978)

CH₃CHOH

2 [benzene ring]

DMSO
───────→
175°

$CH_3-\overset{\overset{H}{|}}{C}-O-\overset{\overset{H}{|}}{C}-CH_3$

[two benzene rings]

99%

JOC 42, 2012 (1977)

$(CH_3)_3C-OH$

1. EtMgBr
─────────→
2. [tetrahydropyran with O-N-succinimide substituent]

[tetrahydropyran]
$O-C(CH_3)_3$

95%

Chem Lett 817 (1977)

[decalin structure, HO]

H₂, MeOH
─────────────→
Pt(II)-Sn(II)
complex catalyst

[decalin structure, OMe]

~100%

Chem Lett, 835 (1978)

$$Ph-\underset{\underset{OH}{|}}{CH}-\underset{\underset{OH}{|}}{CH}-C\equiv CH \quad \xrightarrow[\text{glyme}]{\text{TsCl, NaOH}} \quad Ph-CH\overset{O}{\overline{}}CH-C\equiv CH \qquad 80\%$$

Synthesis, 706 (1977)

Related methods: Protection of Alcohols and Phenols (Section 45A)

Section 124   Ethers and Epoxides from Aldehydes

$$\underset{Pr\quad H}{\overset{MeO\quad OMe}{\diagdown \underset{|}{C}\diagup}} \quad \xrightarrow[\text{Me}_3\text{SiOTf}]{\text{Me}_3\text{SiH}} \quad Pr-CH_2OMe \qquad 100\%$$

Tetr Lett, 4679 (1979)

$$CH_3(CH_2)_8CH(OMe)_2 \quad \xrightarrow[\text{HCl(g), MeOH}]{\text{NaCNBH}_3} \quad CH_3(CH_2)_8CH_2OMe \qquad 83\%$$

Tetr Lett, 1357 (1978)

$$Ph-CH_2CH_2CH(OMe)_2 \quad \xrightarrow{\text{2LiAlH}_4\text{-TiCl}_4} \quad PhCH_2CH_2CH_2OMe \quad 87\%$$

BCS Japan 51, 2059 (1978)

$$PhCHO \xrightarrow{py \cdot BH_3/CF_3COOH} PhCH_2-O-CH_2Ph \qquad 87\%$$

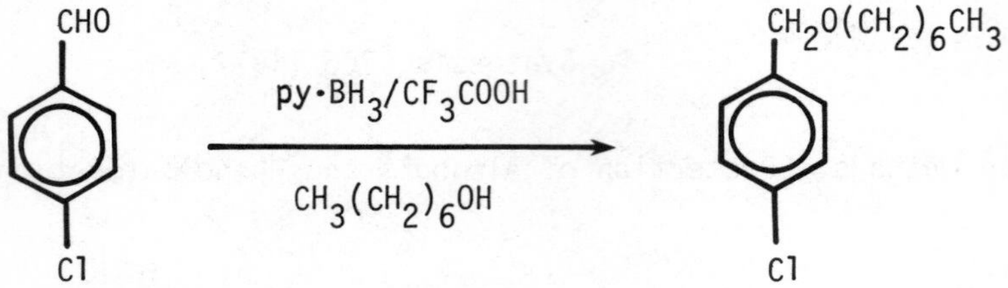

Chem Pharm Bull 27, 2405 (1979)

$$2PhCHO \xrightarrow[\text{2) } H_3O^{\oplus}]{\text{1) } py \cdot BH_3, \text{ TFA}} PhCH_2OCH_2Ph \qquad 87\%$$

Chem Lett, 415 (1979)

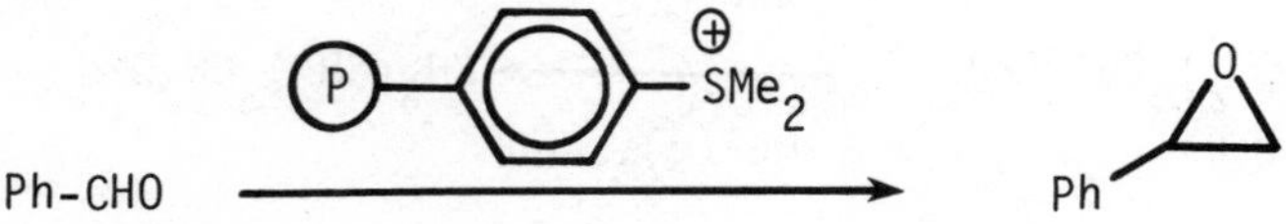

Tetr Lett, 203 (1979)

Related methods: Ethers and Epoxides from Ketones (Section 132)

Section 125   Ethers and Epoxides from Alkyls, Methylenes and Aryls

No examples of the preparation of ethers and epoxides by replace-
ment of alkyl, methylene and aryl groups occur in the literature.
For the conversion of RH → ROR' (R,R'=alkyl) see Section 131
(Ethers from Hydrides)

Section 126   Ethers and Epoxides from Amides

No additional examples

Section 127   Ethers and Epoxides from Amines

No additional examples

Section 128   Ethers and Epoxides from Esters

Chem and Ind, 230 (1977)

Section 129   Ethers and Epoxides from Ethers and Epoxides

JCS Chem Comm, 382 (1979)

Section 130   Ethers from Halides and Sulfonates

Resin (Ph-O$^{\ominus}$)

+ $\xrightarrow{\text{benzene}}$ Bu-O-Ph   100%

Bu-Br

Synthesis, 113 (1977)

Ph-OH

+ $\xrightarrow[\text{DMF}]{\text{Et}_4\text{NF}}$ Ph-O-Et   75%

EtI

Can J Chem 57, 1887 (1979)

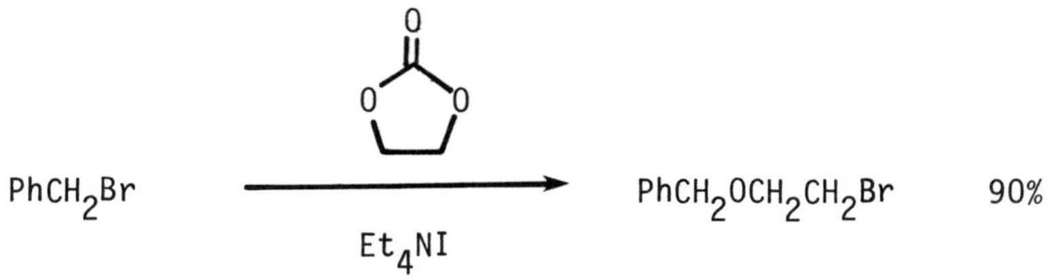

PhCH$_2$Br $\xrightarrow[\text{Et}_4\text{NI}]{}$ PhCH$_2$OCH$_2$CH$_2$Br   90%

Tetr Lett, 2639 (1979)

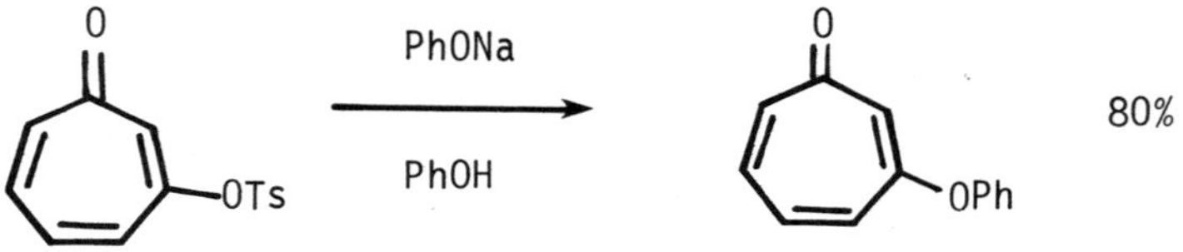

$\xrightarrow[\text{PhOH}]{\text{PhONa}}$ 80%

Synthesis, 298 (1977)

$\xrightarrow{\text{PhO}^{\ominus}, \text{DMF}}$ 85-95%

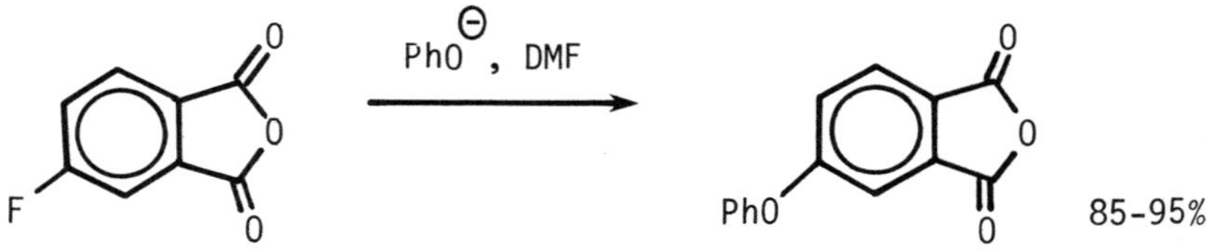

JOC $\underline{42}$, 3425 (1977)

Related methods: Ethers from Alcohols (Section 123)

Section 131   Ethers from Hydrides

No additional examples

Section 132   Ethers and Epoxides from Ketones

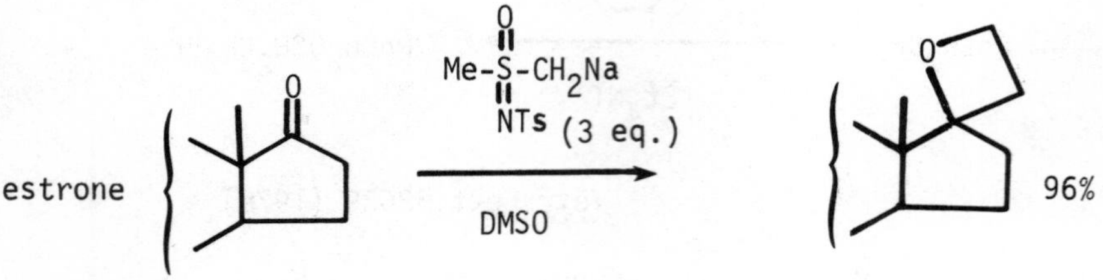

estrone

JACS **101**, 6135 (1979)

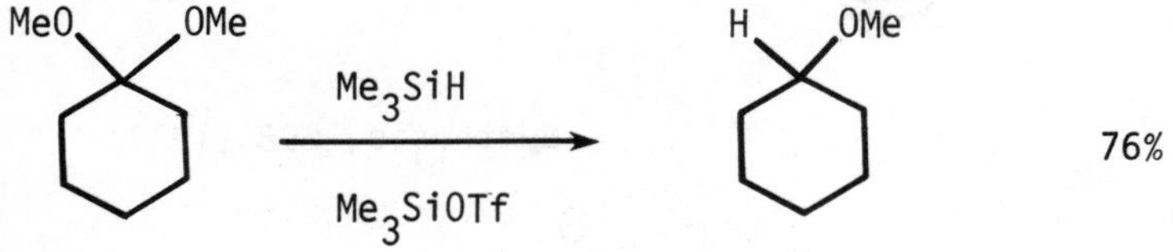

Tetr Lett, 4679 (1979)

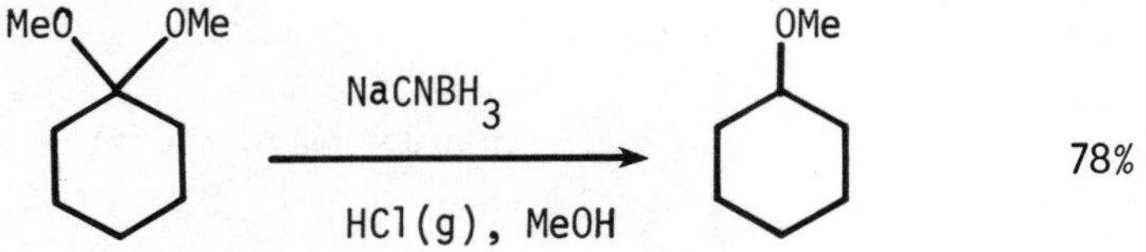

Tetr Lett, 1357 (1978)

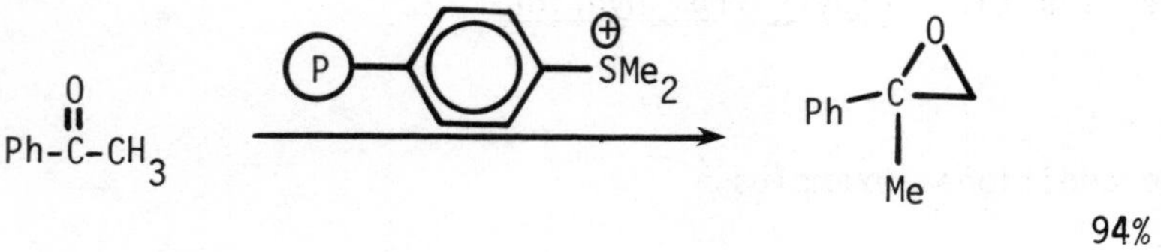

Tetr Lett, 203 (1979)

$$Ph-\overset{\overset{\displaystyle O}{\|}}{C}-C(Me_2)SMe \quad \xrightarrow[\text{2) MeI}]{\text{1) NaBH}_4}$$

Ph ⟨epoxide⟩ Me, Me   68%

JCS Chem Comm, 785 (1978)

Related methods: Ethers and Epoxides from Aldehydes (Section 124)

Section 133   Ethers and Epoxides from Nitriles

No additional examples

Section 134   Ethers and Epoxides from Olefins

Ph ⟨CH=CH⟩ Ph $\xrightarrow[\text{Fe}^{3+}(acac)_3]{\text{H}_2\text{O}_2}$ H, Ph ⟨epoxide⟩ Ph, H   68%

JCS Chem Comm, 948 (1977)

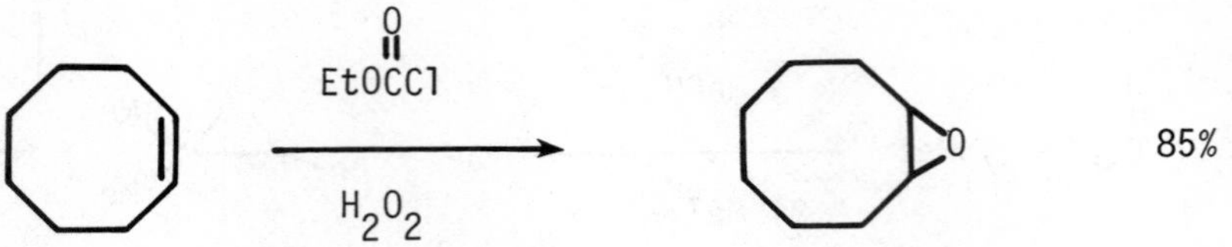

85%

JOC **44**, 2569 (1979)

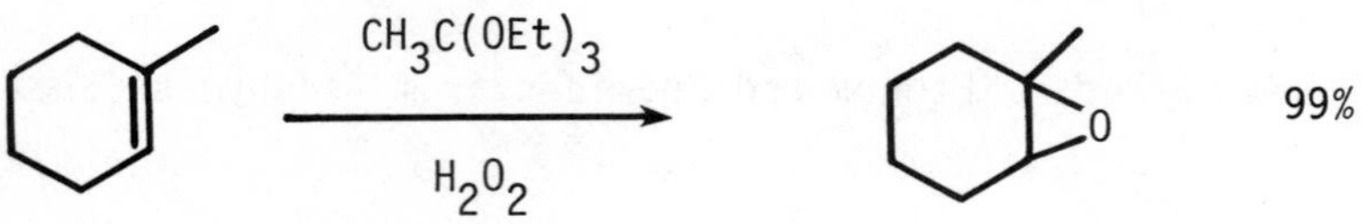

99%

Tetr Lett, 1001 (1979)

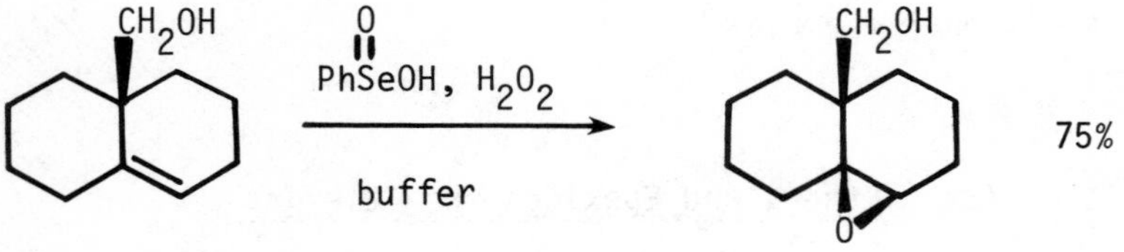

75%

JOC **42**, 2034 (1977)

91%

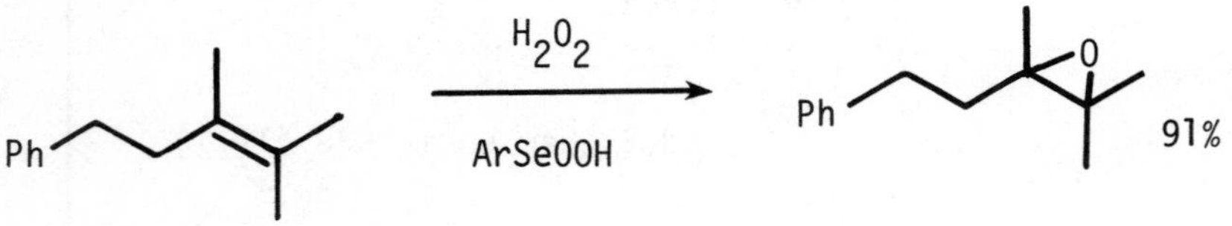

Synthesis, 299 (1978)

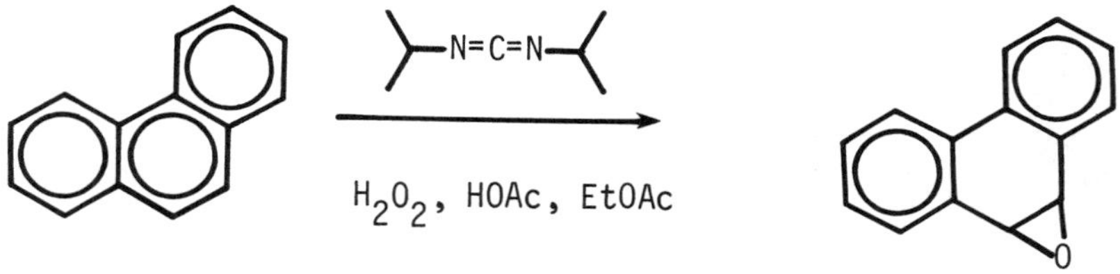

>-N=C=N-<

H_2O_2, HOAc, EtOAc

28%

Tetr Lett, 1369 (1977)

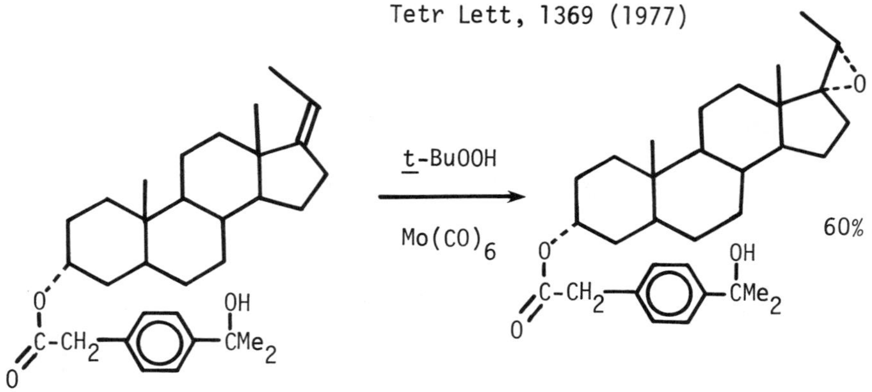

t-BuOOH

$Mo(CO)_6$

60%

Tetr Lett, 623 (1977)

t-BuOOH, trifluoroacetylcamphor

$PhCH_3$, $MoO_2(acac)_2$

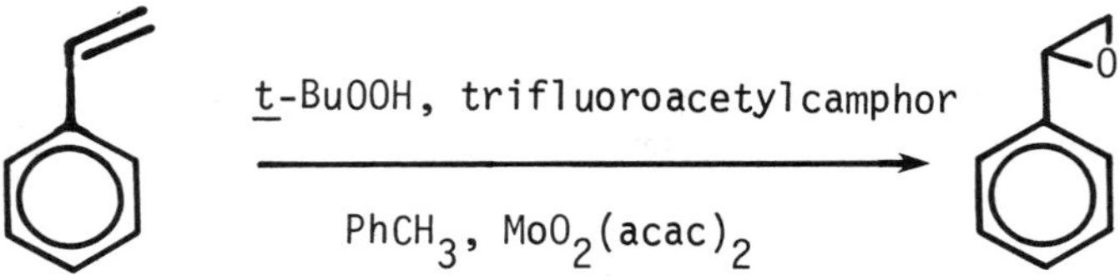

78%

82%ee

Z Chem 18, 218 (1978)

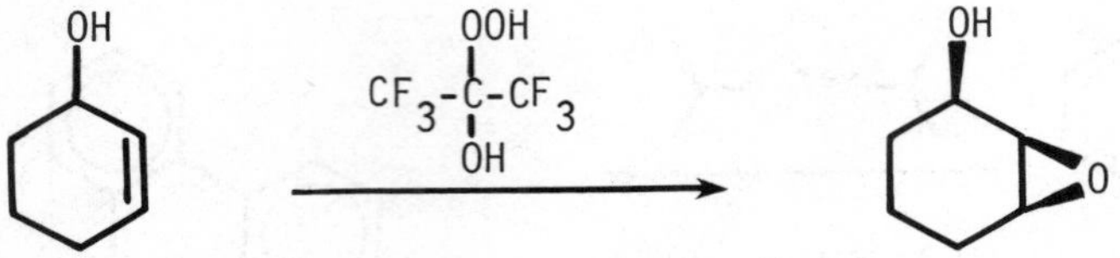

JACS 101, 2484 (1979)

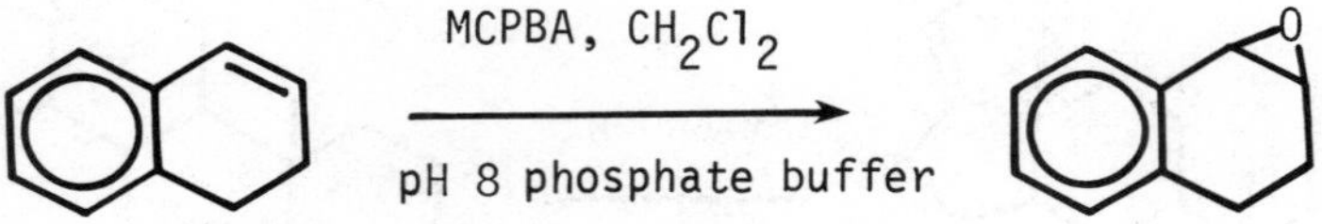

JOC 44, 1351 (1979)

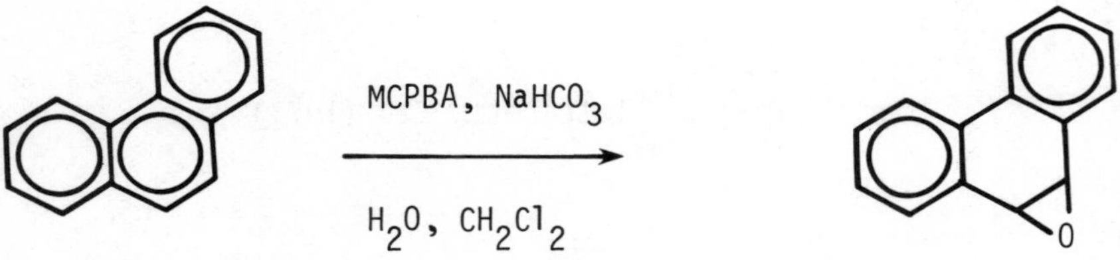

Tetr Lett, 427 (1977)

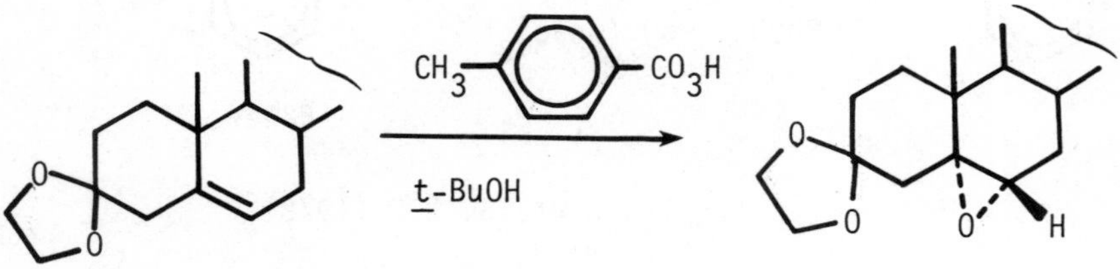

Bull Akad USSR Chem 27, 387 (1978)

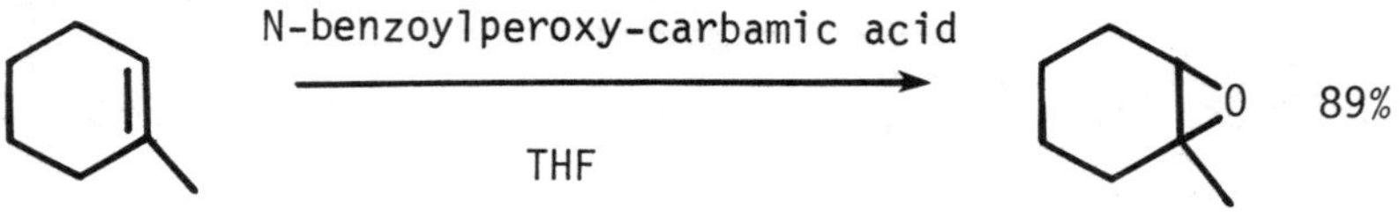

N-benzoylperoxy-carbamic acid

THF

89%

JOC 44, 1485 (1979)

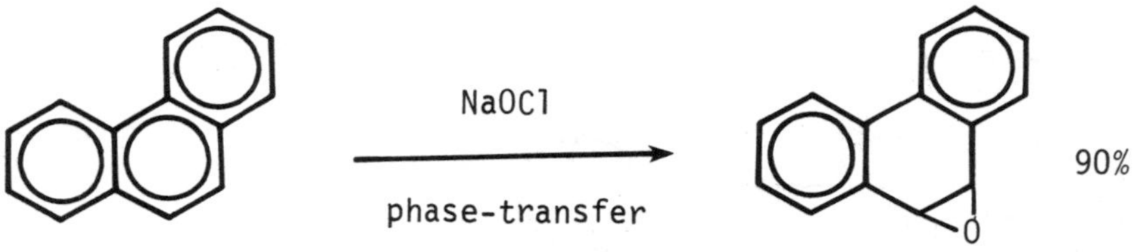

NaOCl

phase-transfer

90%

JACS 99, 8121 (1977)

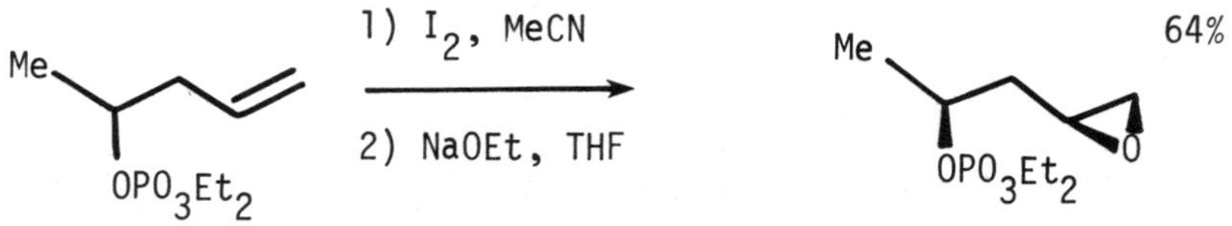

1) I$_2$, MeCN

2) NaOEt, THF

64%

JACS 99, 4829 (1977)

CH$_3$CHO, O$_2$

hν, PhCl

100%

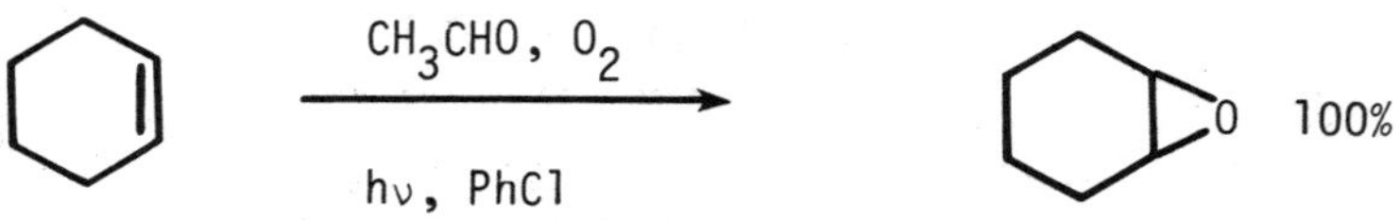

Synthesis, 711 (1977)

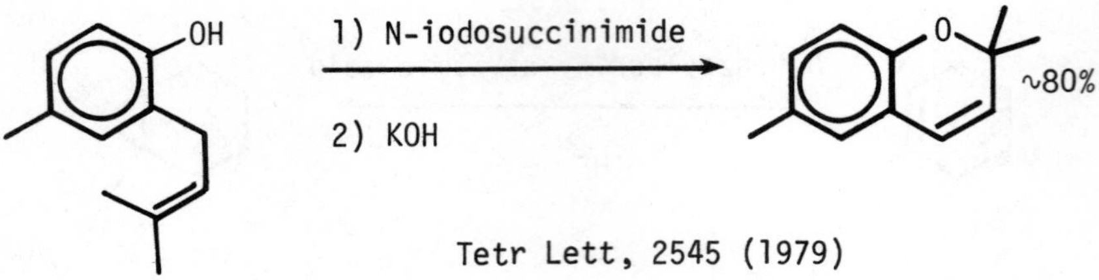

Tetr Lett, 2545 (1979)

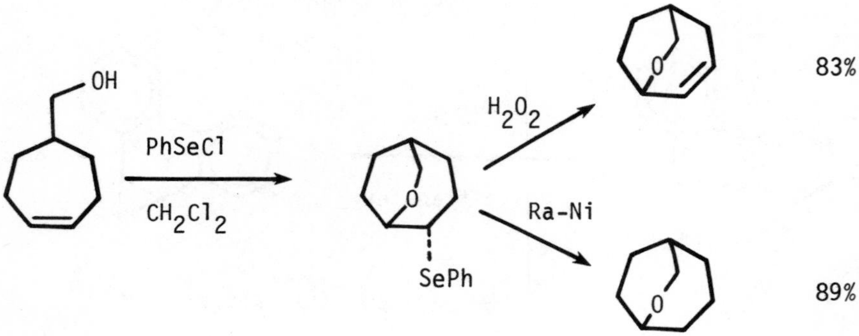

Tetr Lett, 1257 (1977)

JACS 102, 3784 (1978)

Section 135   Ethers and Epoxides from Miscellaneous Compounds

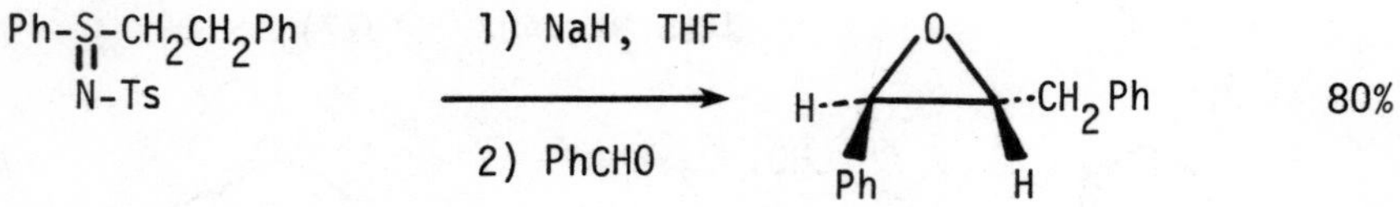

Ph-S-CH$_2$CH$_2$Ph
‖
N-Ts

1) NaH, THF

2) PhCHO

80%

Synthesis, 693 (1977)

CHAPTER 10
PREPARATION OF HALIDES AND SULFONATES

Section 136   <u>Halides from Acetylenes</u>

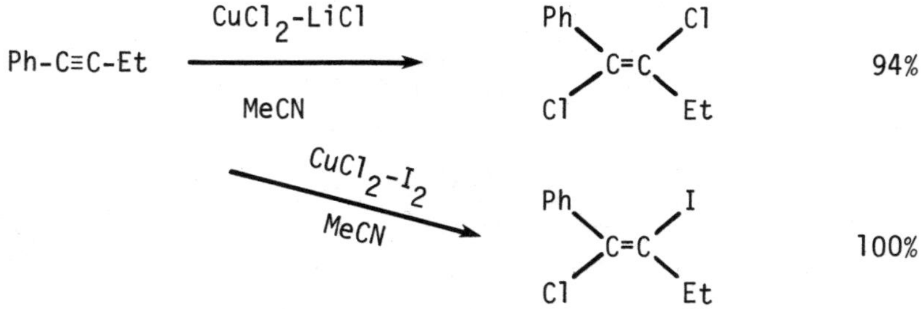

Ph-C≡C-Et
$\xrightarrow[\text{MeCN}]{\text{CuCl}_2\text{-LiCl}}$
Ph, Cl
C=C
Cl, Et
94%

$\xrightarrow[\text{MeCN}]{\text{CuCl}_2\text{-I}_2}$
Ph, I
C=C
Cl, Et
100%

JCS Perkin I, 676 (1977)

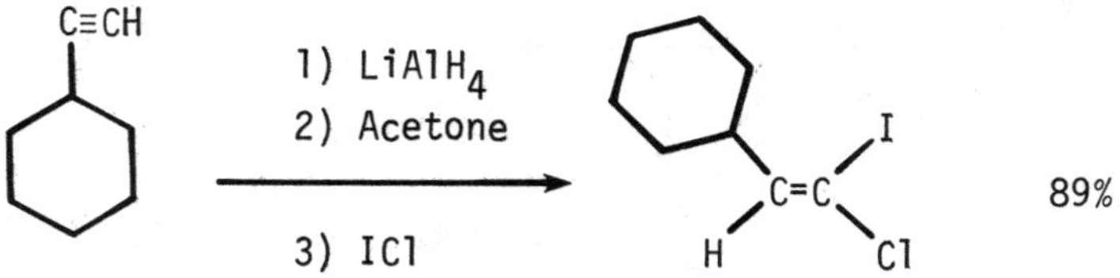

C≡CH

1) LiAlH$_4$
2) Acetone

3) ICl

I
C=C
H, Cl
89%

JACS <u>101</u>, 5101 (1979)

Ph-C≡CH $\xrightarrow{\overset{\oplus}{Et_3}NH\ \overset{\ominus}{HCl_2}}$ $\underset{Cl}{\overset{Ph}{>}}C=CH_2$ 73%

JCS Perkin I, 1797 (1977)

Ph-C≡C-Me $\xrightarrow[\text{2) }I_2\text{, MeCN}]{\text{1) }TeCl_4}$ $\underset{Cl}{\overset{Ph}{>}}C=C\underset{I}{\overset{Me}{<}}$ 85%

Chem Lett, 1357 (1979)

Bu-C≡C-H $\xrightarrow[\text{2) }I_2\text{, THF}]{\text{1) }Cp_2ZrCl_2\text{, }Me_3Al}$ $\underset{Me}{\overset{Bu}{>}}C=C\underset{I}{\overset{H}{<}}$ 85%

Synthesis, 501 (1979)

Ph-C≡CH $\xrightarrow[\text{2) NBS}]{\text{1) [EtCuBr]MgBr}}$ $\underset{Et}{\overset{Ph}{>}}C=C\underset{Br}{\overset{H}{<}}$ >90%

Rec Trav Chim 96, 168 (1977)

Et-C≡C-Et $\xrightarrow[\text{THF}]{\text{pyridine} \cdot (HF)_n}$ 75%

JOC 44, 3872 (1979)

Section 137   Halides from Carboxylic Acids

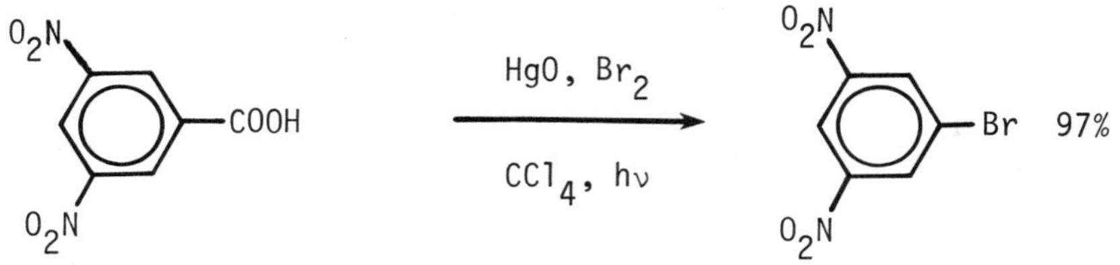

(0% yield without light)

JOC 44, 3405 (1979)

Section 138   Halides and Sulfonates from Alcohols

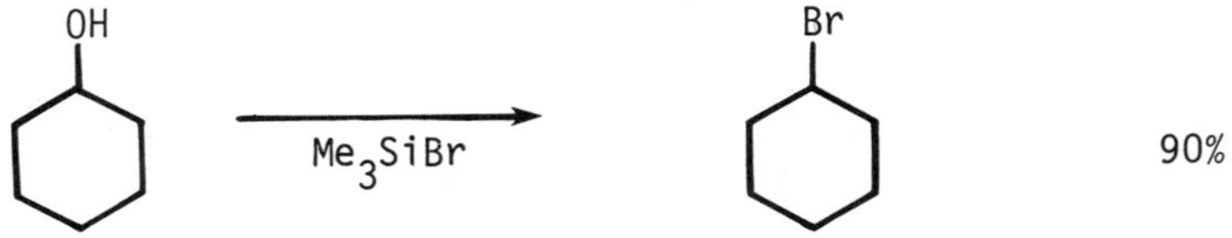

90%

Tetr Lett, 4483 (1978)

95%

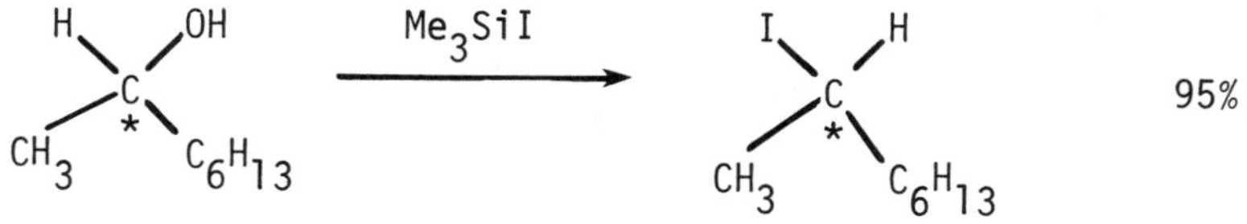

>90% inversion

Tetr Lett, 2659 (1977)

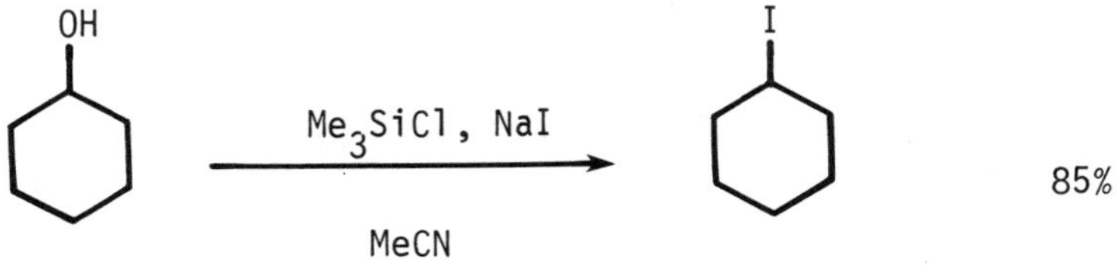

85%

Synthesis, 379 (1979)

OH → I

Me_3SiCl / NaI

98%

JOC **44**, 1247 (1979)

OH → I

1) $Me_3SiSiMe_3/I_2$

2) H_2O

86%

Angew Int Ed **18**, 612 (1979)

$Ph-\underset{\underset{OH}{|}}{C}H-CH_2COOEt$

1) Me_3SiCl, pyridine

2) $PhPF_4$

$Ph-\underset{\underset{F}{|}}{C}H-CH_2COOEt$

92%

Tetr Lett, 4507 (1978)

$CH_2=CH-\underset{\underset{OH}{|}}{C}H-CH_3$

PPh_3

$Cl_3C\underset{\underset{O}{||}}{C}CCl_3$

$CH_2=CH-\underset{\underset{Cl}{|}}{C}H-CH_3$

94%

Tetr Lett, 2999 (1977)

$$PhCH_2OCH_2CH_2OH \xrightarrow{\text{DMPADC}^*} PhCH_2OCH_2CH_2Cl \qquad 92\%$$

*N, N-Dimethylphosphoramidic dichloride

Chem Lett, 923 (1978)

$$R-OH \xrightarrow[\text{trinonyl borate}]{\text{HCl, ZnCl}_2} R-Cl$$

R = 1° alkyl

JOC (USSR) 13, 604 (1977)

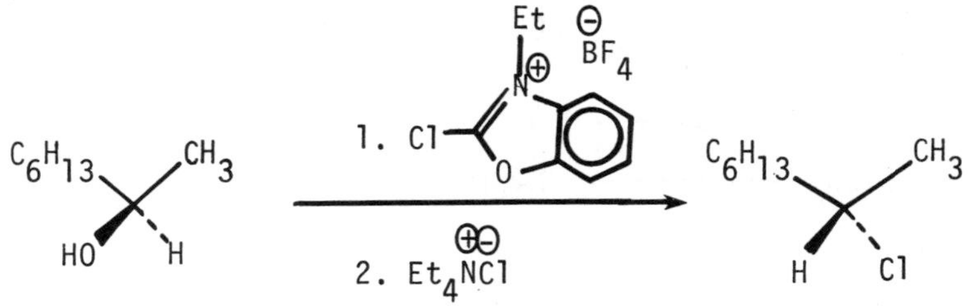

76%

>90% inversion

Chem Lett, 383 (1977)

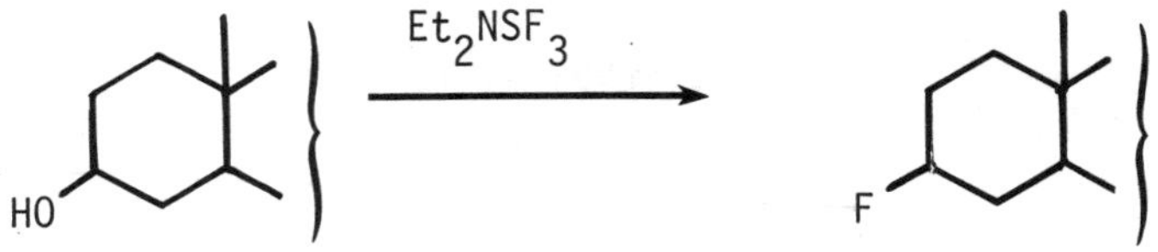

~80%

Tetr Lett, 1823 (1979)

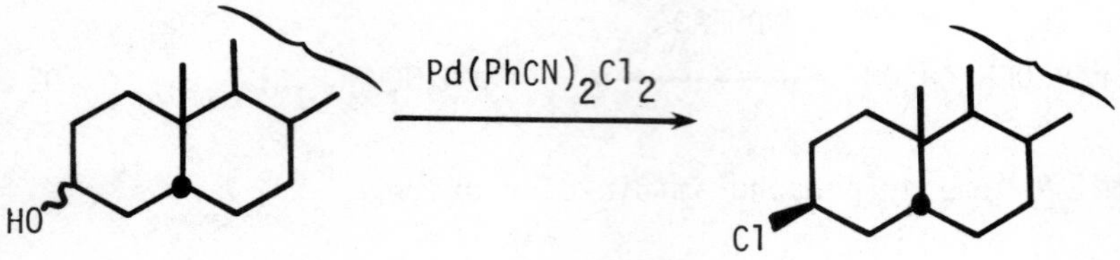

Tetr Lett, 4575 (1978)

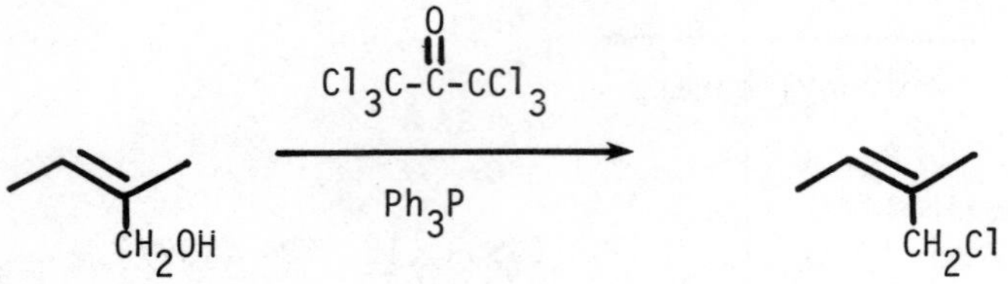

100%

JOC 44, 359 (1979)

TfOTf

R-OH $\xrightarrow{\text{pyridine}}$ R-OTf   28-89%

JOC 42, 3109 (1977)

PhSO$_2$Cl, benzene

Bu-CH$_2$OH $\longrightarrow$ Bu-CH$_2$OSO$_2$Ph   85%

Synthesis, 822 (1979)

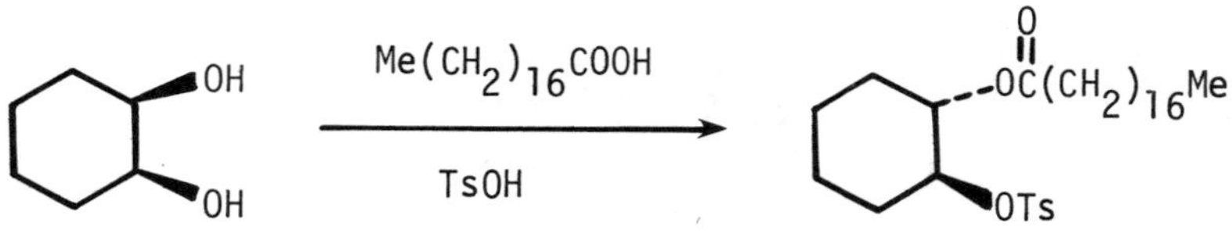

Aust J Chem <u>30</u>, 2479 and 2487 (1977)

Section 139   <u>Halides from Aldehydes</u>

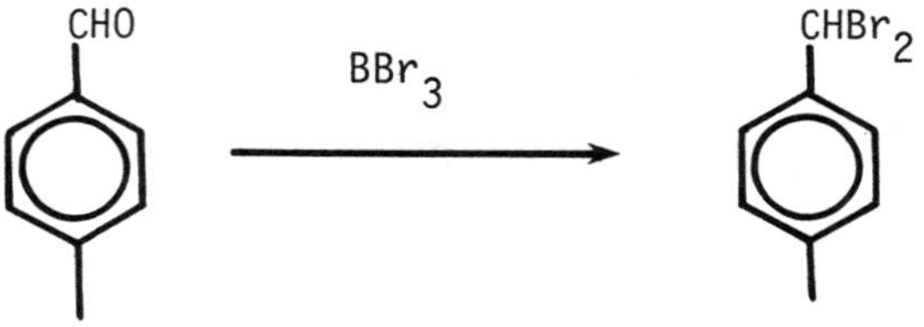

72%

Synth Comm <u>9</u>, 341 (1979)

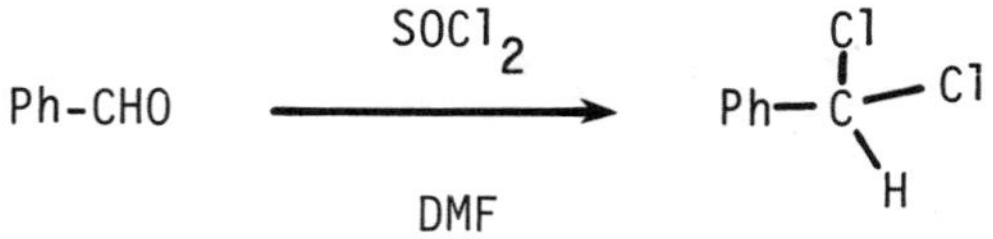

90%

JOC <u>43</u>, 4367 (1978)

Section 140   Halides from Alkyls

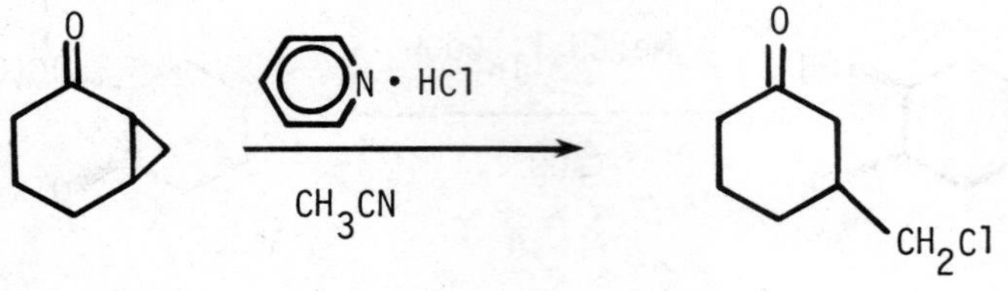

Synthesis, 227 (1978)

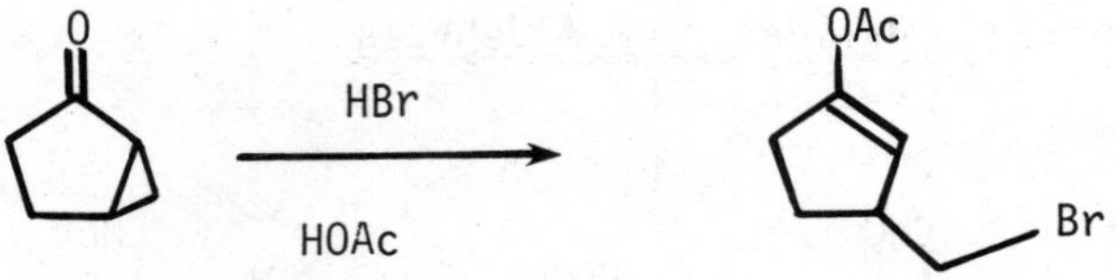

Helv Chim Acta 62, 2338 (1979)

For the conversion RH → RHal see Section 146 (Halides from Hydrides)

Section 141   Halides and Sulfonates from Amides

No additional examples

Section 142   Halides from Amines

Bu–NH$_2$ $\xrightarrow{\Delta}$ Bu–I   69%

Synthesis, 634 (1977)

$\xrightarrow[\text{MeCN, CuCl}_2]{\underline{t}\text{-BuSNO}}$ 98%

Tetr Lett, 4519 (1978)

$\xrightarrow[\text{MeCN}]{\underline{t}\text{-BuONO, CuCl}_2}$ 92%

JOC 42, 2426 (1977)

$\xrightarrow[\text{I}_2\text{, benzene}]{\underline{t}\text{-BuSNO}_2}$ 80%

Chem Lett, 939 (1979)

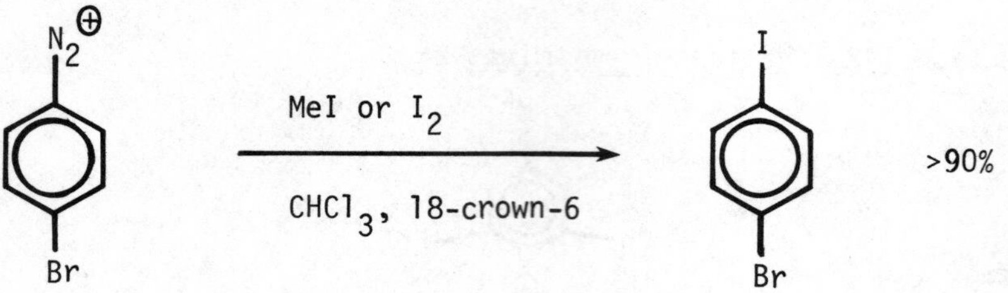

Tetr Lett, 3519 (1977)

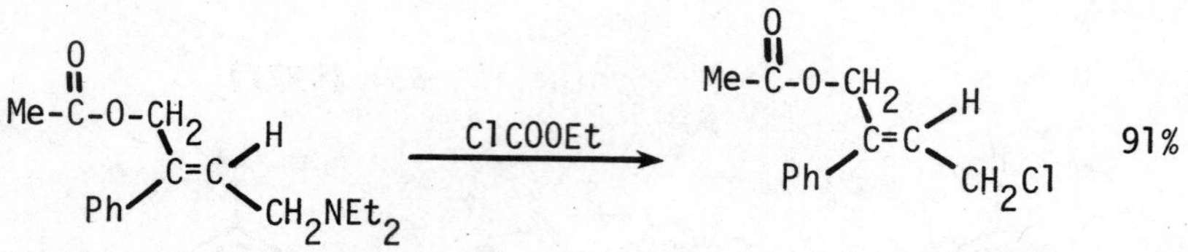

91%

Synthesis, 786 (1977)

30%

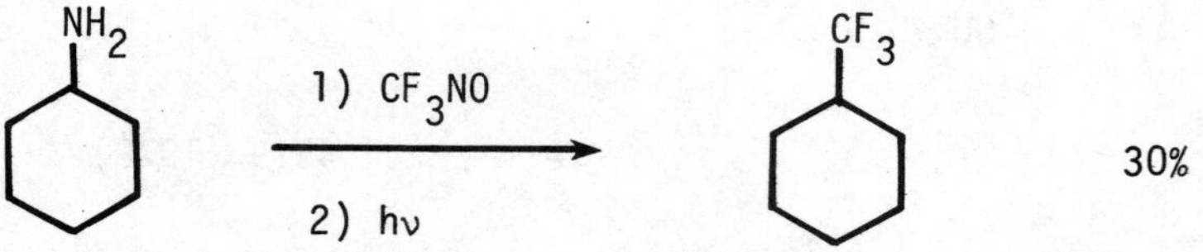

Angew Int Ed $\underline{16}$, 854 (1977)

$$Ph_2CN_2 \xrightarrow[\text{Bu}_4\text{NClO}_4, \text{ CH}_2\text{Cl}_2]{\text{KHF}_2, \text{ H}_2\text{O}} Ph_2CHF$$

60%

Tetr Lett, 1447 (1977)

Section 143   <u>Halides and Sulfonates from Esters and Epoxides</u>

No additional examples

Section 144   <u>Halides from Ethers</u>

$$\text{cyclohexyl-OEt} \xrightarrow[\text{2) } H_2O]{\text{1) } Me_3SiSiMe_3/I_2} \text{cyclohexyl-I} \quad 98\%$$

Angew Int Ed <u>18</u>, 612 (1979)

$$\underline{n}\text{-}C_{16}H_{33}\text{-}OCH_3 \xrightarrow[\text{R}_4\text{PBr}]{\text{HBr}} \underline{n}\text{-}C_{16}H_{33}Br \quad 88\%$$

Synthesis, 771 (1978)

$$\xrightarrow{PBr_3\text{-DMF}} \quad 87\%$$

Chem Lett, 891 (1977)

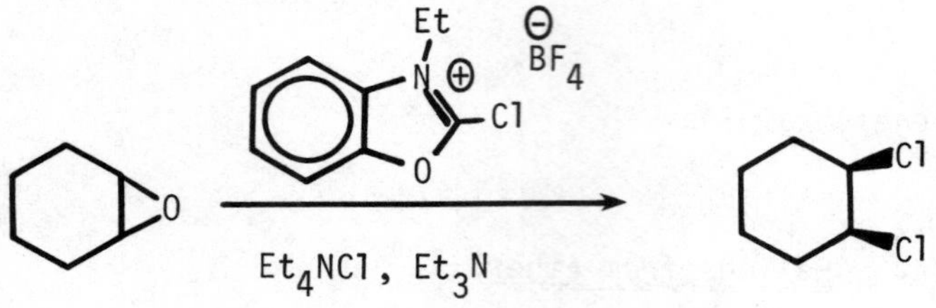

58%

Chem Lett, 1013 (1977)

Section 145   Halides from Halides and Sulfonates

$$BuBr \xrightarrow[\text{silica gel column}]{KI, Bu_4PI} BuI \qquad 96\%$$

Synthesis, 952 (1979)

$$Ph\text{-}CCl_3 \xrightarrow{PhSbF_4} PhCF_3 \qquad 91\%$$

JOC (USSR) 13, 561 (1977)

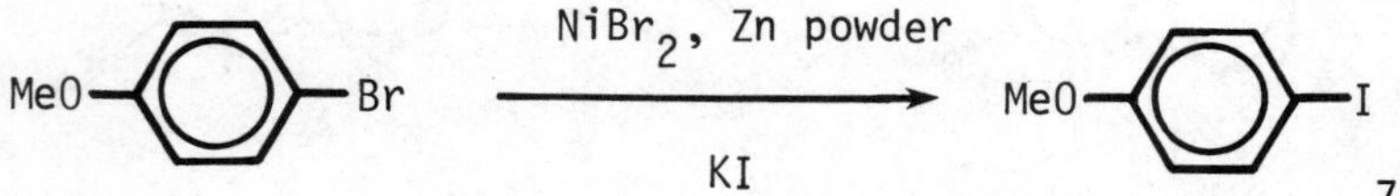

MeO—⟨ ⟩—Br $\xrightarrow[\text{KI}]{NiBr_2, \text{Zn powder}}$ MeO—⟨ ⟩—I   78%

Chem Lett, 191 (1978)

$$\text{BuBr} \quad \xrightarrow[\text{2) Br}_2]{\overset{\displaystyle 1)\ Ph_2\overset{\textstyle O}{\overset{\|}{As}}CH_2Li}{}} \quad \text{BuCH}_2\text{Br} \qquad 79\%$$

Angew Int Ed <u>16</u>, 53 (1977)

$$\begin{array}{c} \text{BuLi} \\[4pt] + \\[4pt] \text{PhN(SO}_2\text{CF}_3)_2 \end{array} \quad \xrightarrow[\text{2) H}_2\text{O}]{\text{1) Et}_2\text{O}} \quad \text{BuSO}_2\text{CF}_3 \qquad 74\%$$

JOC <u>42</u>, 3875 (1977)

Section 146   <u>Halides from Hydrides</u>

α-Halogenations of ketones, esters, and acids are found in Sections 369 (Haloketones), 359 (Haloesters), and 319 (Haloacids).

$$\underset{\text{Et}_2}{\overset{\text{Me}}{|}}\text{C-H} \quad \xrightarrow[\text{cat.}(C_6H_{13})_3B]{\text{PhICl}_2} \quad \underset{\text{Et}_2}{\overset{\text{Me}}{|}}\text{C-Cl} \qquad 82\%$$

Chem Lett, 961 (1979)

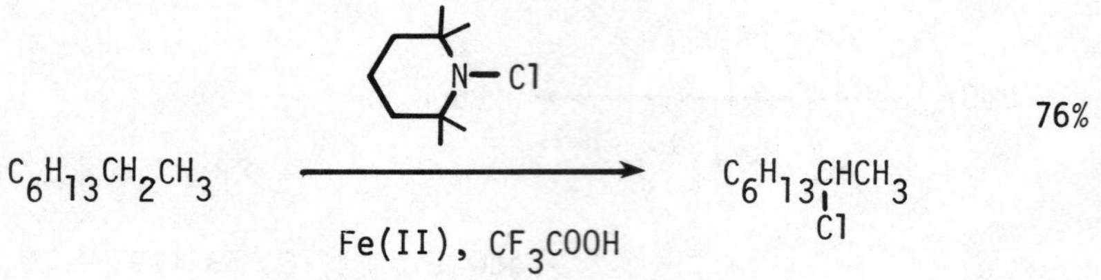

$C_6H_{13}CH_2CH_3$ $\xrightarrow[\text{Fe(II), CF}_3\text{COOH}]{}$ $C_6H_{13}\underset{\underset{Cl}{|}}{C}HCH_3$ 76%

JOC 44, 3728 (1979)

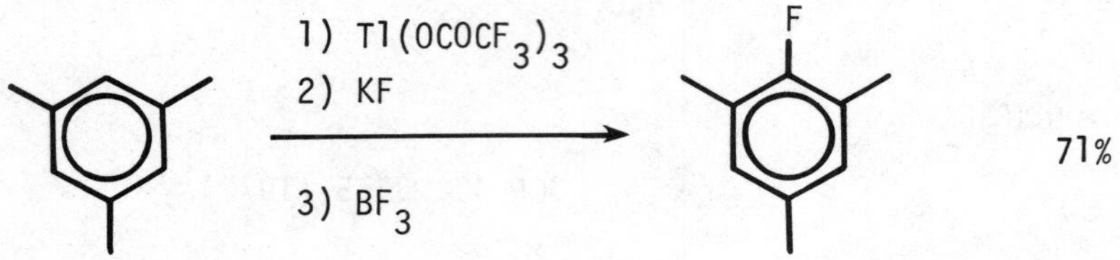

1) Tl(OCOCF$_3$)$_3$
2) KF
3) BF$_3$

71%

JOC 42, 362 (1977)

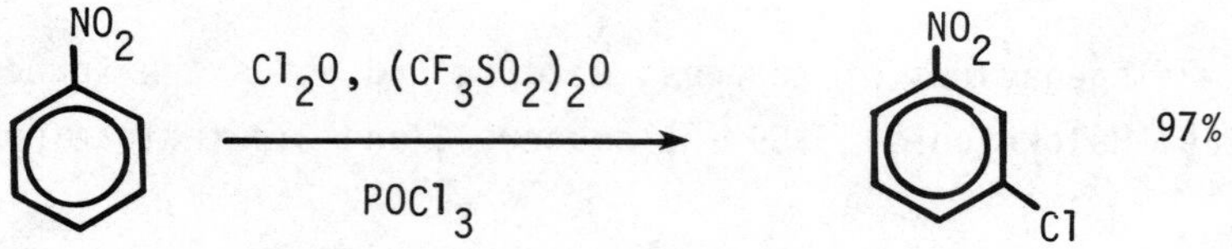

Cl$_2$O, (CF$_3$SO$_2$)$_2$O

POCl$_3$

80% meta

97%

Chem Ber 112, 1677 (1979)

NBS-DMF

93%

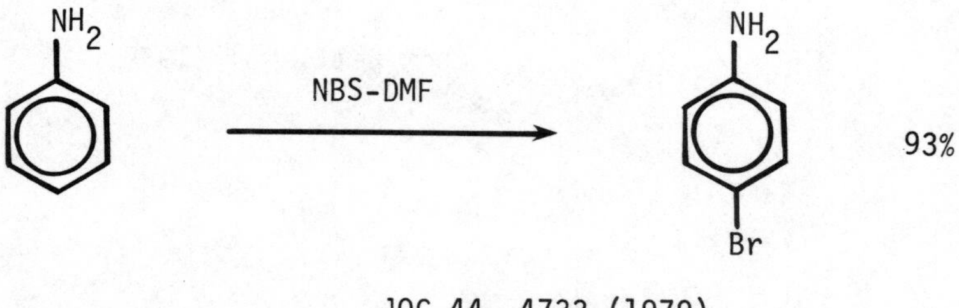

JOC 44, 4733 (1979)

CH_3—⟨benzene⟩—CH_3 $\xrightarrow[\text{Pd(PPh}_3)_4]{\text{SO}_2\text{Cl}_2}$ CH_3—⟨benzene⟩—CH_2Cl 84%

Chem Lett, 223 (1978)

⟨fluorene⟩ $\xrightarrow{\text{DBU-BrCCl}_3}$ ⟨9,9-dibromofluorene⟩ Br Br 99%

same conditions

$PhC{\equiv}CH$ $\xrightarrow{\hspace{3cm}}$ $PhC{\equiv}CBr$ 99%

Chem Lett, 73 (1978)

⟨anisole, OCH_3⟩ $\xrightarrow[\text{CCl}_4]{\text{Cl}_2,\ h\nu}$ ⟨OCH_2Cl benzene⟩ ∼50%

Chem and Ind, 127 (1977)

$C_7H_{15}\text{-}CH_2NO_2$ $\xrightarrow[\text{2) Cl}_2,\ \text{CH}_2\text{Cl}_2]{\text{1) KOH, H}_2\text{O}}$ $C_7H_{15}\text{-}\overset{\text{Cl}}{\underset{}{\text{CHNO}_2}}$ 92%

JOC 42, 3764 (1977)

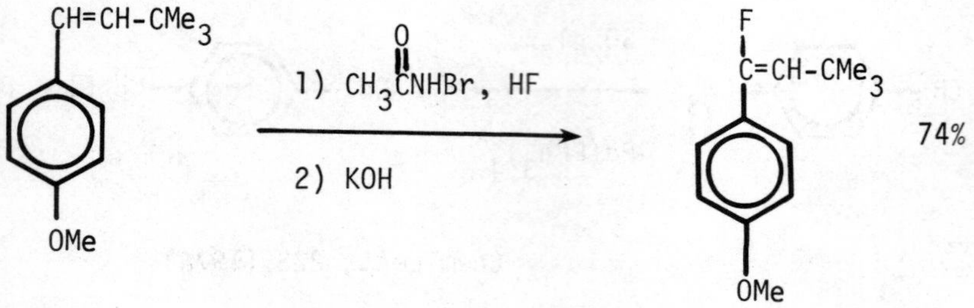

Synthesis, 217 (1978)

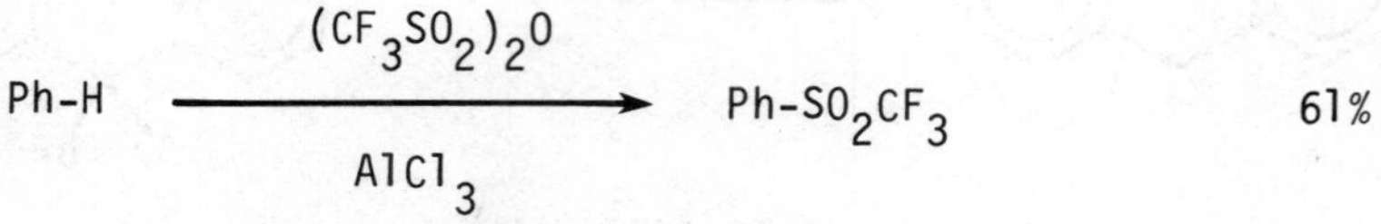

$$Ph-H \xrightarrow[AlCl_3]{(CF_3SO_2)_2O} Ph-SO_2CF_3 \qquad 61\%$$

JOC 42, 3875 (1977)

"Selective Halogenation of Steroids Using Attached Aryl Iodide Templates"

JACS 99, 905 (1977)

Section 147   Halides from Ketones

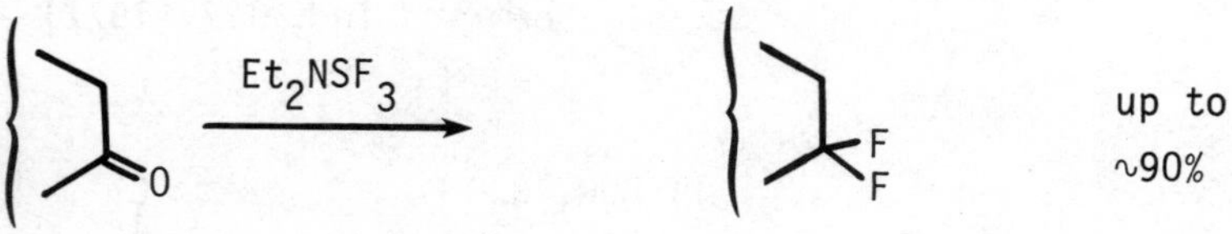

up to
~90%

JCS Perkin I, 1354 (1979)

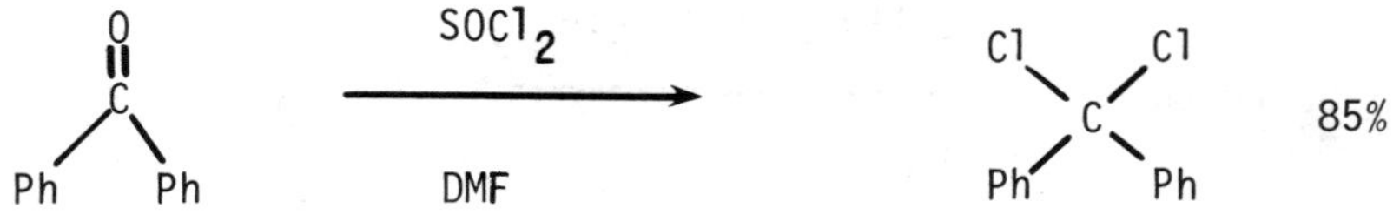

JOC <u>43</u>, 4367 (1978)

Section 148   <u>Halides and Sulfonates from Nitriles</u>

No examples

Section 149   <u>Halides from Olefins</u>

For halocyclopropanations see Section 74 (Alkyls from Olefins)

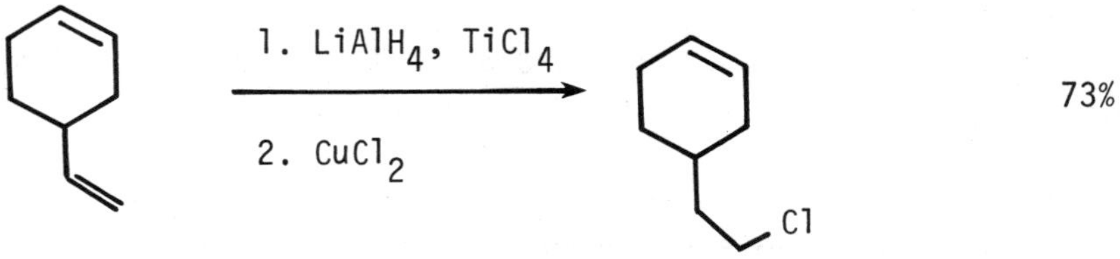

Chem Lett, **833** (1978)

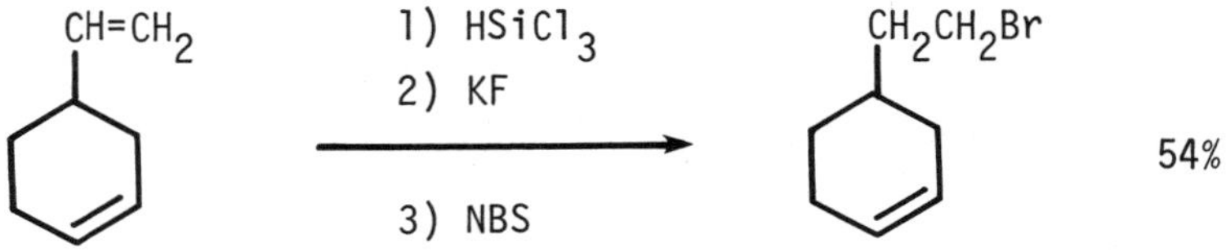

JACS <u>100</u>, 290 (1978)

$$\text{Bu-CH=CH}_2 \xrightarrow[\text{2. Br}_2]{\text{1. LiAlH}_4, \text{ Cp}_2\text{Ti(AlH}_3\text{)}_2} \text{Bu-CH}_2\text{CH}_2\text{Br} \qquad 79\%$$

Chem Lett, 1117 (1977)

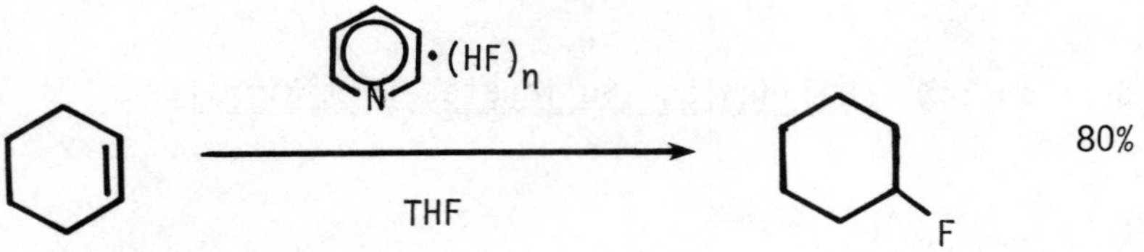

80%

JOC 44, 3872 (1979)

79%

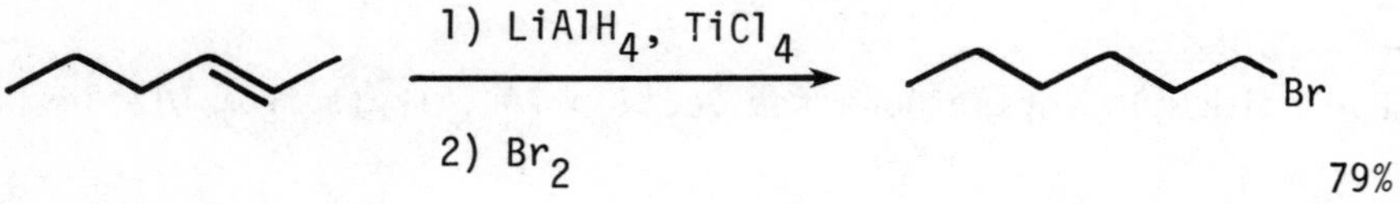

J Organometal Chem 142, 71 (1977)

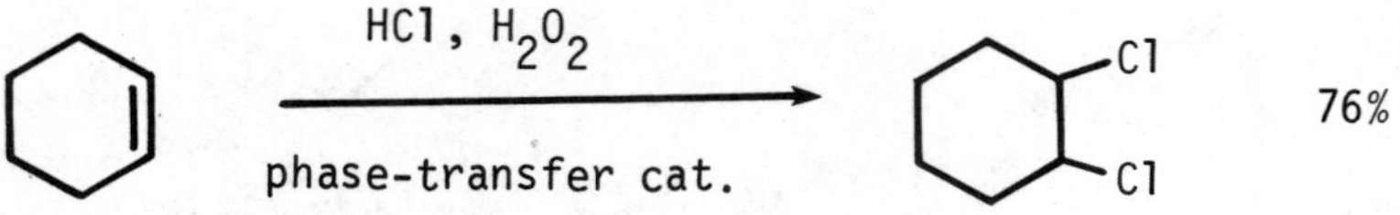

76%

Synthesis, 676 (1977)

Ph-CH=CHCH$_3$ $\xrightarrow{\text{XeF}_2}$ Ph-CH-CH-CH$_3$ 60%
                                              | |
                                              F   F

JOC <u>42</u>, 1559 (1977)

$\xrightarrow[\text{Ru(II) complex}]{\text{CCl}_4}$ 83%

Chem Lett, 115 (1978)

Use of a heterogeneous solvent suspension method for benchtop fluorination of alkenes using XeF$_2$.

Tetr Lett, 363 (1977)

Section 150   <u>Halides from Miscellaneous Compounds</u>

Ph-N=N-NMe$_2$ $\xrightarrow{\text{HF, pyridine}}$ Ph-F   97%

JCS Chem Comm, 914 (1979)

Review: "Introduction of Fluorine into Organic Molecules: Why and How"

Tetrahedron <u>34</u>, 3 (1978)

Review: "The Invention of Reactions Useful for the Synthesis
of Specifically Fluorinated Natural Products"

Pure and Appl Chem <u>49</u>, 1241 (1977)

CHAPTER 11
PREPARATION OF HYDRIDES

This chapter lists hydrogenolysis and related reactions by which functional groups are replaced by hydrogen, e.g. $RCH_2X \rightarrow RCH_2-H$ or R-H

Section 151   Hydrides from Acetylenes

No examples of the reaction $RC{\equiv}CR \rightarrow RH$ occur in the literature.

Section 152   Hydrides from Acid Halides

JCS Perkin I, 1137 (1979)

Section 153   Hydrides from Alcohols and Phenols

This section lists examples of the hydrogenolysis of alcohols and phenols, ROH → RH

1) CH_3SO_2Cl

2) electrolysis, DMF, Et_4NOTs

63%

Tetr Lett, 2157 (1979)

steroid

K, 18-crown-6

THF

86%

JCS Chem Comm, 1175 (1979)

$\underline{n}$-$C_{12}H_{25}OH \longrightarrow \underline{n}$-$C_{12}H_{25}SePh \xrightarrow{Ph_3SnH} \underline{n}$-$C_{12}H_{26}$ 73%

JCS Chem Comm, 41 (1978)

1-naphthyl—$\overset{\overset{\displaystyle H}{|}}{\underset{\underset{\displaystyle Ph}{|}}{C}}$—OH $\xrightarrow{\text{NaBH}_4/\text{CF}_3\text{COOH}}$ 1-naphthyl—$\overset{\overset{\displaystyle H}{|}}{\underset{\underset{\displaystyle Ph}{|}}{C}}$—H   97%

Synthesis, 172 (1977)

$C_5H_{11}-\overset{\overset{\displaystyle O}{\|}}{C}-NHCH_2OH$ $\xrightarrow[\text{TFA}]{\text{NaCNBH}_3}$ $C_5H_{11}-\overset{\overset{\displaystyle O}{\|}}{C}-NHMe$ 86%

Synth Comm <u>7</u>, 549 (1977)

Ph_2CHOH $\xrightarrow[\text{CH}_2\text{Cl}_2]{\text{Me}_2\text{SiI}_2}$ Ph_2CH_2 96%

$Ph-\overset{\overset{\displaystyle O}{\|}}{C}-\underset{\underset{\displaystyle OH}{|}}{C}HPh$ $\xrightarrow[\text{CH}_2\text{Cl}_2]{\text{Me}_2\text{SiI}_2}$ $Ph-\overset{\overset{\displaystyle O}{\|}}{C}CH_2Ph$ 45%

Tetr Lett, 4941 (1979)

$Ph-\overset{\overset{\displaystyle O}{\|}}{C}-\underset{\underset{\displaystyle OH}{|}}{C}H-Ph$ $\xrightarrow[\text{2) sodium thiosulfate}]{\text{1) Me}_3\text{SiI, CH}_2\text{Cl}_2}$ $Ph-\overset{\overset{\displaystyle O}{\|}}{C}-CH_2Ph$ 95%

Synth Comm <u>9</u>, 665 (1979)

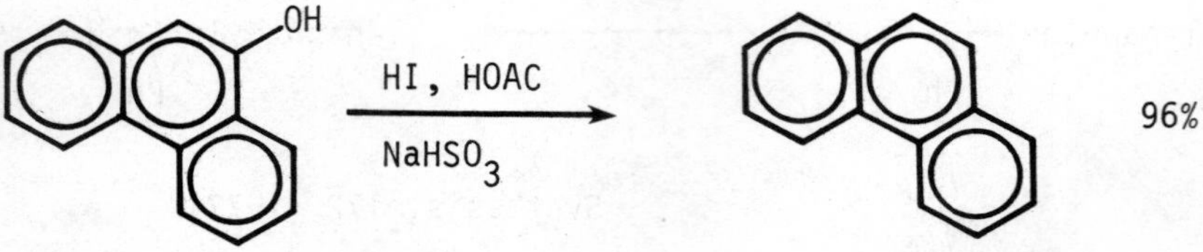

HI, HOAC

NaHSO₃

96%

JOC 44, 4813 (1979)

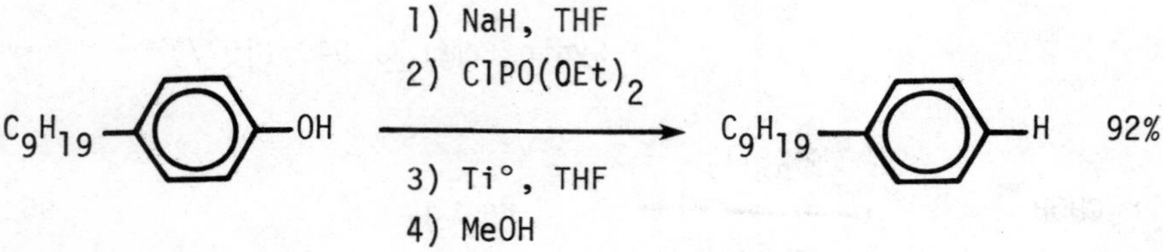

C_9H_{19}—⟨ ⟩—OH

1) NaH, THF
2) ClPO(OEt)₂

3) Ti°, THF
4) MeOH

C_9H_{19}—⟨ ⟩—H   92%

JOC 43, 4797 (1978)

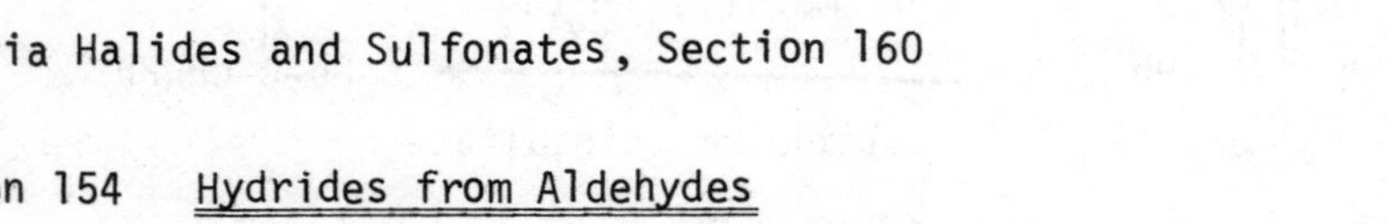

Zn, HCl

ether

97%

Tetrahedron 33, 511 (1977)

Also via Halides and Sulfonates, Section 160

Section 154   Hydrides from Aldehydes

No additional examples

For the conversion RCHO → RMe etc. see Section 64 (Alkyls from Aldehydes)

Section 155   Hydrides from Alkyls

AlCl$_3$, CS$_2$

90%

JCS Perkin I, 176 (1979)

Section 156   Hydrides from Amides

No additional examples

Section 157   Hydrides from Amines

This section lists examples of the conversion RNH$_2$ → RH

NaBH$_3$CN

HMPA

91%

JCS Chem Comm, 1089 (1978)

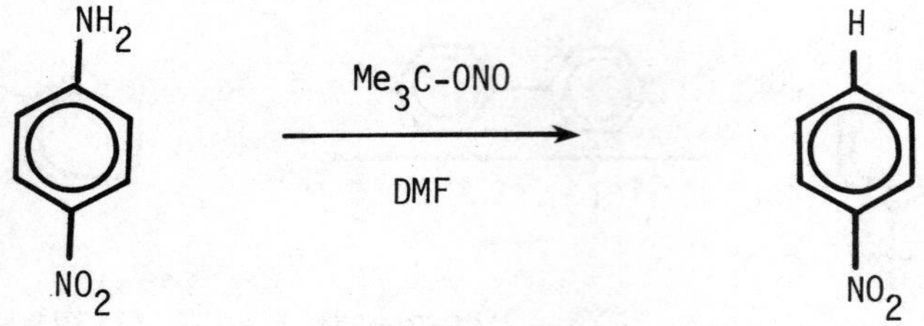

JOC <u>42</u>, 3494 (1977)

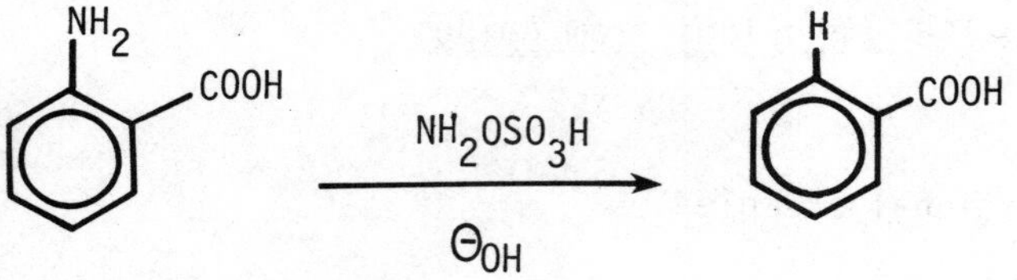

JACS <u>100</u>, 341 (1978)

Ph-NHNH$_2$ $\xrightarrow[\substack{\text{benzene/H}_2\text{O} \\ \text{Et}_4\text{NBr}}]{\text{TosN}_3, \text{ NaOH}}$ Ph-H   40%

Tetr Lett, 3059 (1978)

Ar-N$_2^{\oplus}$BF$_4^{\ominus}$ $\xrightarrow{\text{PhSH}}$ Ar-H   84-100%

Ar = subst. Ph

Chem Lett, 1051 (1979)

Section 158 Hydrides from Esters

This section lists examples of the reactions RCOOR' → RH and RCOOR' → R'H

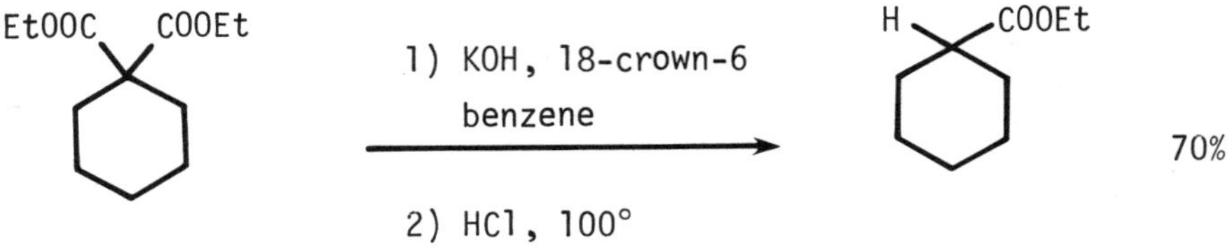

EtOOC
COOEt
1) KOH, 18-crown-6
benzene
2) HCl, 100°
H
COOEt
70%

Synthesis, 37 (1977)

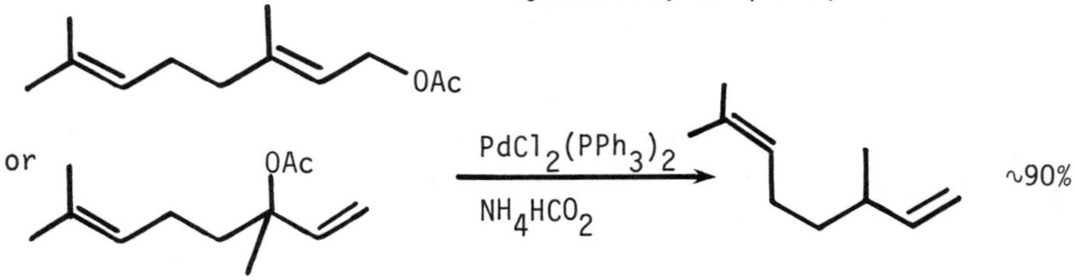

OAc
or
OAc
PdCl2(PPh3)2
NH4HCO2
~90%

Tetr Lett, 613 (1979)

OAc
O
O
O
O
O
HMPT, hν
CH3
O
O
O
O
O
85%

Synthesis, 774 (1977)

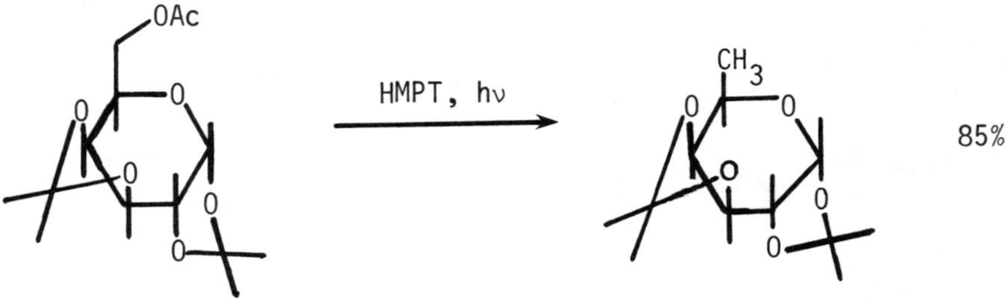

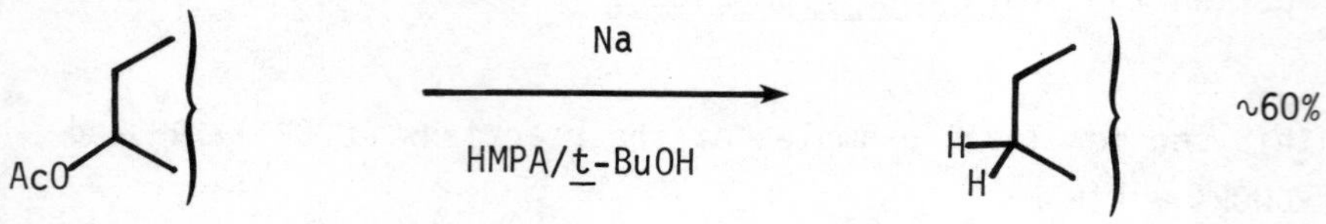

JCS Chem Comm, 567 (1978)

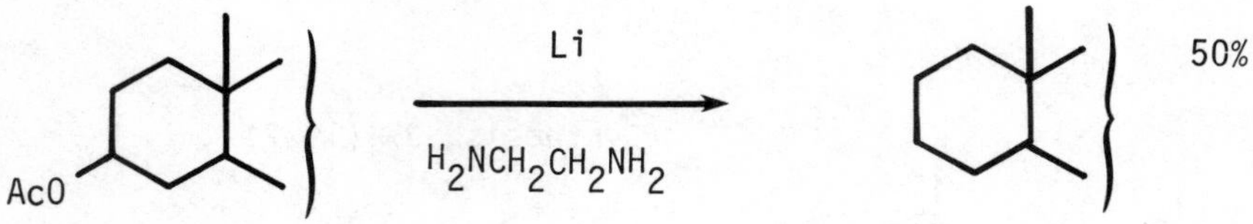

Indian J Chem 18B, 179 (1979)

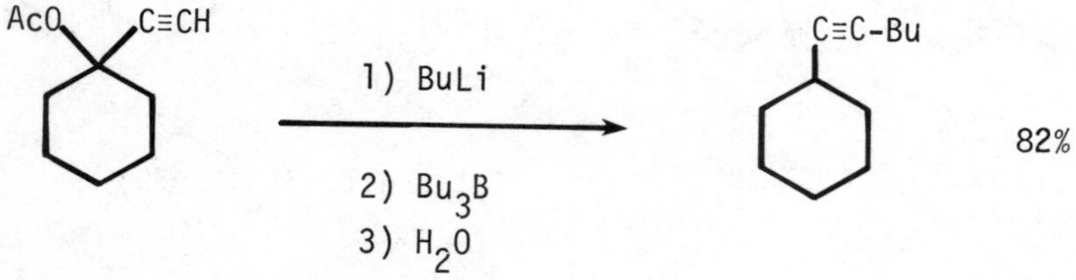

JOC 42, 2650 (1977)

Section 159   Hydrides from Ethers

This section lists examples of the reaction R-X-R' → RH, where
X=0 or S.

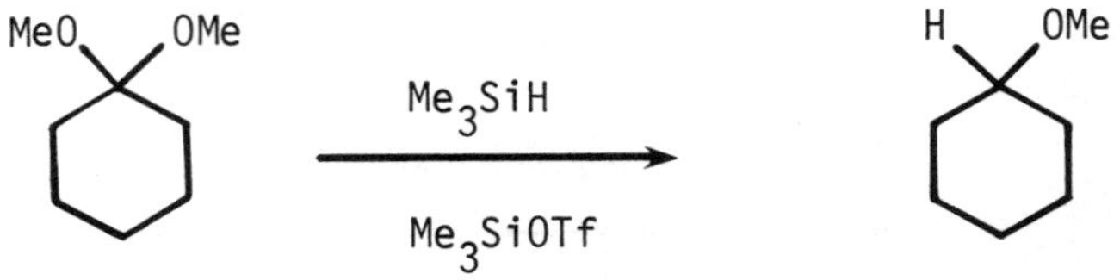

MeO
OMe
Me_3SiH
Me_3SiOTf
H
OMe

76%

Tetr Lett, 4679 (1979)

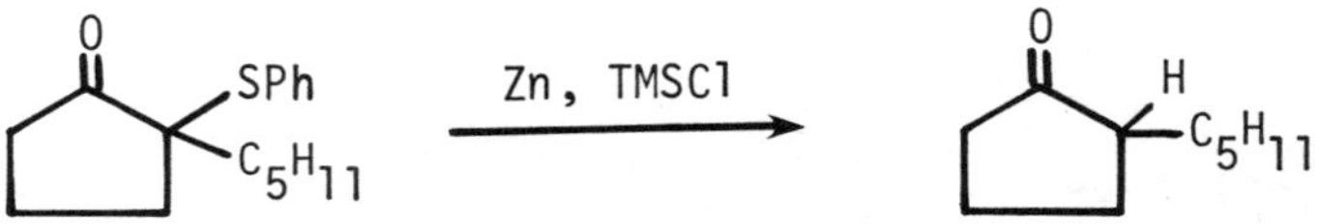

O
SPh
C_5H_{11}
Zn, TMSCl
O
H
C_5H_{11}

84%

Synth Comm 7, 427 (1977)

Section 160 Hydrides from Halides and Sulfonates

This section lists the reduction of halides and sulfonates RX→RH

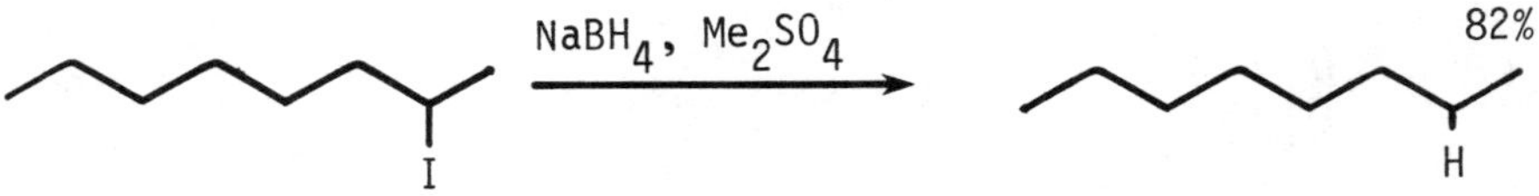

$NaBH_4$, Me_2SO_4
I
82%
H

Full paper, many examples.

JOC 43, 2259 (1978)

$$\text{R-X} \xrightarrow[\text{HMPA}]{\begin{array}{c}\text{NaBH}_3\text{CN,}\\ \text{9-BBNCN, or}\\ \text{polymeric cyanoborane}\end{array}} \text{R-H}$$

R = alkyl

X = halide, tosylate

JOC 42, 82 (1977)

JOC 42, 2166 (1977)

$$\underset{\text{Br}}{\overset{|}{\text{C}_3\text{H}_7\text{-CH-C}_6\text{H}_{13}}} \xrightarrow{\text{LiAlH}_4\text{-CoCl}_2} \text{decane} \quad 98\%$$

Tetr Lett, 4481 (1977)

89%

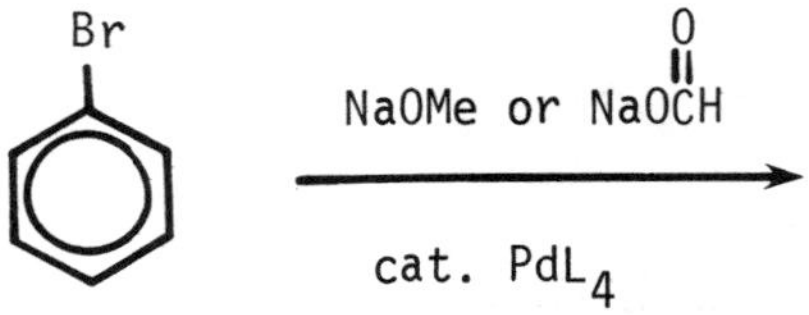

JOC 43, 1619 (1978)
Tetr Lett, 1913 (1978)

100%

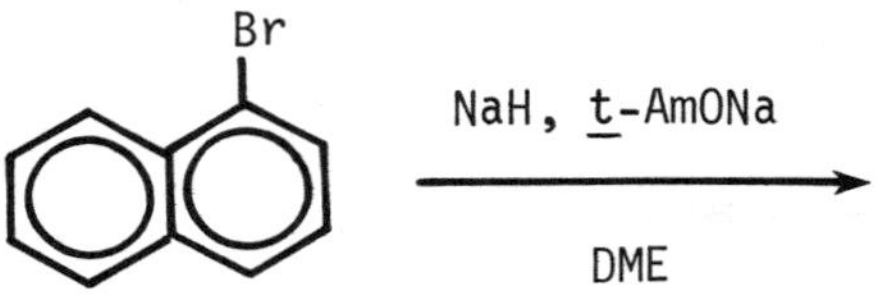

Tetr Lett, 3951 (1977)

Ph-Br —($Et_3N \cdot HCOOH$, Pt/C)→ Ph-H 92%

JOC 42, 3491 (1977)

83%

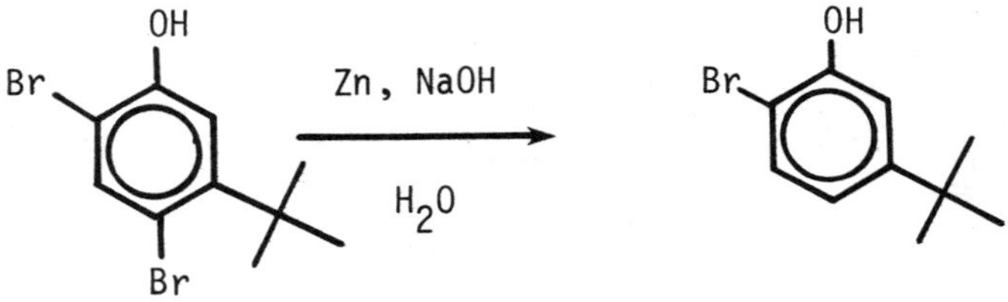

JOC 42, 835 (1977)

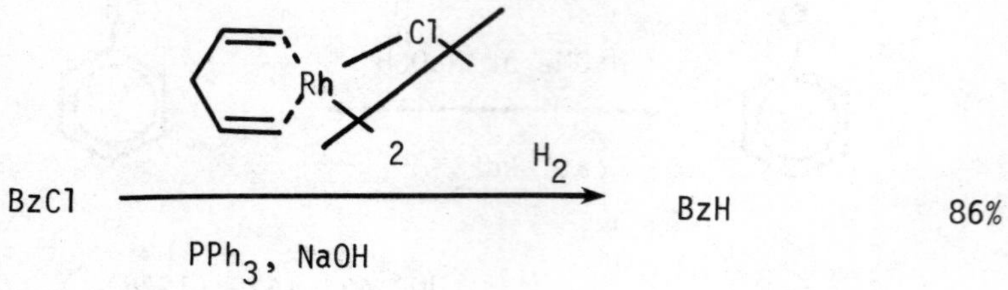

BzCl $\xrightarrow[\text{PPh}_3, \text{ NaOH}]{}$ BzH   86%

(Rh catalyst structure with Cl, subscript 2, H_2 above arrow)

J Organometal Chem 148, 311 (1978)

$CH_3(CH_2)_8CH_2I \xrightarrow{\text{MgH}_2} CH_3(CH_2)_8CH_3$ 100%

JOC 43, 1557 (1978)

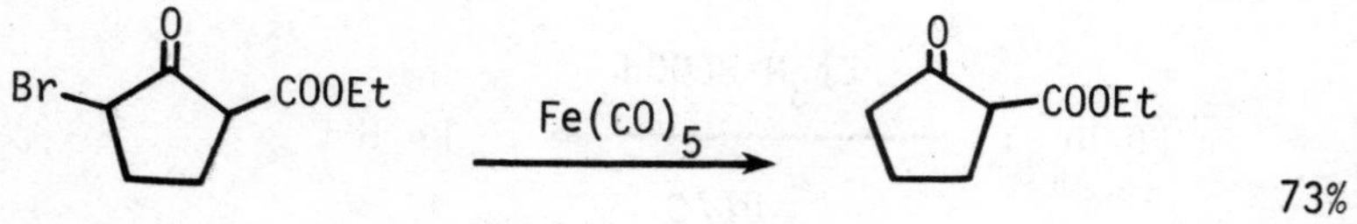

Br—(cyclopentanone with COOEt) $\xrightarrow{\text{Fe(CO)}_5}$ (cyclopentanone with COOEt)

73%

JOC 44, 641 (1979)

Ph—(chain with Br)—CBr$_3$ $\xrightarrow[\text{THF}]{\text{Ni(CO)}_4}$ Ph—(chain with Br)—CHBr$_2$ 60%

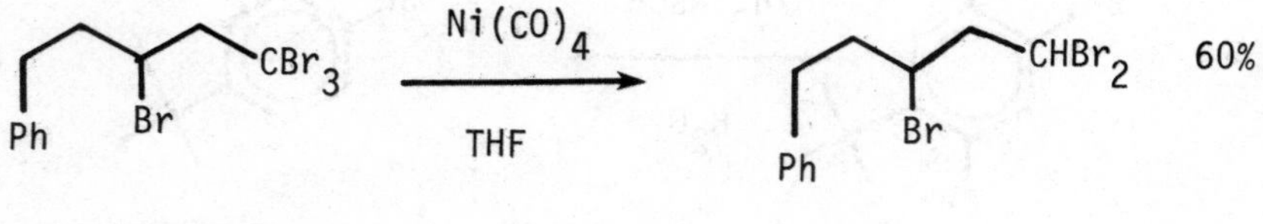

Chem Pharm 25, 1749 (1977)

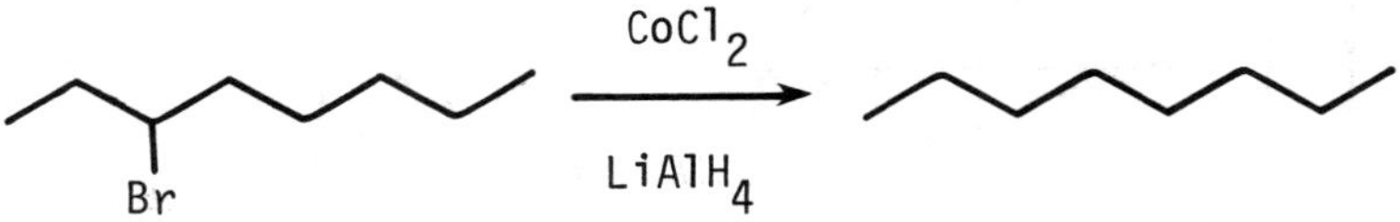

JOC **43**, 1263 (1978)

$$\underline{n}\text{-}C_{10}H_{21}I \xrightarrow{\text{LiCuH}_2} \underline{n}\text{-}C_{10}H_{22} \qquad 100\%$$

JOC **43**, 183 (1978)

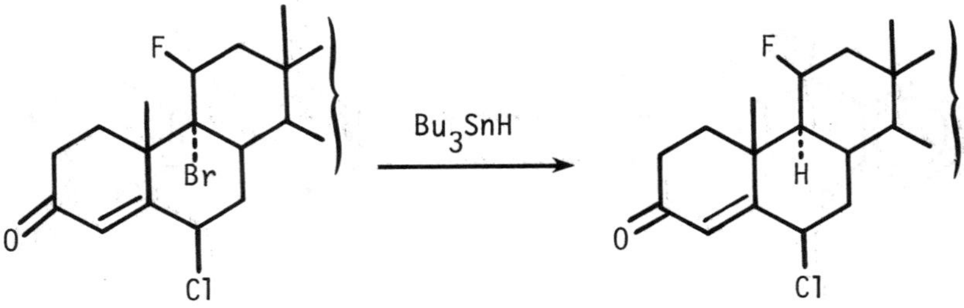

JOC **44**, 151 (1979)

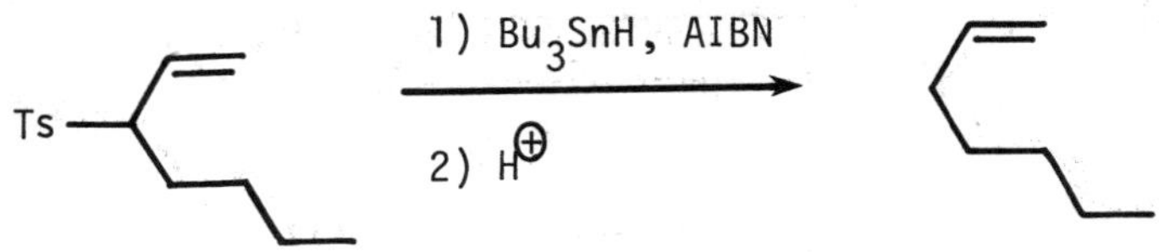

JACS **101**, 5414 (1979)

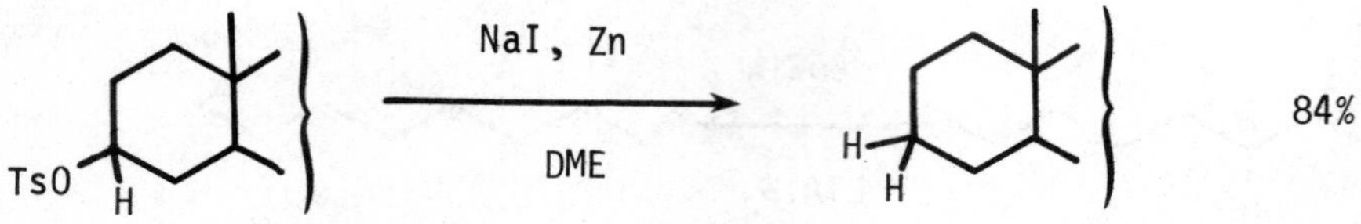

84%

Coll Czech Chem Comm 44, 246 (1979)

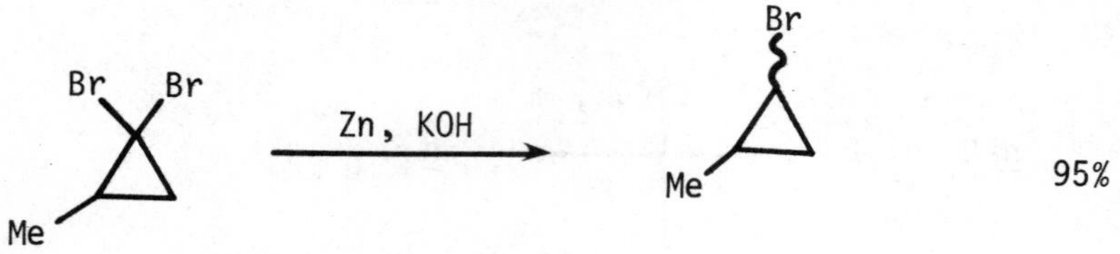

95%

JOC 43, 3500 (1978)

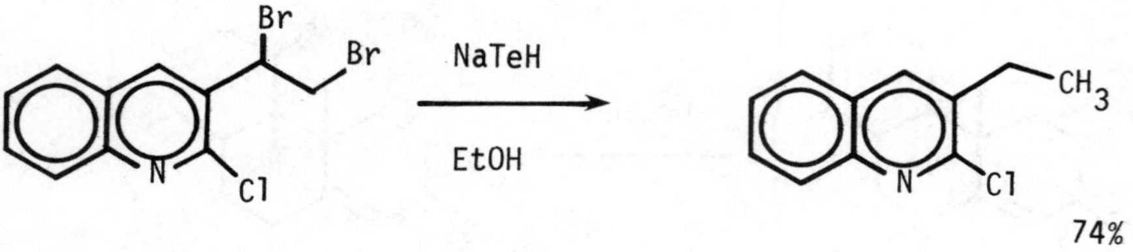

74%

Synthesis, 545 (1978)

80%

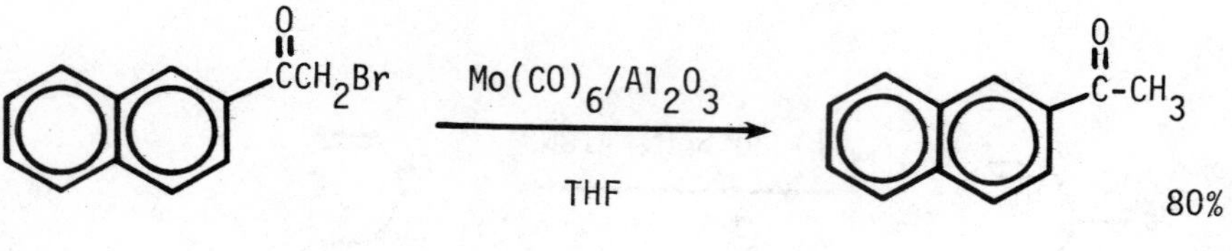

JOC 44, 2568 (1979)

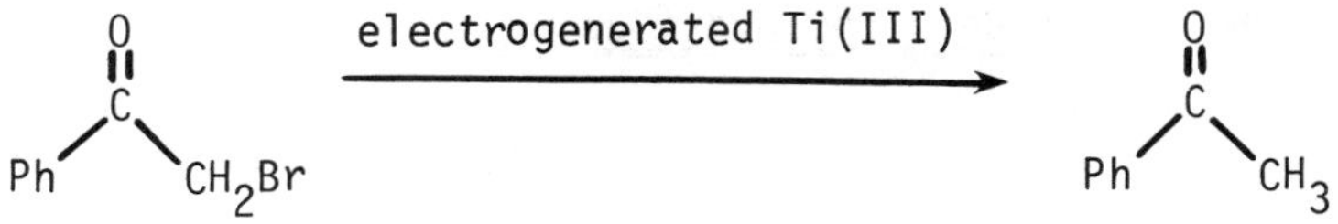

electrogenerated Ti(III)
O
C
Ph
CH2Br
O
C
Ph
CH3

58%

Can J Chem 56, 2269 (1978)

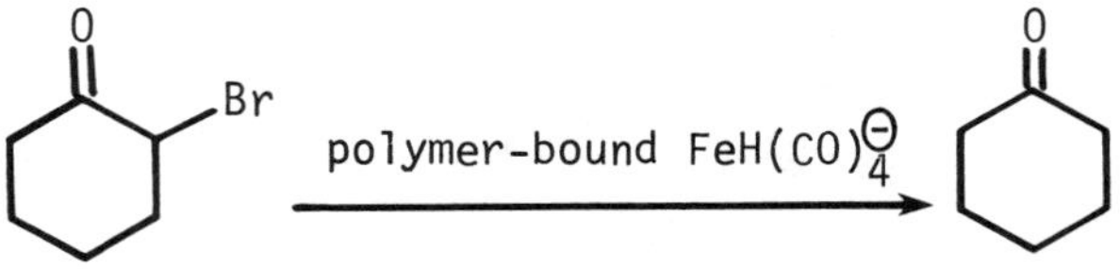

O
Br
polymer-bound FeH(CO)4
O

92%

JOC 43, 1598 (1978)

Ce(III), NaI
O
Br
THF
O

90%

Synth Comm 9, 241 (1979)

O
Cl
NaI
pyridine·SO3
O

86%

Synthesis, 59 (1979)

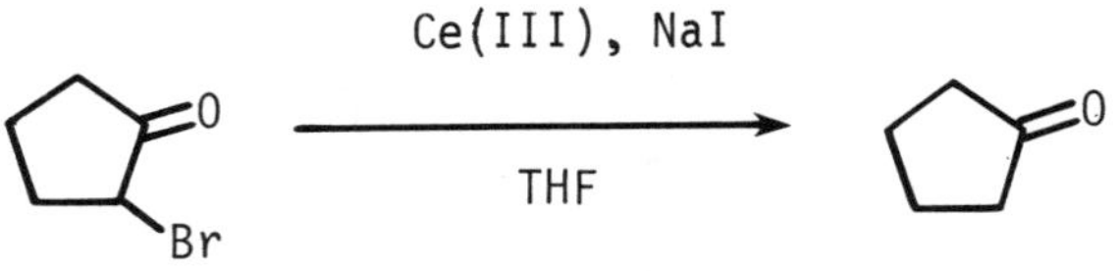

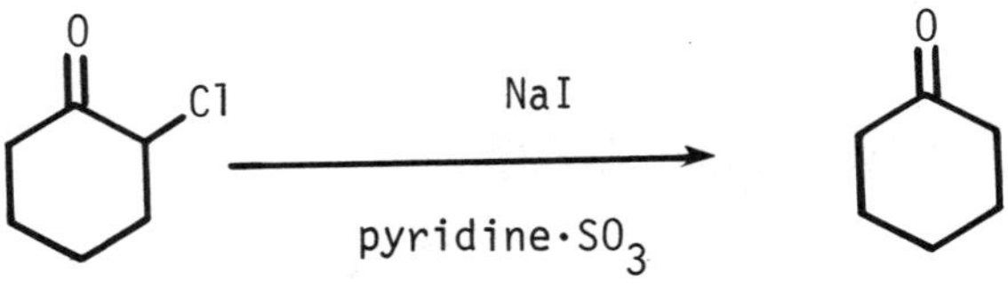

Section 161   Hydrides from Hydrides

No additional examples

Section 162   Hydrides from Ketones

No additional examples

For the conversion $R_2CO \rightarrow R_2CH_2$ or R_2CHR' see Section 72 (Alkyls and Methylenes from Ketones)

Section 163   Hydrides from Nitriles

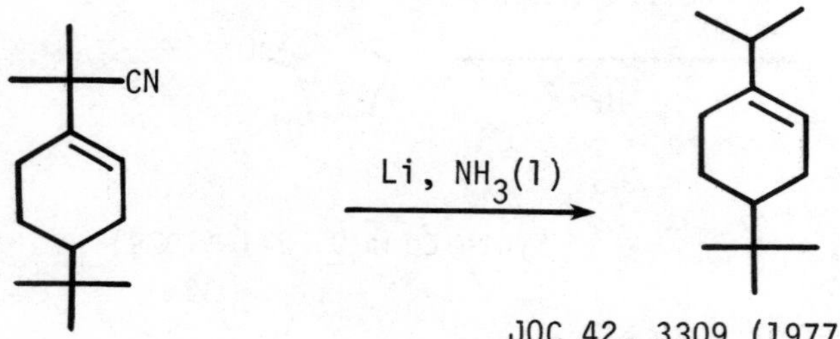

Li, NH$_3$(l)

JOC 42, 3309 (1977)

80%

Section 164   Hydrides from Olefins

No additional examples

Section 165 Hydrides from Miscellaneous Compounds

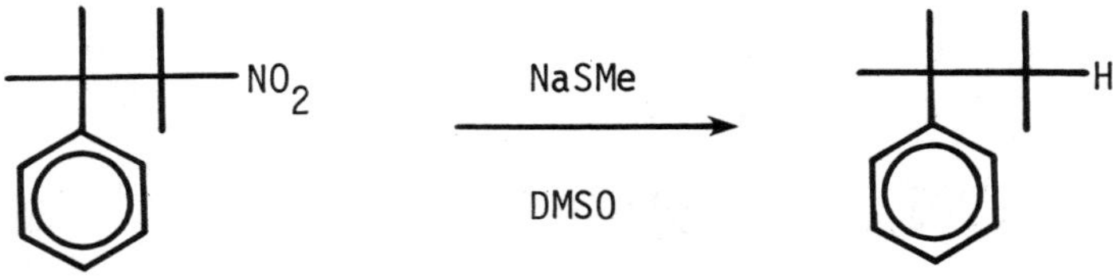

94%

Many examples (must be tertiary).

JACS 101, 647 (1979)

$$\text{R-N=C} \xrightarrow[\text{or } Na^{\oplus}\ Napth^{\ominus}]{Na,\ NH_3(l)} \text{R-H} \qquad \sim 90\%$$

JOC 43, 2396 (1978)

CHAPTER 12
PREPARATION OF KETONES

Section 166   <u>Ketones from Acetylenes</u>

$$Bu-C{\equiv}C-H \xrightarrow[\text{EtOH, H}_2\text{O}]{\text{Hg/Nafion-H}} Bu-\overset{\overset{\displaystyle O}{\|}}{C}-CH_3 \qquad 90\%$$

Synthesis, 671 (1978)

$$Ph-C{\equiv}C-Pr \xrightarrow[\text{CH}_2\text{Cl}_2\text{, Adogen}]{\text{KMnO}_4\text{, H}_2\text{O}} Ph-\overset{\overset{\displaystyle O}{\|}}{C}-\overset{\overset{\displaystyle O}{\|}}{C}-Pr \qquad 81\%$$

Synthesis, 462 (1978)

$$CH_3(CH_2)_6-C{\equiv}C-(CH_2)_6CH_3 \xrightarrow[\text{acetone}]{\text{KMnO}_4} CH_3(CH_2)_6-\overset{\overset{\displaystyle O}{\|}}{C}-\overset{\overset{\displaystyle O}{\|}}{C}-(CH_2)_6CH_3$$
$$81\%$$

JOC <u>44</u>, 1574 (1979)

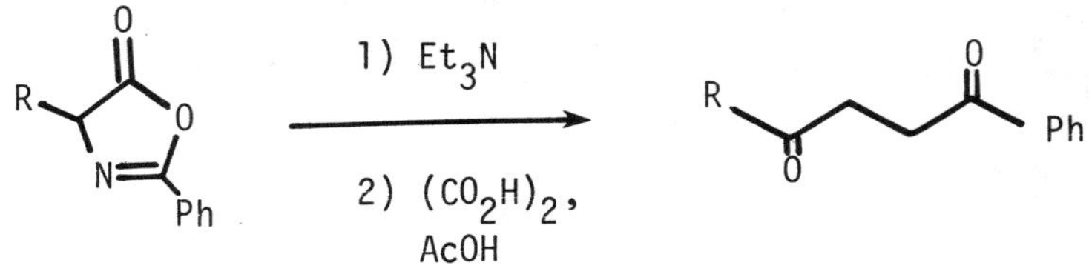

1) Et$_3$N

2) (CO$_2$H)$_2$, AcOH

+

HC≡C-COPh   Chem Ber 112, 322 (1979)

Section 167   Ketones from Carboxylic Acids and Acid Halides

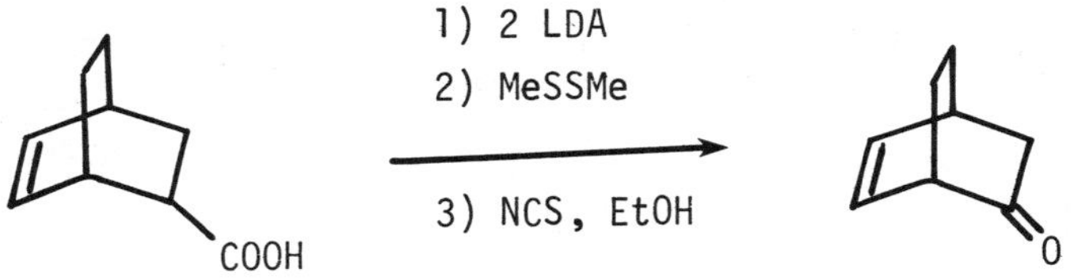

1) 2 LDA
2) MeSSMe

3) NCS, EtOH

JACS 99, 3101 (1977)

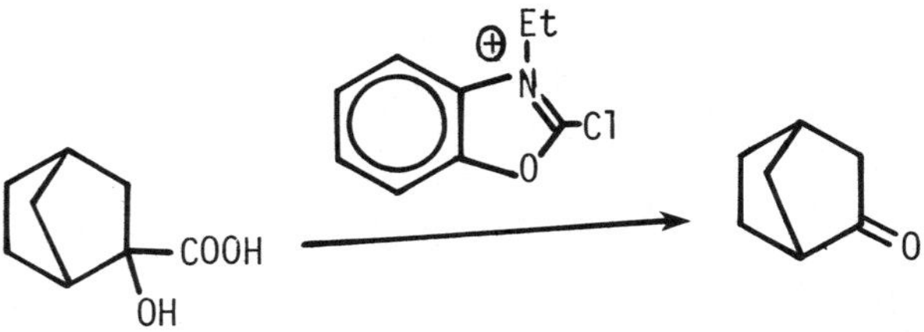

66%

Chem Lett, 49 (1978)

HOOC COOH
      \ /
C_6H_{13}-C-$(CH_2)_2COCH_3$

$\xrightarrow[\text{ether/water, NaOH}]{\text{electrolysis}}$

C_6H_{13}
        \
         C=O   63%
        /
$CH_3CO(CH_2)_2$

Tetr. Lett, 1047 (1979)

$\xrightarrow[170°]{NaCl \cdot AlCl_3}$

96%

JOC **44**, 3724 (1979)

Cl—⬡—C-Cl (O)

+

Ph-CH$_3$

$\xrightarrow{\text{Nafion-H}}$

82%

Synthesis, 672 (1978)

+ Ph-Cl

$\xrightarrow{Fe_2(SO_4)_3}$

79%

Synthesis, 54 (1977)

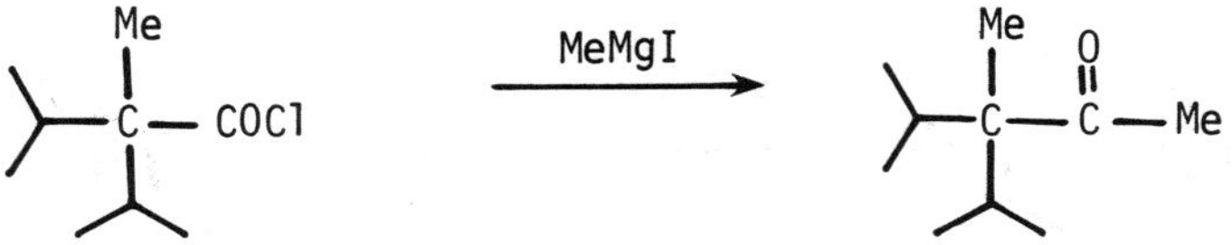

J Chem Research (S), 46 (1978)

$PrMgBr$ + $Ph-\overset{O}{\underset{\|}{C}}-Cl$ $\xrightarrow{THF}$ $Pr-\overset{O}{\underset{\|}{C}}-Ph$ 86%

Tetr Lett, 4303 (1979)

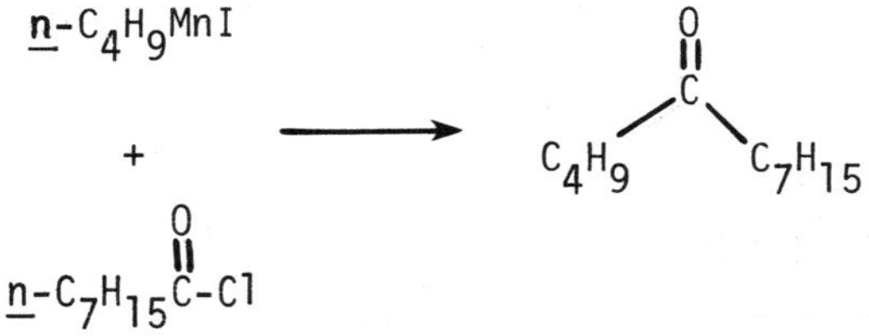

Synthesis, 130 (1977)

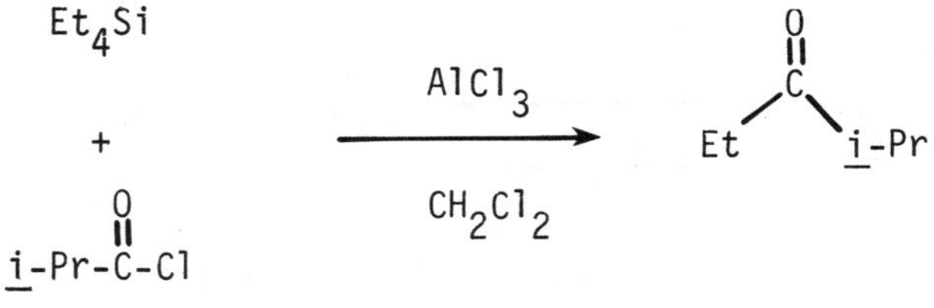

Synthesis, 677 (1977)

$$C_{11}H_{23}-\overset{\overset{\displaystyle O}{\|}}{C}-Cl \xrightarrow[\text{2) } H_2O]{\text{1) } Me_3SiCH_2AlCl_2} C_{11}H_{23}\overset{\overset{\displaystyle O}{\|}}{C}CH_3 \quad 70\%$$

J Prakt Chem <u>320</u>, 341 (1978)

$$PhCH_2CH_2\overset{\overset{\displaystyle O}{\|}}{C}-Cl \xrightarrow[\text{BzPdL}_2Cl]{Me_4Sn} PhCH_2CH_2\overset{\overset{\displaystyle O}{\|}}{C}CH_3 \quad 99\%$$

JOC <u>44</u>, 1613 (1979)

$$Ph-\overset{\overset{\displaystyle O}{\|}}{C}-Cl \xrightarrow{\begin{array}{c} Bu_3B, MeLi, \\ CuCl \cdot COD \end{array}} Ph-\overset{\overset{\displaystyle O}{\|}}{C}-Bu \quad 85\%$$

Tetr Lett, 173 (1977)

$$2 \ C_7H_{15}-\overset{\overset{\displaystyle O}{\|}}{C}-Cl \xrightarrow[\text{2) } CF_3COOH]{\text{1) } Fe_2(CO)_9} C_7H_{15}\overset{\overset{\displaystyle O}{\|}}{C}C_7H_{15} \quad 54\%$$

Tetr Lett, 3861 (1977)

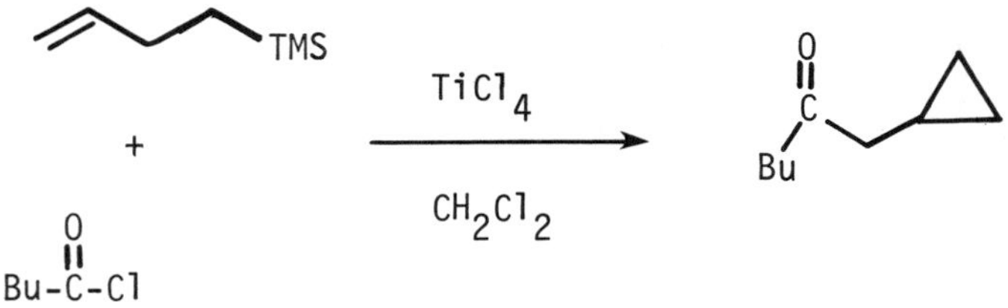
TMS
+
O
Bu-C-Cl
TiCl4
CH2Cl2
O
C
Bu

65%

Tetr Lett, 4045 (1977)

Section 168 Ketones from Alcohols and Phenols

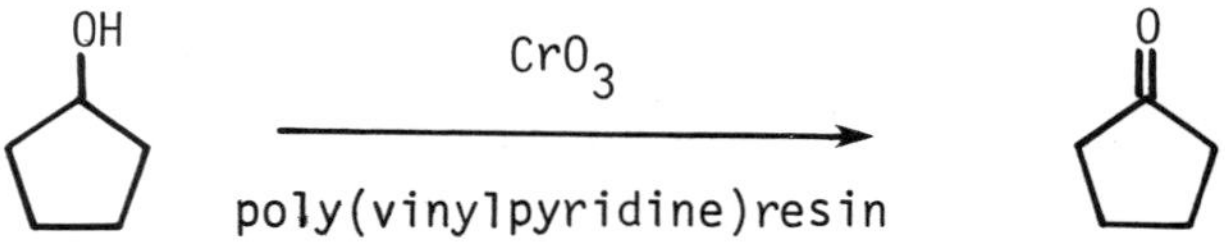
OH
CrO3
poly(vinylpyridine)resin
O

100%

JOC 43, 2618 (1978)

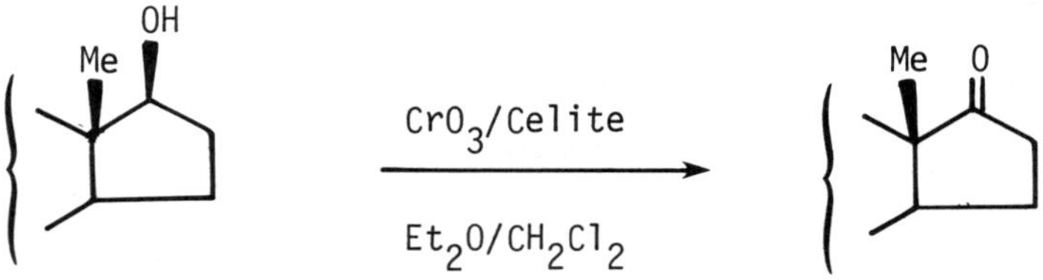
OH
Me
CrO3/Celite
Et2O/CH2Cl2
Me
O

71%

Synthesis, 815 (1979)

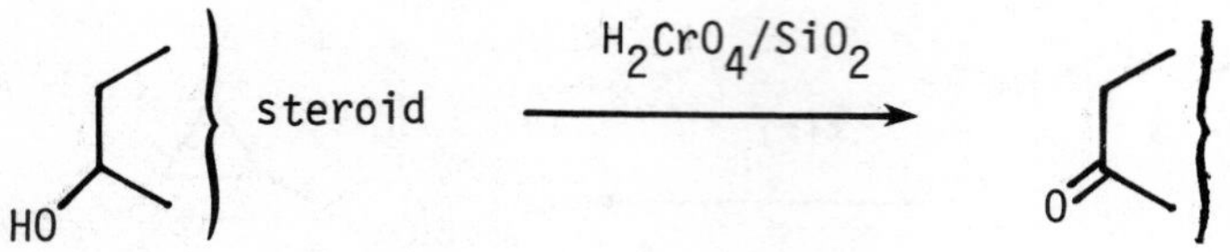

H_2CrO_4/SiO_2

steroid

88%

Synthesis, 534 (1978)

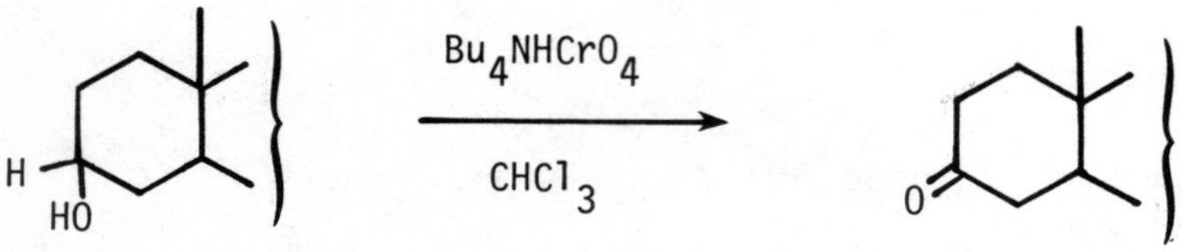

Bu_4NHCrO_4

$CHCl_3$

70%

Synthesis, 356 (1979)

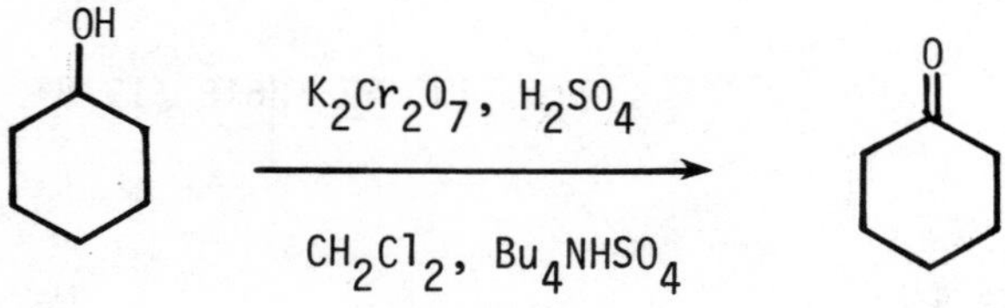

$K_2Cr_2O_7$, H_2SO_4

CH_2Cl_2, Bu_4NHSO_4

65%

Tetr Lett, 1601 (1978)

chromic acid

silica gel

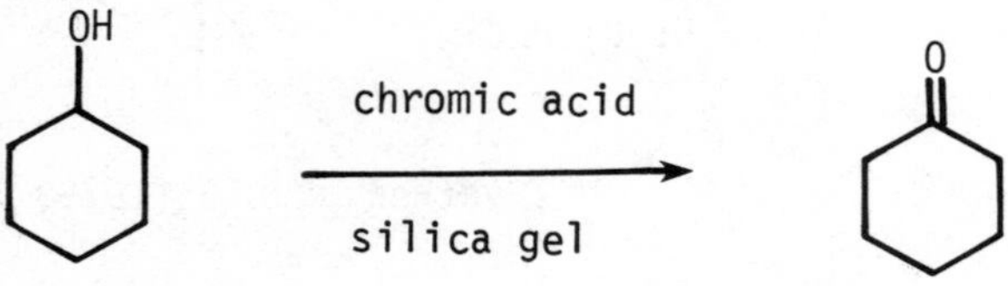

76%

Tetrahedron 35, 1789 (1979)

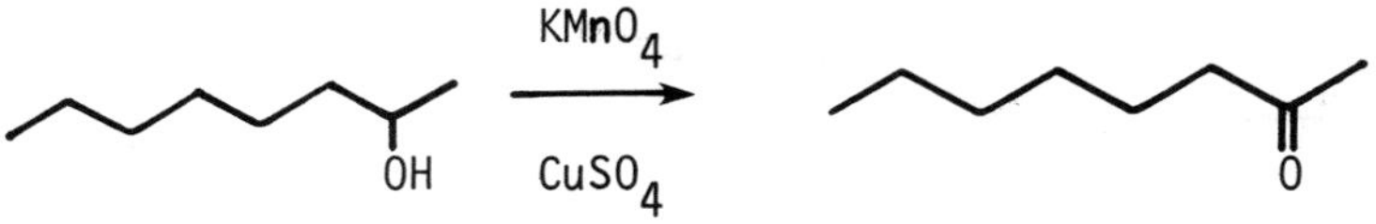

KMnO$_4$

CuSO$_4$

96%

JOC $\underline{44}$, 3446 (1979)

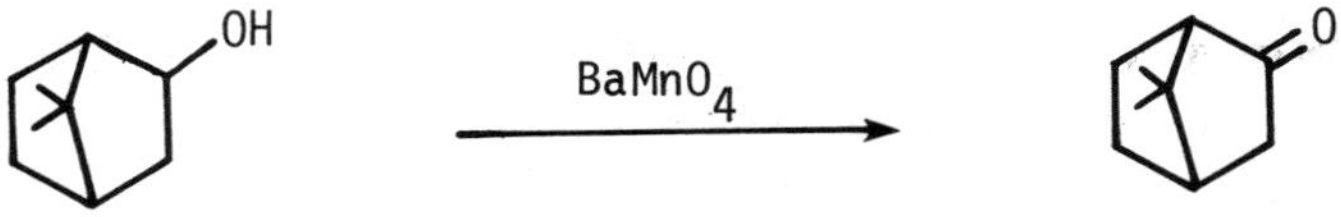

BaMnO$_4$

90%

Tetr Lett, 839 (1978)

1) DMSO, (COCl)$_2$

2) Et$_3$N

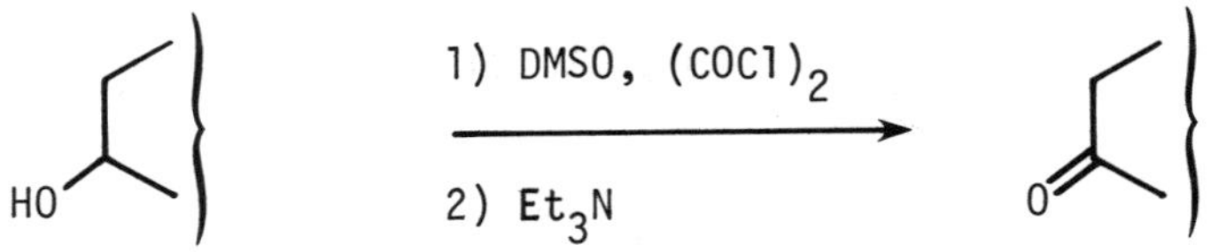

95%

Synthesis, 297 (1978)
Tetrahedron $\underline{34}$, 1651 (1978)
JOC $\underline{43}$, 2480 (1978)

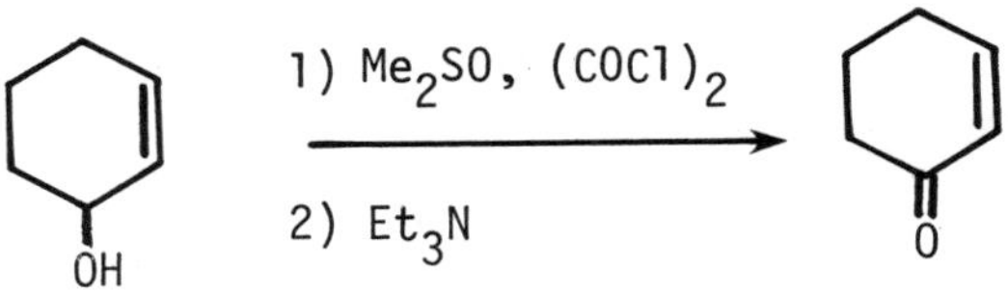

1) Me$_2$SO, (COCl)$_2$

2) Et$_3$N

98%

JOC $\underline{44}$, 4148 (1979)

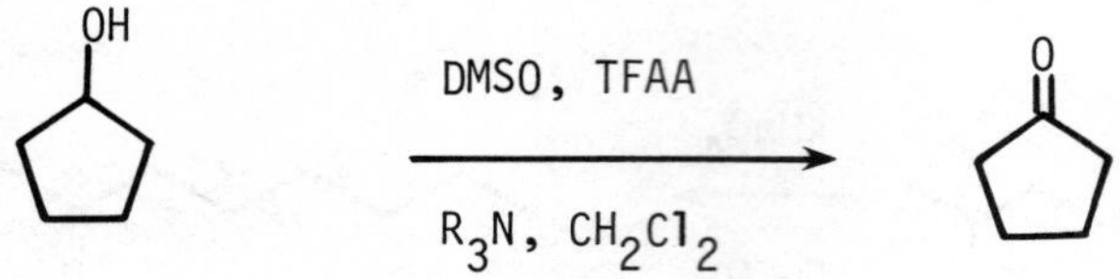

Synthesis, 297 (1978)

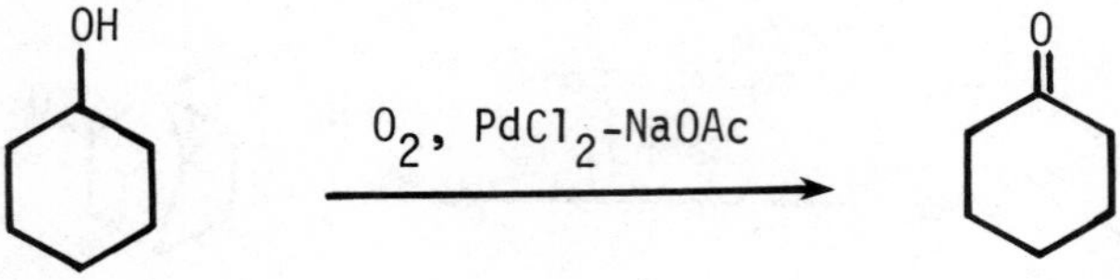

JCS Chem Comm, 157 (1977)

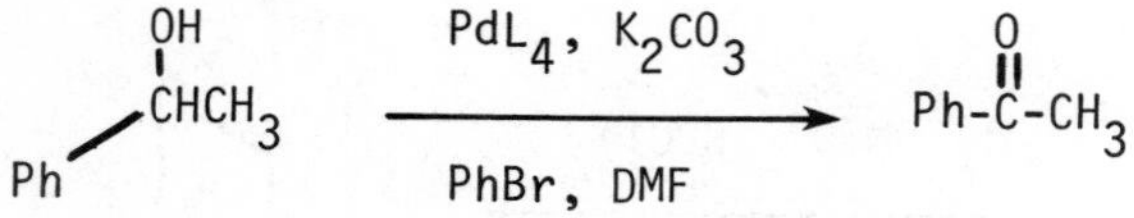

Tetr Lett, 1401 (1979)

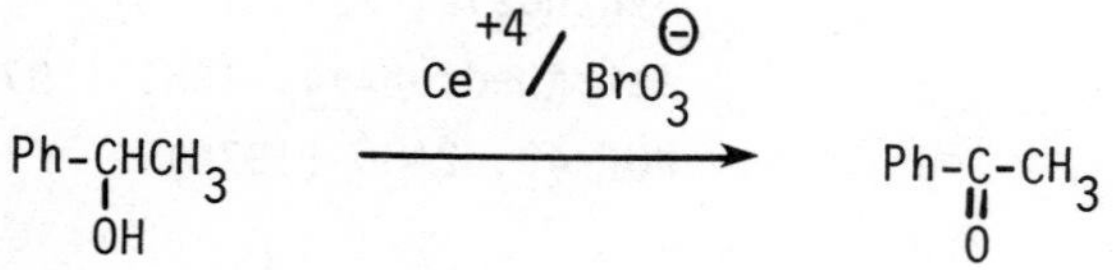

Synthesis, 936 (1978)

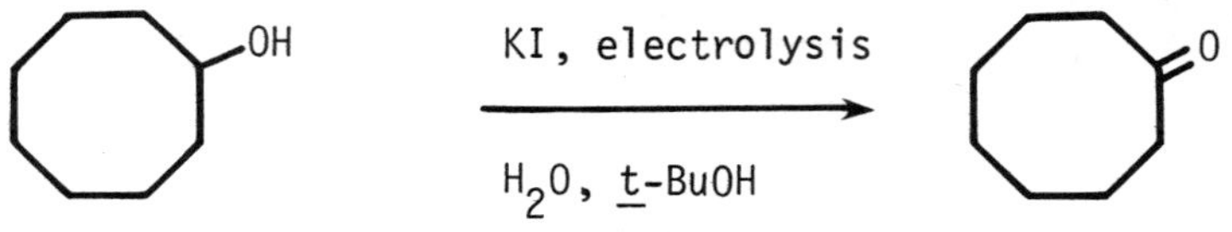
OH
KI, electrolysis
H2O, t-BuOH
O

59%

Tetr Lett, 165 (1979)

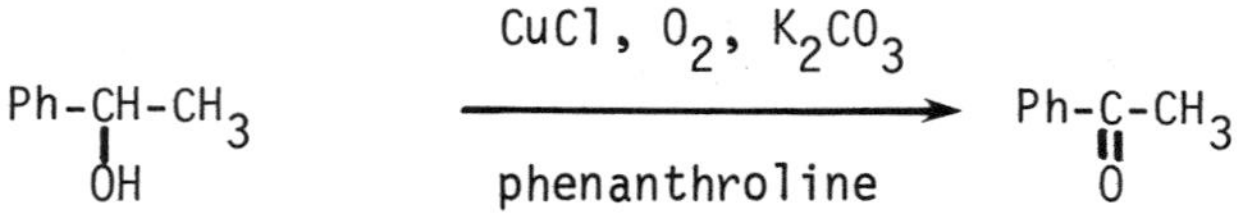
Ph-CH-CH3
OH
CuCl, O2, K2CO3
phenanthroline
Ph-C-CH3
O

Tetr Lett, 1215 (1977)

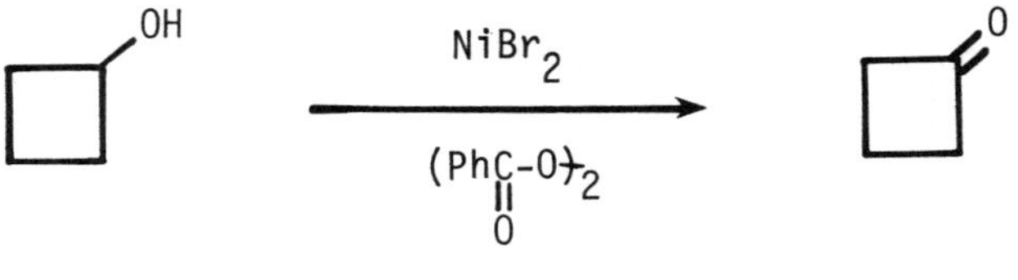
OH
NiBr2
(PhC-O)2
O
O

93%

JOC 44, 2955 (1979)

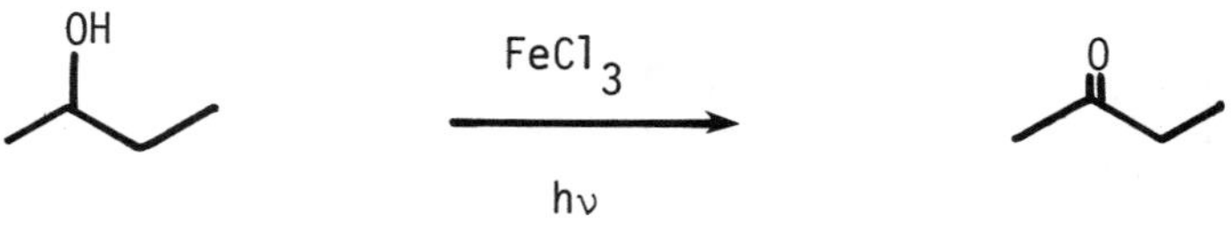
OH
FeCl3
hν
O

75%

JOC 42, 171 (1977)

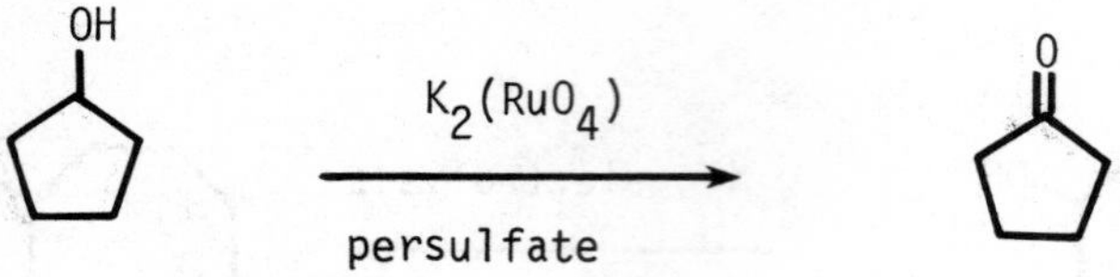

$K_2(RuO_4)$

persulfate

71%

JCS Chem Comm, 58 (1979)

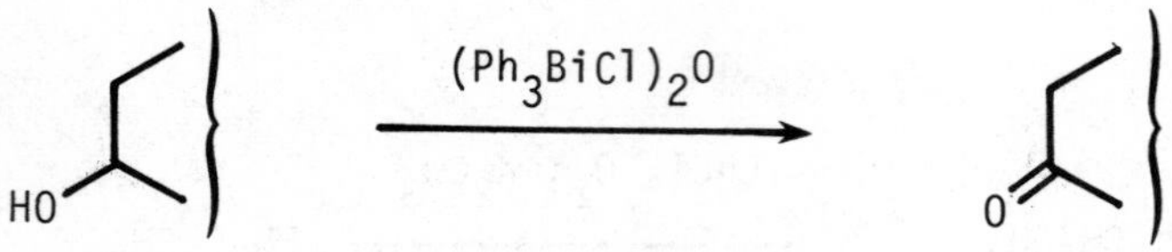

$(Ph_3BiCl)_2O$

~80%

JCS Chem Comm, 1099 (1978)

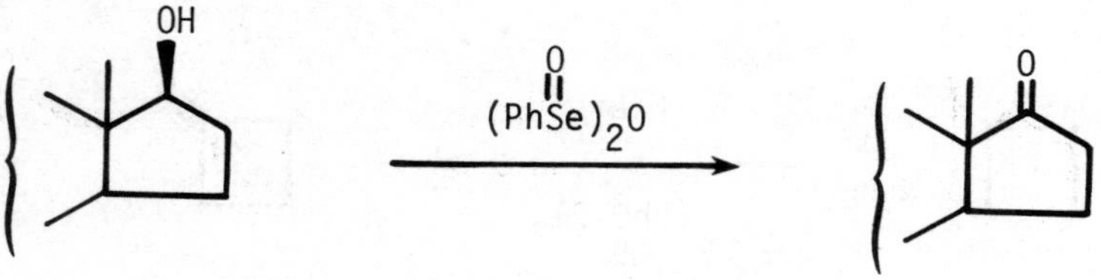

$(PhSe)_2O$

86%

JCS Chem Comm, 952 (1978)

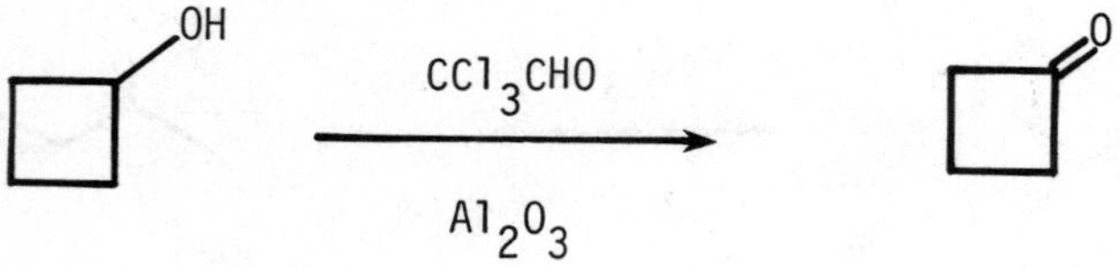

CCl_3CHO

Al_2O_3

99%

Synthesis, 555 (1977)

5-deazaflavin

KOH

82%

JCS Chem Comm, 825 (1977)

cholestanol

1) PrMgBr

2) azodicarbonyl
   dipiperidine

93%

BCS Japan $\underline{50}$, 2773 (1977)

$$\underset{\text{HCl/benzene}}{\overset{\text{DDQ, HIO}_4}{\longrightarrow}}$$

Ph-CH=CH-CH-Me   Ph-CH=CH-C-Me   80%
         | ‖
         OH   O

Synthesis, 848 (1978)

Ph$_3$CBF$_4$

CH$_2$Cl$_2$

59%

Tetr Lett, 2771 (1978)

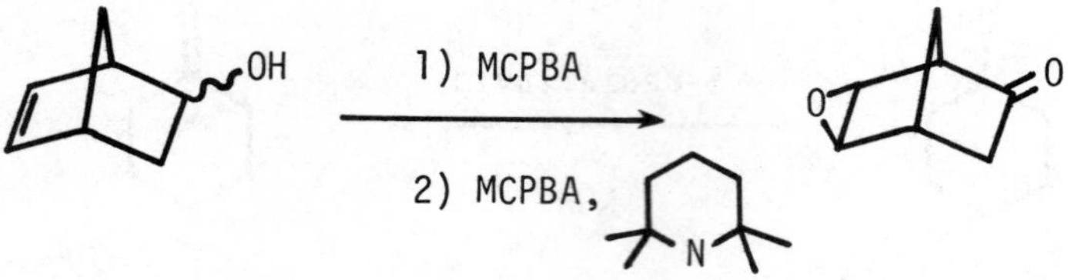

86%

JOC 42, 2077 (1977)

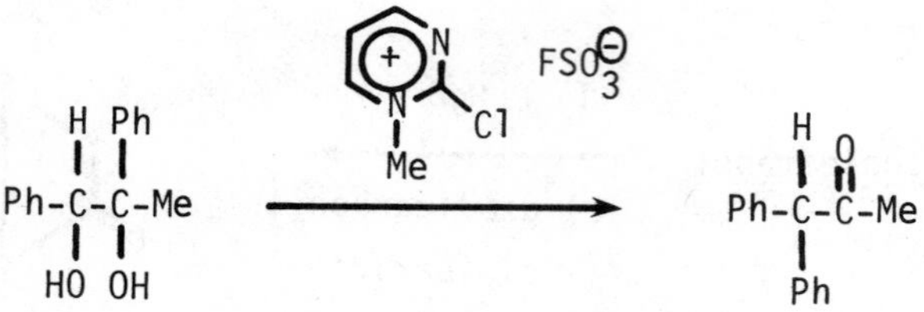

88%

Chem Lett, 179 (1977)

Related Methods: Aldehydes from Alcohols and Phenols (Section 48)

Section 169   Ketones from Aldehydes

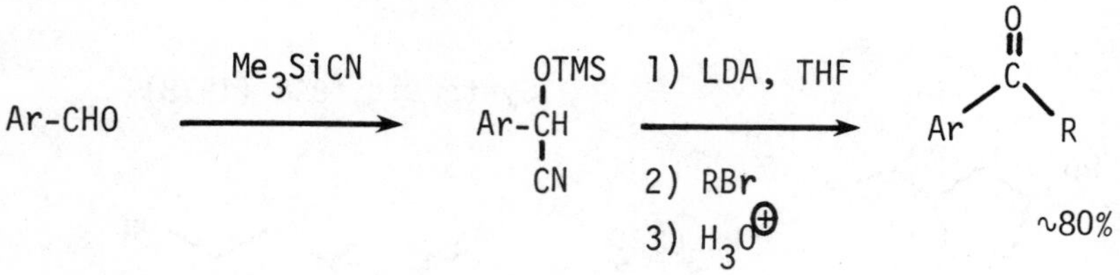

~80%

Ar = subst. Ph, pyridyl, furyl
R = Me, i-Pr, Bz

Chem Ber 112, 2045 (1979)

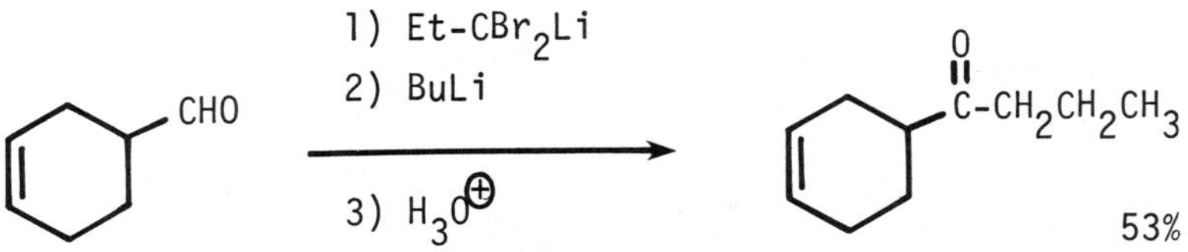

Synthesis, 968 (1979)

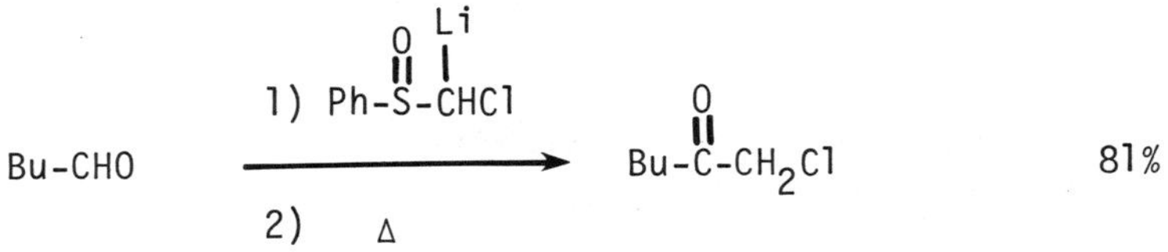

Tetr Lett, 1225 (1977)
Chem Lett, 209 (1979)

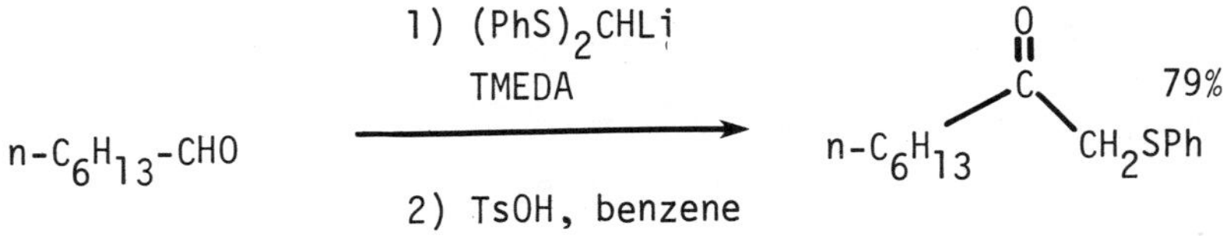

JCS Perkin I, 1074 (1979)

Section 170   Ketones from Alkyls, Methylenes, and Aryls

Synthesis, 144 (1979)

$$Ph-CH_2CH_2CH_2CH_3 \xrightarrow{BzEt_3NMnO_4} Ph\overset{\overset{\displaystyle O}{\|}}{-C}-CH_2CH_2CH_3 \quad 44\%$$

Angew Int Ed 18, 68 and 69 (1979)

$$PhCH_2Ph \xrightarrow{\underset{PhSe-O-SePh}{\overset{\overset{\displaystyle O}{\|}\;\overset{\displaystyle O}{\|}}{}}} \underset{Ph\qquad Ph}{\overset{\overset{\displaystyle O}{\|}}{C}} \quad 90\%$$

Tetr Lett, 3331 (1979)

$$C_7H_{15}-CH_2-CH=CH_2 \xrightarrow[\underline{t}-BuOOH]{SeO_2} C_7H_{15}-\overset{\overset{\displaystyle O}{\|}}{C}-CH=CH_2 \quad 61\%$$

JACS 99, 5526 (1977)

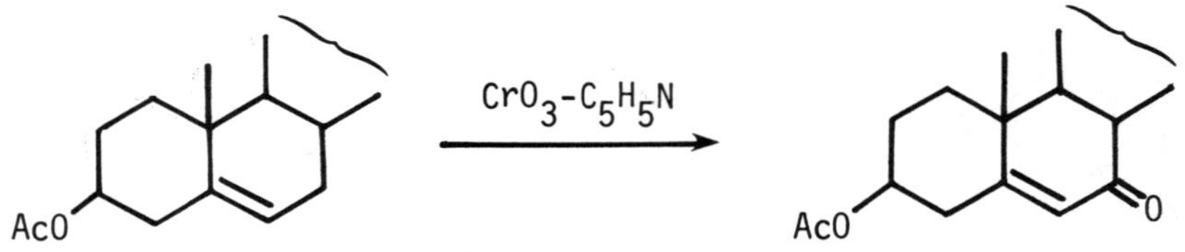

$CrO_3-C_5H_5N$

87%

J Chem Res (S), 42 (1979)

$\underline{n}-C_5H_{11}-CH_2-CH=CH-COOMe$ $\xrightarrow[\text{Ac}_2\text{O, HOAc}]{\text{CrO}_3}$ $\underline{n}-C_5H_{11}-\overset{\overset{\text{O}}{\|}}{C}-CH=CH-COOMe$

86%

Bull Chem Soc Japan <u>52</u>, 184 (1979)

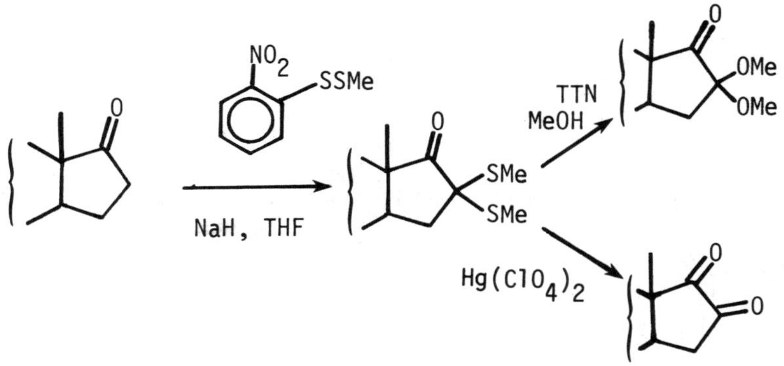

67%

NaH, THF

TTN
MeOH

$Hg(ClO_4)_2$

57%

Tetr Lett, 5021 (1978)

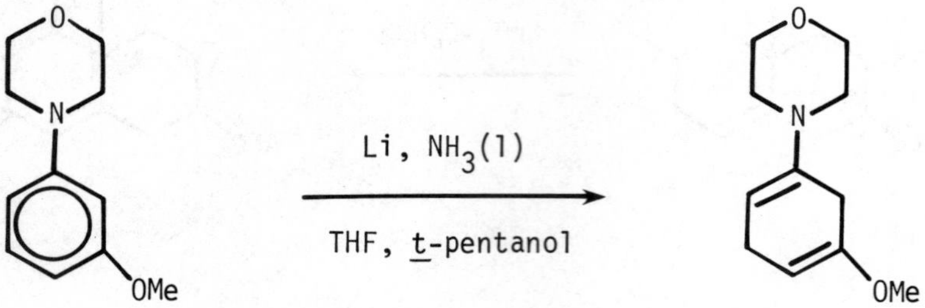
Li, NH_3(l)
THF, t-pentanol
OMe
OMe

Aust J Chem 31, 1625 (1978)

Section 171 Ketones from Amides

No additional examples

Section 172 Ketones from Amines

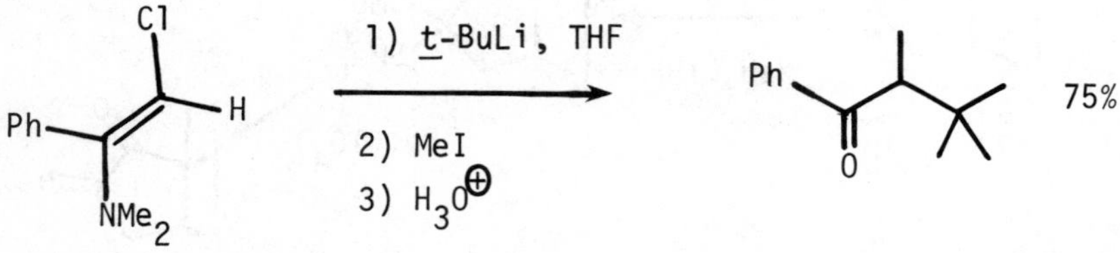
Cl
H
Ph
NMe_2
1) t-BuLi, THF
2) MeI
3) $H_3O^{\oplus}$
Ph
O
75%

JOC 44, 3585 (1979)

Section 173　　Ketones from Esters

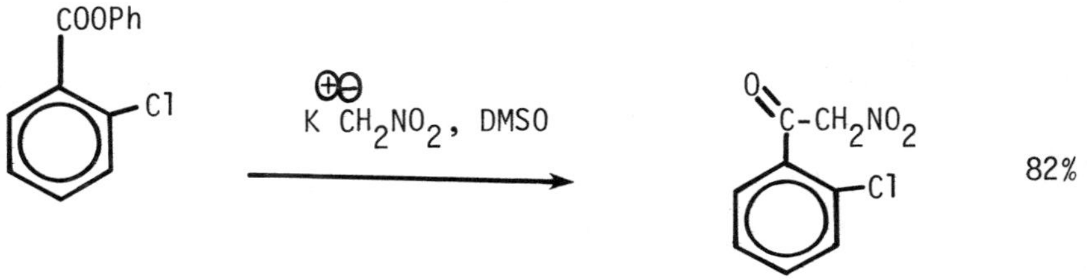

Synthesis, 295 (1979)

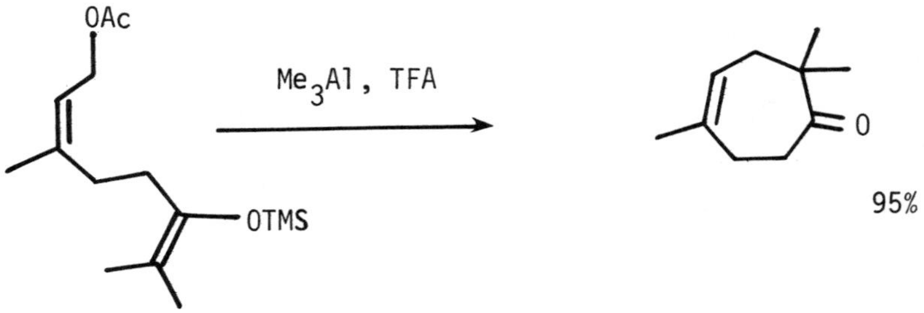

JACS 99, 4192 (1977)

Section 174   Ketones from Ethers and Epoxides

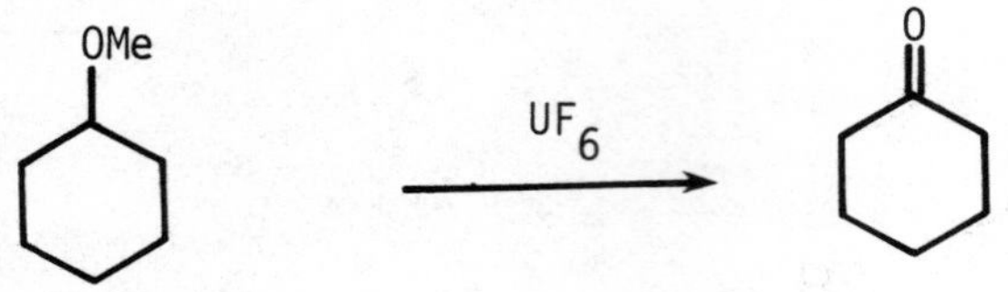

UF_6

86%

JACS 100, 5396 (1978)

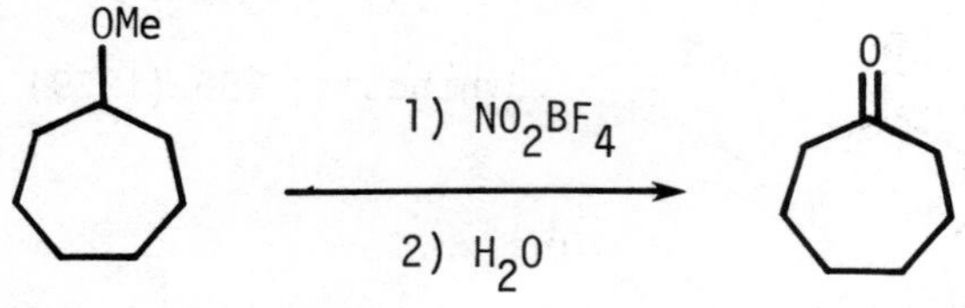

1) NO_2BF_4

2) H_2O

57%

JOC 42, 3097 (1977)

1) PhSeNa, EtOH

2) MCPBA, pyr

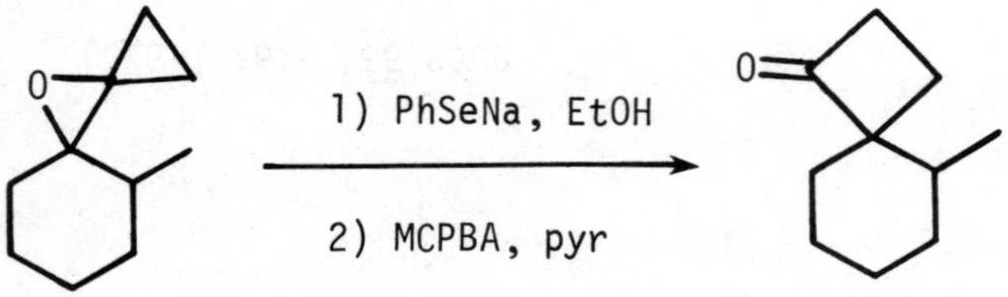

~80%

JACS 99, 7601 (1977)

Section 175   __Ketones from Halides__

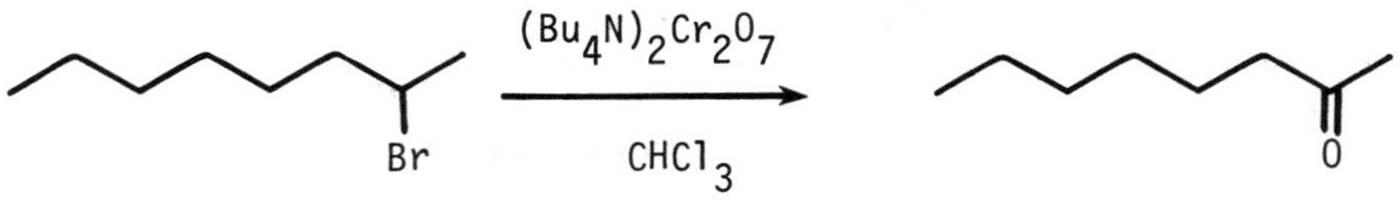

$(Bu_4N)_2Cr_2O_7$

$CHCl_3$

72%

Chem & Ind, 213 (1979)

1)

2)   Δ

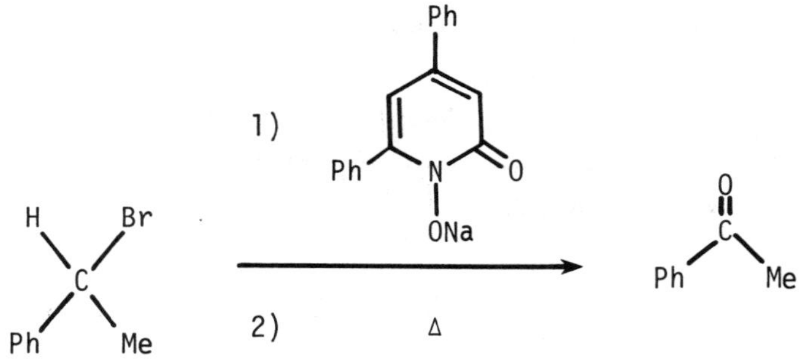

60%

JCS Perkin I, 2493 (1979)

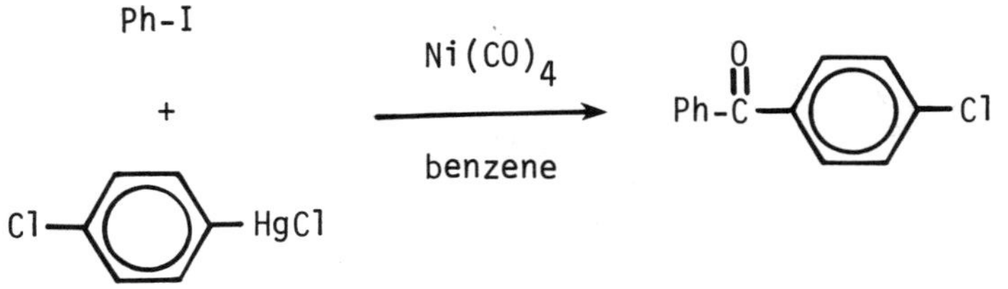

Ph-I

+

$Ni(CO)_4$

benzene

90%

Synthesis, 776 (1977)

Bu-MgBr $\xrightarrow[\text{2) MeI}]{\text{1) Fe(CO)}_5}$

$$\underset{\text{Bu}}{\overset{\overset{\displaystyle O}{\overset{\|}{C}}}{}} \underset{\text{Me}}{}$$

70%

Tetr Lett, 761 (1978)

PhI + Me$_4$Sn $\xrightarrow[\text{PhPd(PPh}_3)_2\text{I}]{\text{CO, HMPA}}$ Ph-$\overset{\overset{\displaystyle O}{\|}}{C}$-Me   123%

Tetr Lett, 2601 (1979)

2 PhCH$_2$Br $\xrightarrow[\text{Bu}_4\text{NBr}]{\text{Fe(CO)}_5, \text{ NaOH}}$ PhCH$_2$-$\overset{\overset{\displaystyle O}{\|}}{C}$-CH$_2$Ph   94%

Chem Lett, 321 (1979)

MeOCH$_2$COOMe $\xrightarrow[\text{2) anodic oxidation}]{\text{1) 2BuMgX}}$

64%

$$\underset{\text{Bu}}{\overset{\overset{\displaystyle O}{\overset{\|}{C}}}{}} \underset{\text{Bu}}{}$$

Tetr Lett, 3625 (1977)

1) Et_2NCH_2CN, LDA
2) LDA, R'X

R-X $\xrightarrow{\hspace{4cm}}$

3) $(COOH)_2$, H_2O/THF

$R-\overset{\overset{\displaystyle O}{\|}}{C}-R'$ ~80%

R, R' = 1° alkyl

Tetr Lett, 5175 (1978)

1) NaOH, Bu_4NI
2) BzBr

$TosCH_2N=C$ $\xrightarrow{\hspace{4cm}}$

3) i-PrI

$\underset{\displaystyle Bz \qquad i\text{-}Pr}{\overset{\overset{\displaystyle O}{\|}}{C}}$ 65%

Tetr Lett, 4229 (1977)

$\xrightarrow[\text{HOAc}]{Hg(OAc)_2,\ BF_3 \cdot Et_2O}$

60%

Tetr Lett, 1943 (1978)

$\xrightarrow[\text{CH}_2\text{Cl}_2]{Hg(OCOCF_3)_2}$

90%

Tetr Lett, 3489 (1979)

Related methods: Ketones from Ketones (Section 177), Aldehydes from Halides (Section 55)

Section 176   Ketones from Hydrides

This section lists examples of the replacement of hydrogen by ketonic groups, RH → RCOR'. For the oxidation of methylenes R_2CH_2 → R_2CO see Section 170 (Ketones from Alkyls and Methylenes)

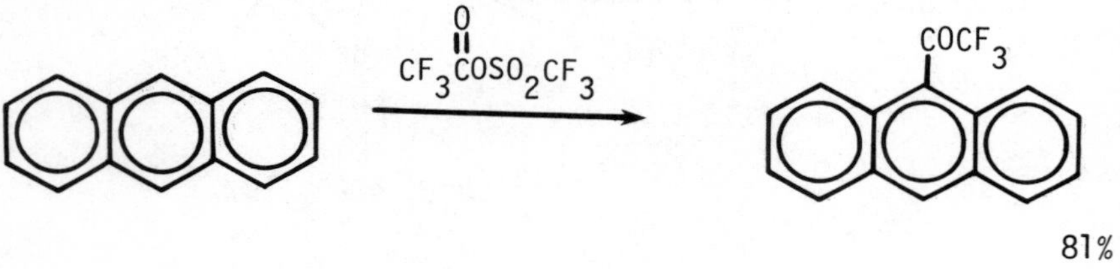

JOC <u>44</u>, 313 (1979)

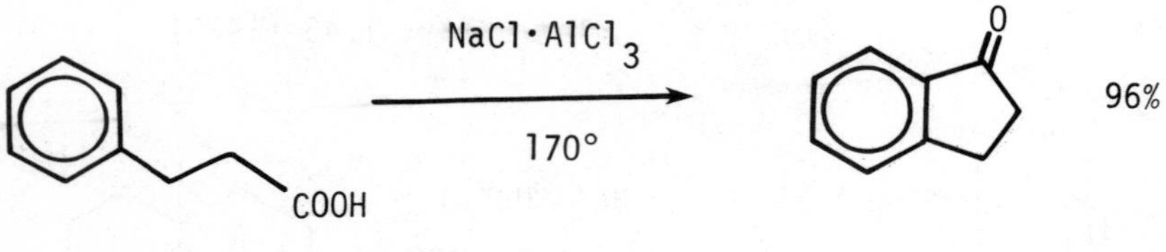

JOC <u>44</u>, 3724 (1979)

PhCH$_3$

+

Ph$_2$C=C=O

CO
———————————→
Rh$_4$(CO)$_{12}$

Ph$_2$CH-C—⟨benzene⟩—CH$_3$ 57%

Chem Lett, 535 (1978)

Section 177   Ketones from Ketones

This section contains alkylations of ketones and protected ketones, ketone transpositions and annelations, ring expansions and ring openings, and dimerizations. Conjugate reductions and Michael alkylations of enones are listed in Section 74 (Alkyls from Olefins).

For the preparation of enamines from ketones see Section 356 (Amine-Olefin).

OBR$_3^{\ominus}$ K$^{\oplus}$ (cyclohexenyl)

MeI
———————————→
THF

(2-methylcyclohexanone)   95%

Tetr Lett, 845 (1979)

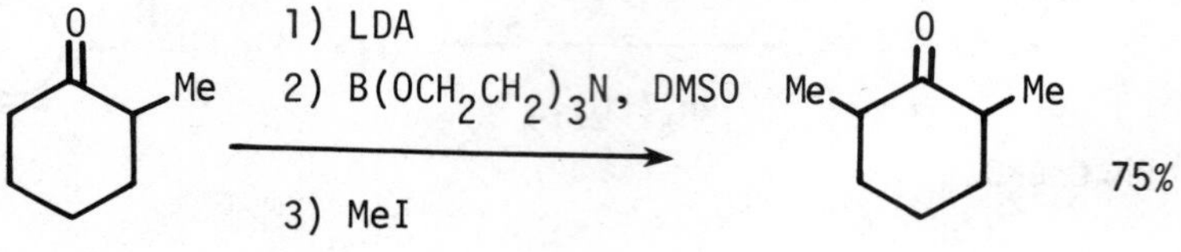

Synth Comm 8, 9 (1978)

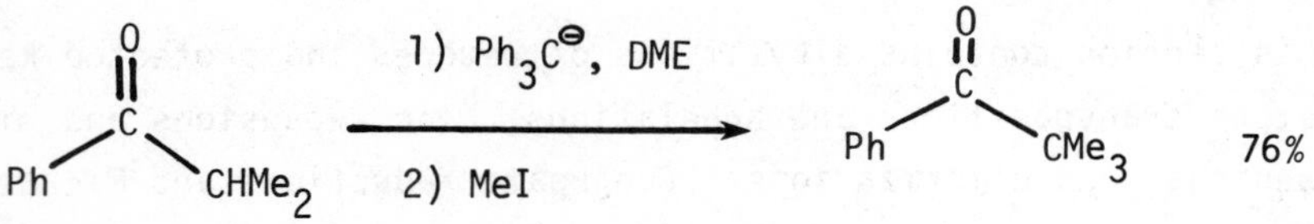

Synth Comm 7, 137 (1977)

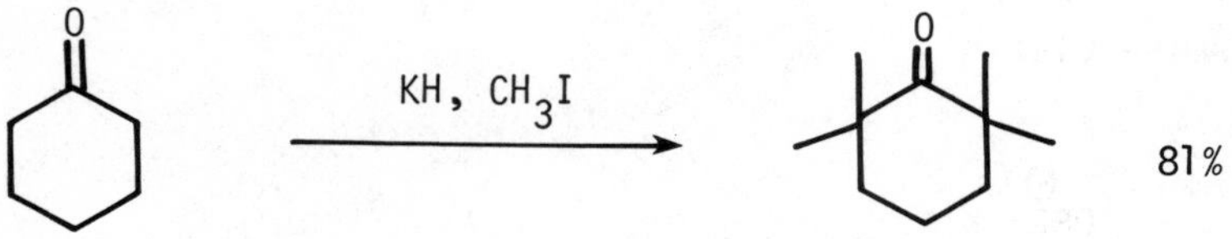

JOC 43, 1834 (1978)

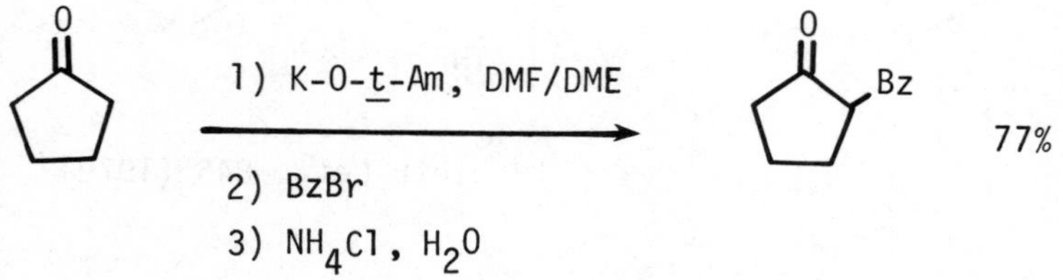

Synth Comm 8, 563 (1978)

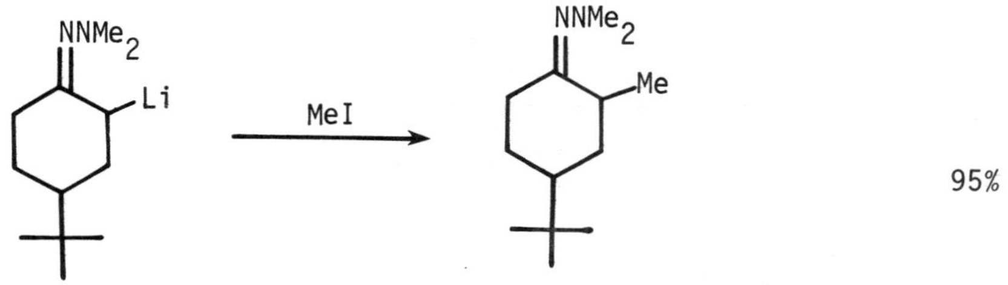
NNMe2
Li
MeI
NNMe2
Me
95%

Chem Ber 111, 1337 (1978)

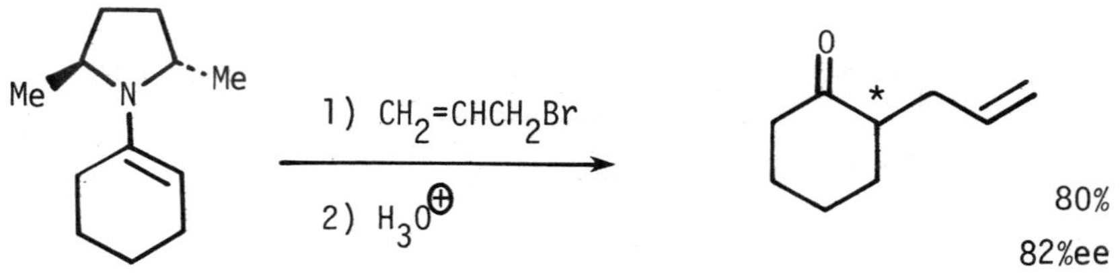
Me
N
Me
1) CH2=CHCH2Br
2) H3O⊕
O
*
80%
82%ee

JOC 42, 1663 (1977)

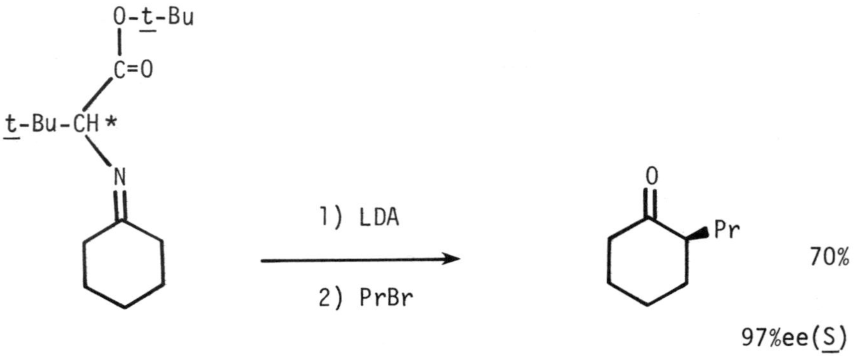
O-t-Bu
C=O
t-Bu-CH *
N
1) LDA
2) PrBr
O
Pr
70%
97%ee(S)

Tetr Lett, 573 (1978)

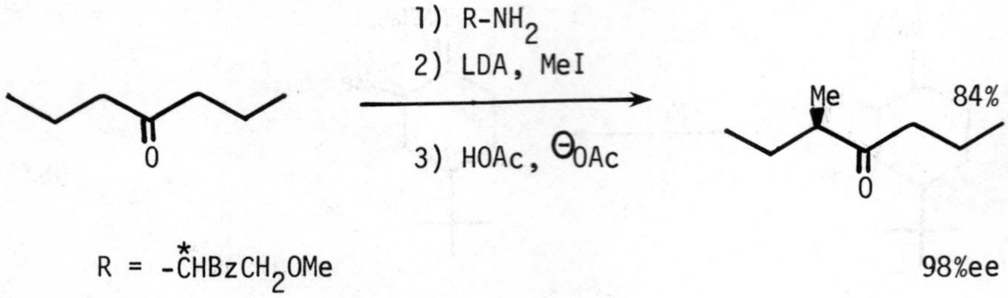

R = -*CHBzCH$_2$OMe

84%

98%ee

JOC <u>43</u>, 3245 (1978)

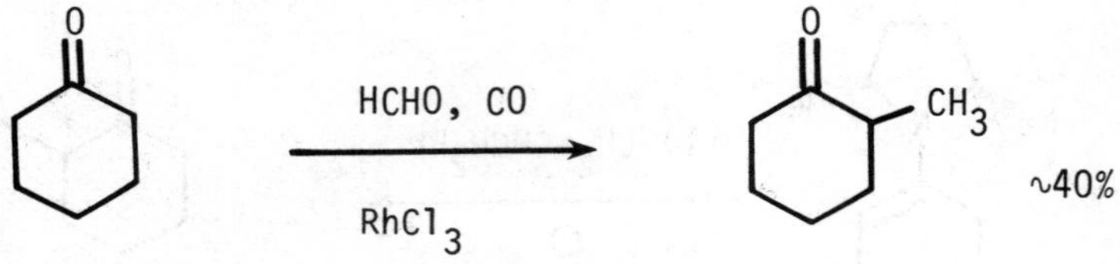

~40%

Chem Lett, 215 (1978)

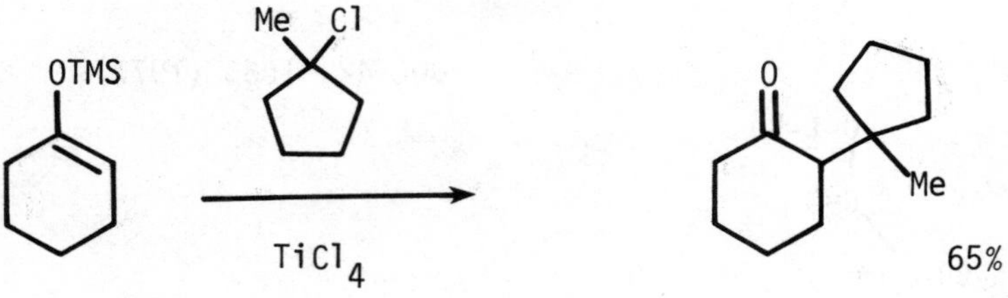

65%

Tetr Lett, 1427 (1979)

Ph-CH=C
  COCH$_3$
  COPh

[FeH(CO)$_4$]$^{\ominus}$

EtOH

PhCH$_2$CH$_2$CPh

69%

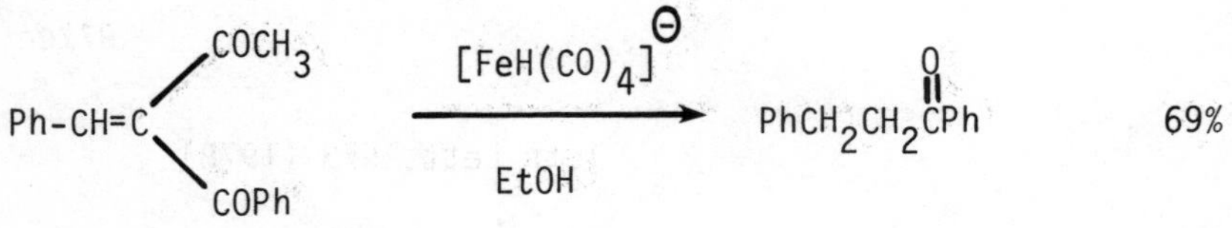

BCS Japan <u>51</u>, 835 (1978)

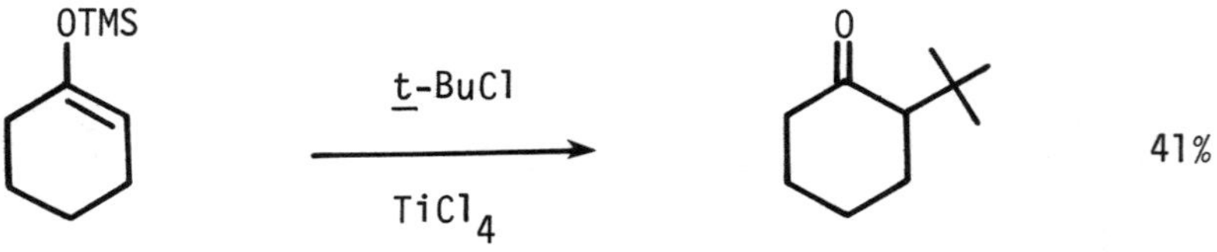

Tetr Lett, 4183 (1977)

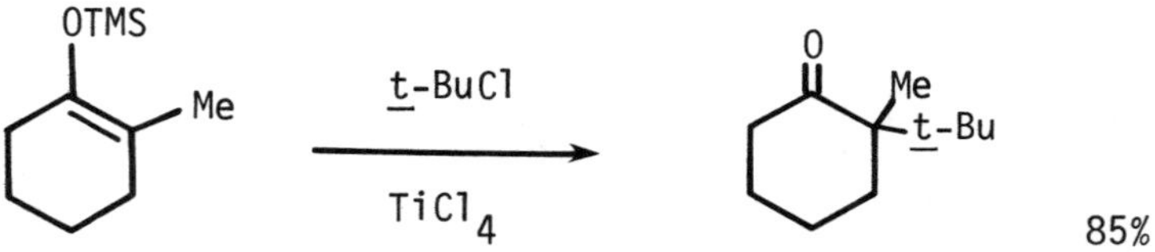

Angew Int Ed 17, 48 (1978)

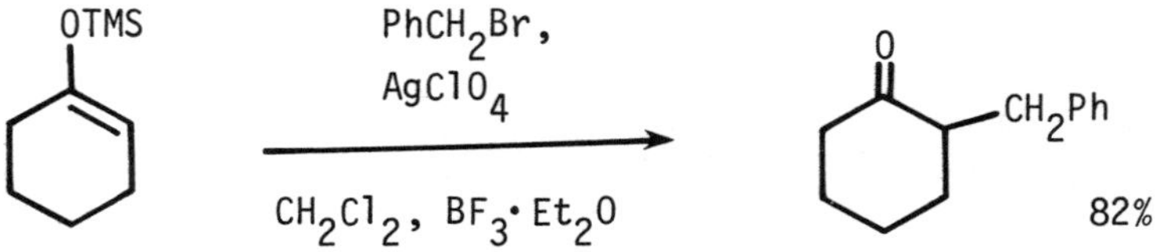

Bull Chem Soc Japan 52, 1241 (1979)

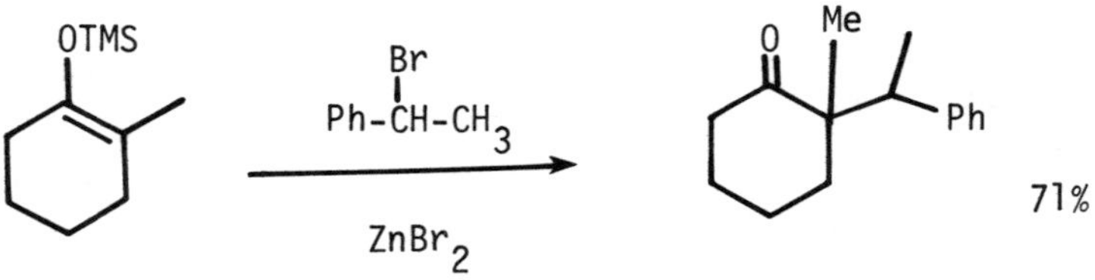

Tetr Lett, 1519 (1979)

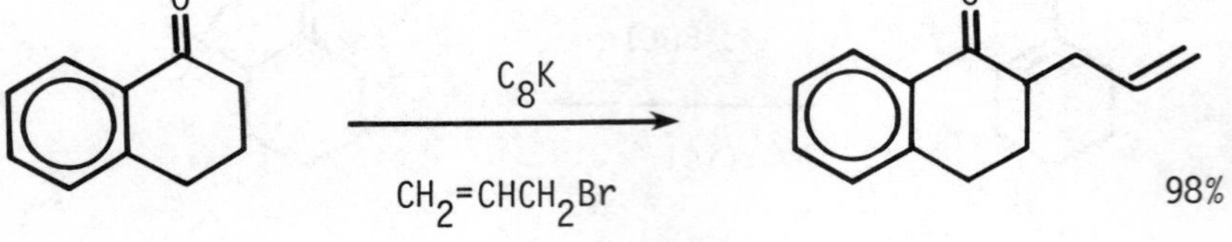

Tetr Lett, 3121 (1977)

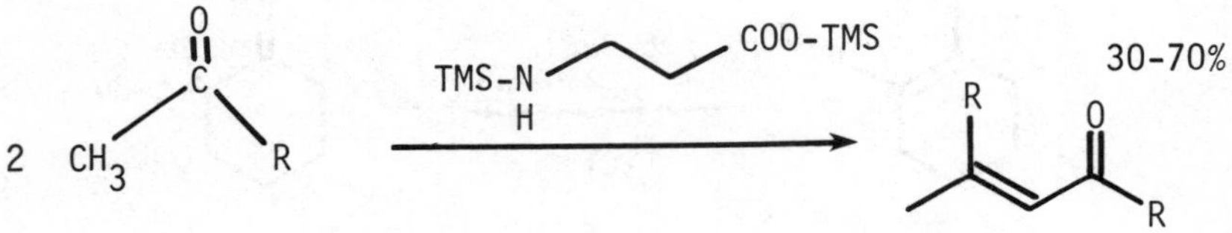

Angew Int Ed 16, 251 (1977)

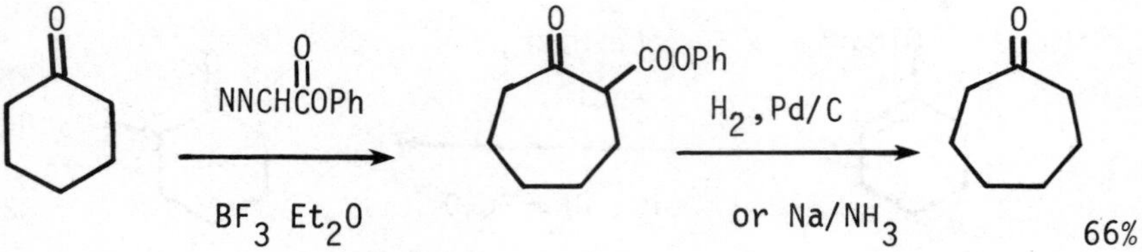

Synth Comm 8 413 (1978)

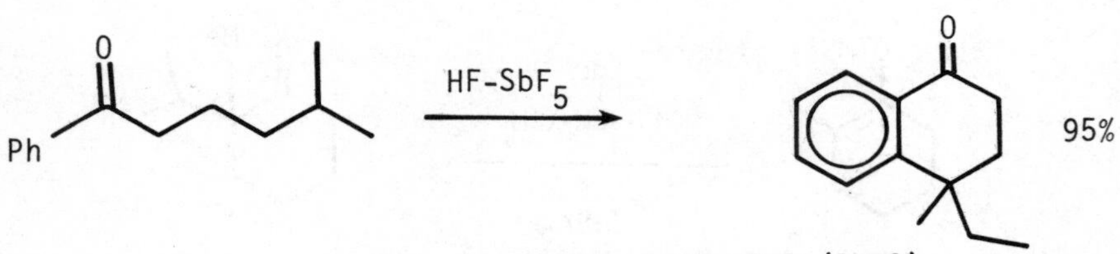

Chem Lett, 231 (1978)

Also works with styrene etc.

JACS <u>100</u>, 1791 and 1799 (1978)

$$Ph-CH-CN \quad \xrightarrow[\text{2) BuBr}]{\text{1) LDA, THF}} \quad Ph-\overset{\overset{\displaystyle O}{\|}}{C}-CH_2-Bu \quad 63\%$$

(Ph-CH-CN with NMe$_2$ substituent)

3) $H_3O^{\oplus}$

Synthesis, 127 (1979)

Review: "Synthesis of Aldehydes, Ketones, and Carboxylic Acids from Lower Carbonyl Compounds by C-C Coupling Reactions"

Synthesis, 633 (1979)

Ketones may also be alkylated and homologated via olefinic ketones (Section 374)

Related methods: Aldehydes from Aldehydes (Section 49)

Section 178   Ketones from Nitriles

No additional examples

Section 179   Ketones from Olefins

$$\text{~~~~} \xrightarrow[\substack{2) \text{ } \text{CH}_2\text{=CH-C(=O)-CH}_3}]{1) \text{ LiAlH}_4, \text{ TiCl}_4} \text{C}_8\text{H}_{17}\text{-C(=O)-CH}_3 \qquad 54\%$$

Chem Lett, 167 (1979)

$$\text{Bu-CH=CH}_2 \xrightarrow[\substack{2) \text{ PhCOCl}}]{1) \text{ LiAlH}_4, \text{ TiCl}_4} \text{BuCH}_2\text{CH}_2\text{C(=O)-Ph} \qquad 71\%$$

Chem Lett, 623 (1979)

$$\xrightarrow{\text{Pb(OAc)}_4} \qquad 52\%$$

Indian J Chem 14B, 704 (1976)

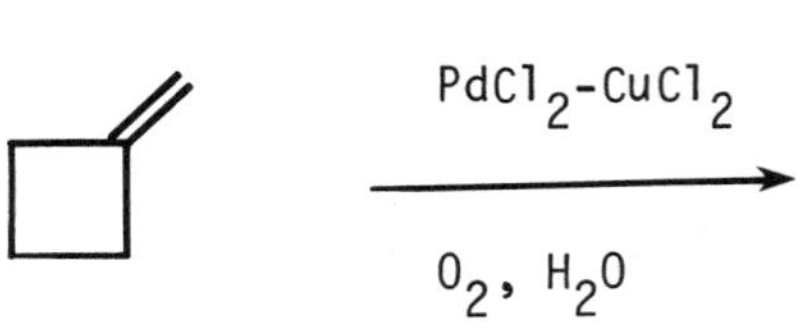

PdCl$_2$-CuCl$_2$

O$_2$, H$_2$O

75%

JCS Chem Comm, 583 (1977)

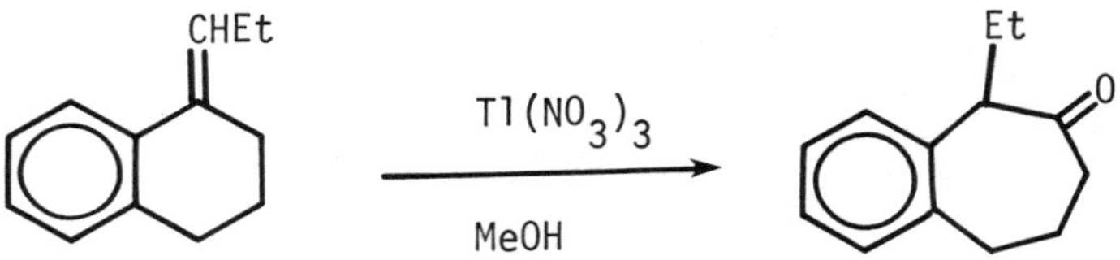

Tl(NO$_3$)$_3$

MeOH

96%

Tetr Lett, 1827 (1977)

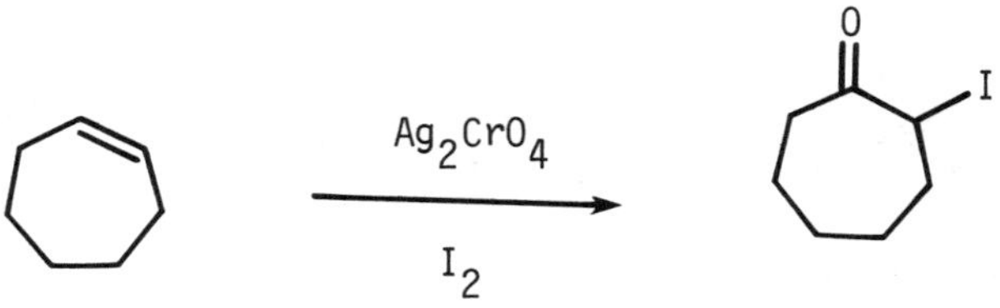

Ag$_2$CrO$_4$

I$_2$

65%

JOC _42_, 4268 (1977)

hν, O$_2$

pyridine, FeCl$_3$

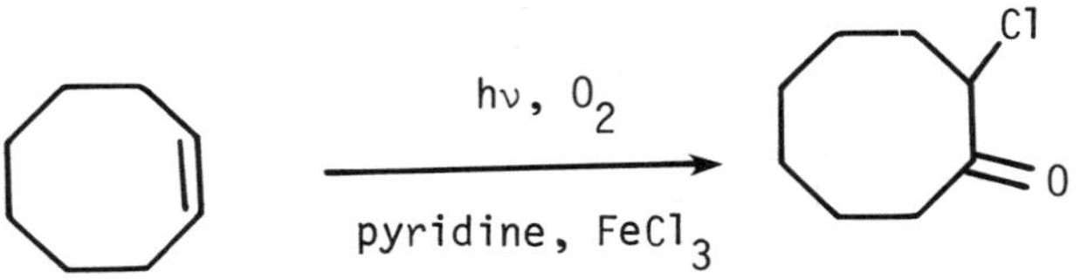

40%

Chem Lett, 161 (1978)

See also Section 134 (Ethers and Epoxides from Olefins) and
Section 174 (Ketones from Ethers and Epoxides).

Section 180   Ketones from Miscellaneous Compounds

Conjugate reductions and reductive alkylations of enones are
listed in Section 74 (Alkyls from Olefins).

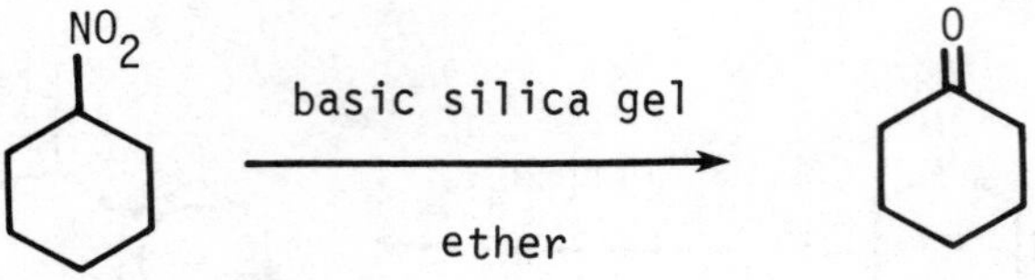

JACS 99, 3861 (1977)

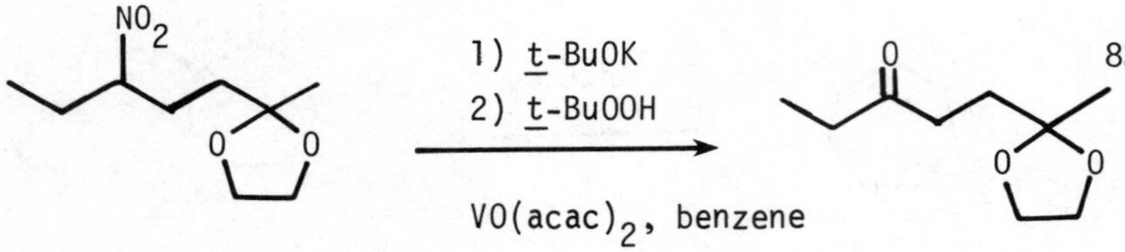

Tetr Lett, 331 (1977)

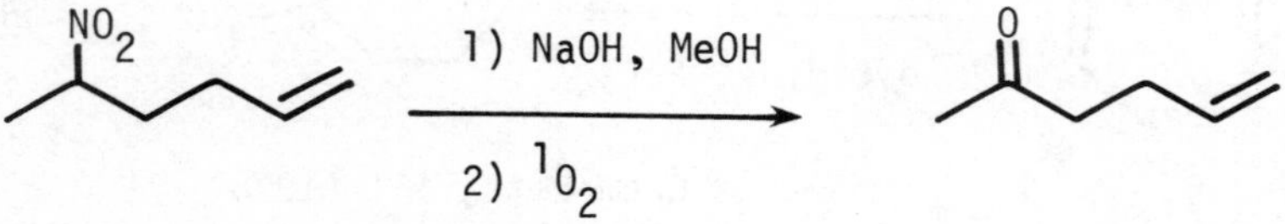

JOC 43, 1271 (1978)

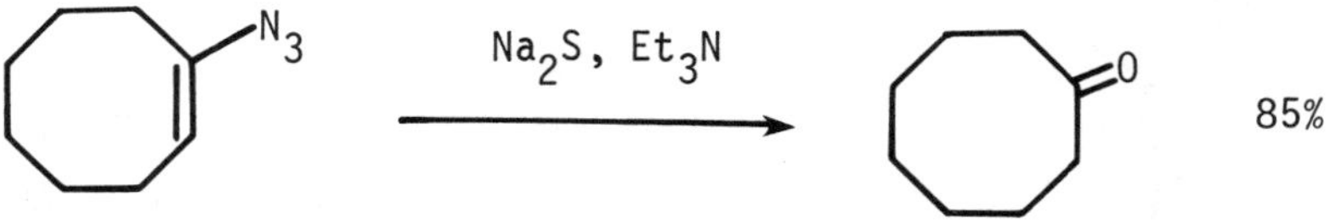

JOC **44**, 4712 (1979)

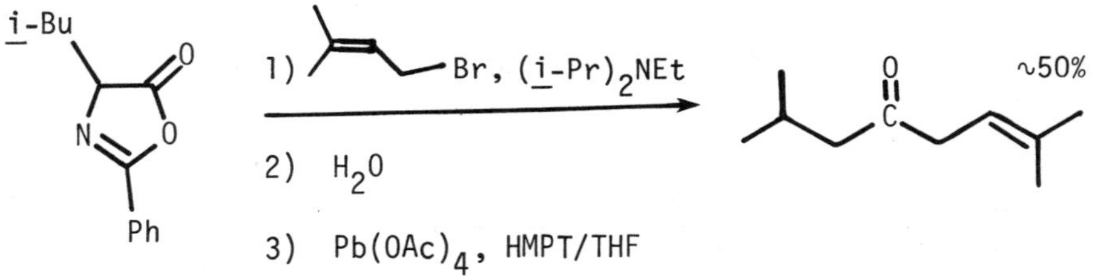

Angew Int Ed **17**, 450 (1978)

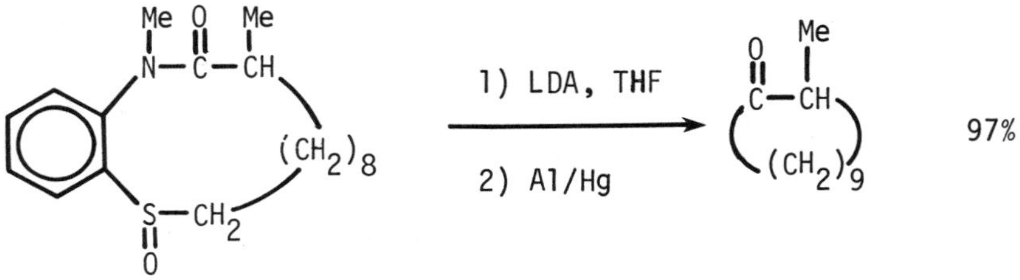

Tetr Lett, 4487 (1979)

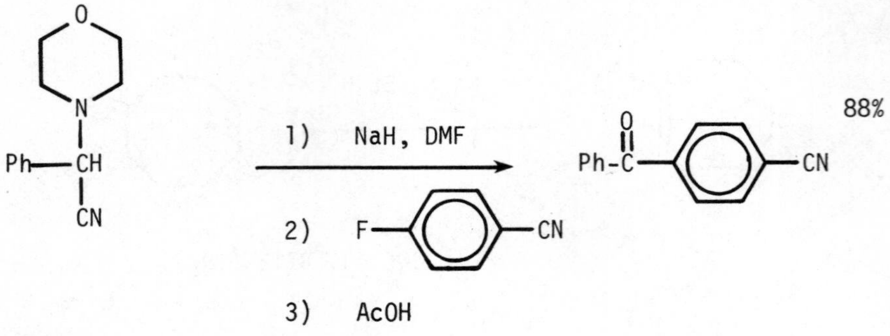

JOC 44, 4597 (1979)

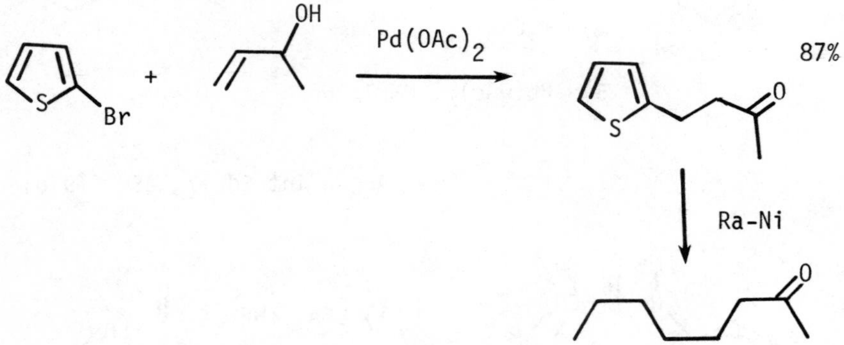

Chem Lett, 423 (1977)

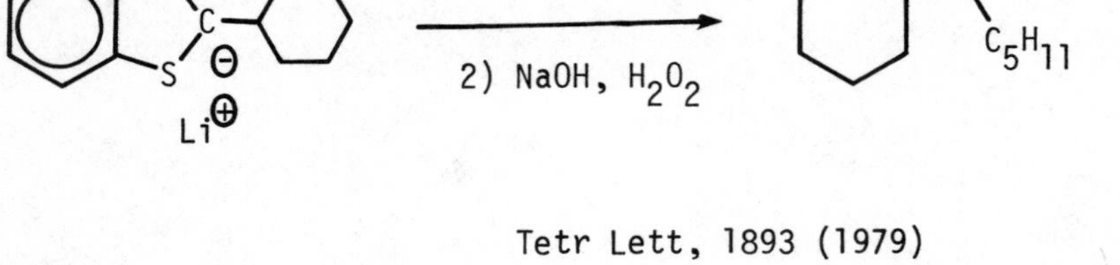

Tetr Lett, 1893 (1979)

Section 180A   Protection of Ketones

See Section 367 (Ether-Olefin) for the formation of enol ethers.
Many of the methods in Section 60A (Protection of Aldehydes) are
also applicable to ketones.

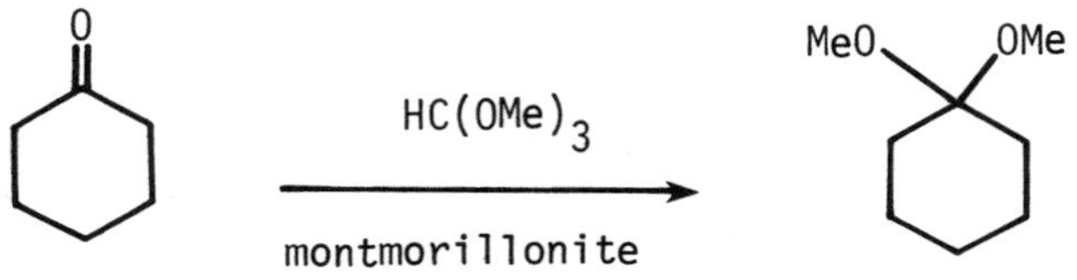

Synthesis, 467 (1977)

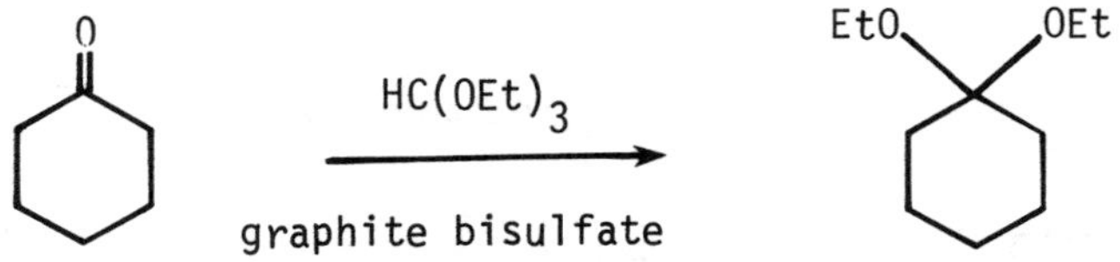

BSC France, 499 (1977)

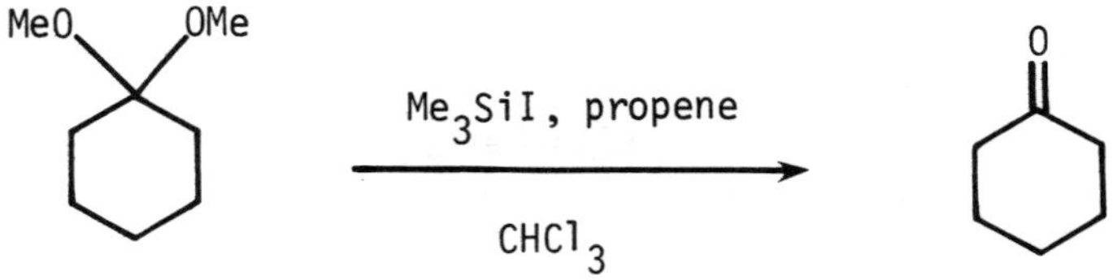

Tetr Lett, 4175 (1977)

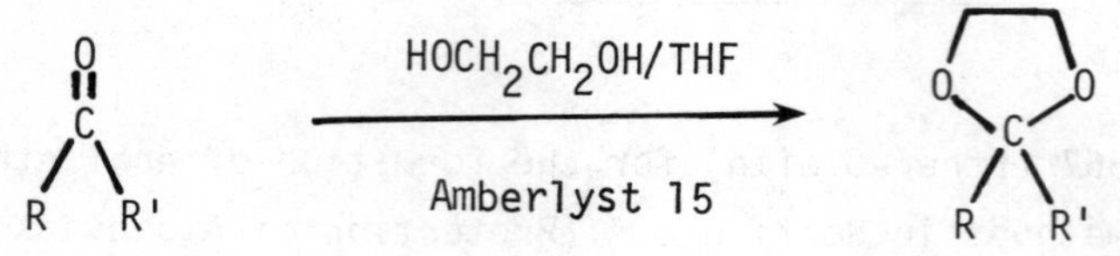

JCS Perkin I, 158 (1979)

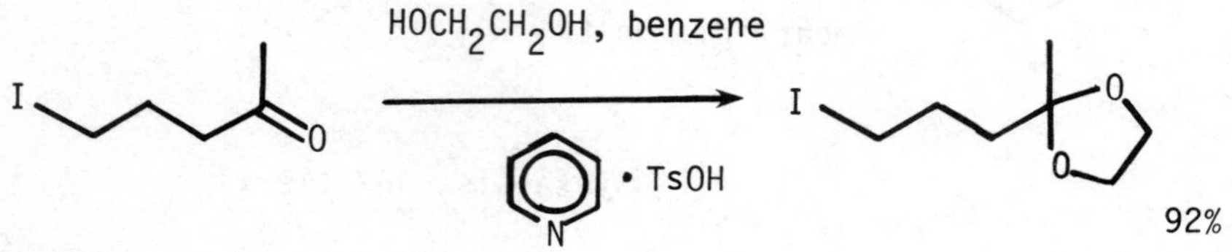

92%

Synthesis, 724 (1979)

65%

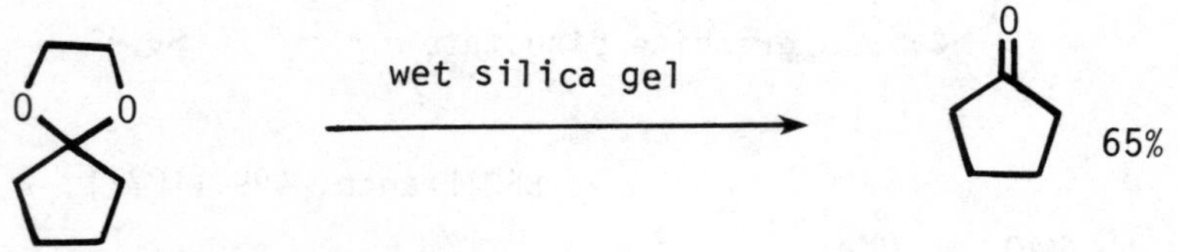

Synthesis, 63 (1978)

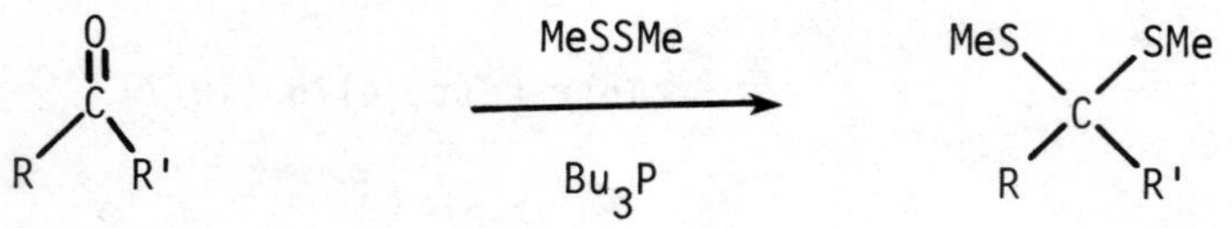

Chem Lett, 767 (1979)

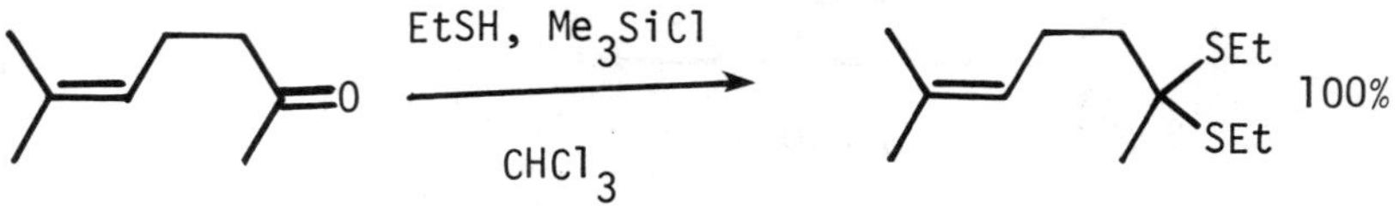

Synth Comm <u>7</u>, 283 (1977)

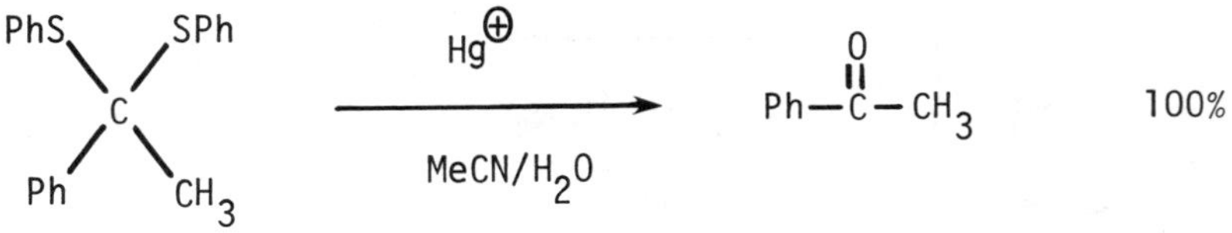

Tetr Lett, 675 (1978)

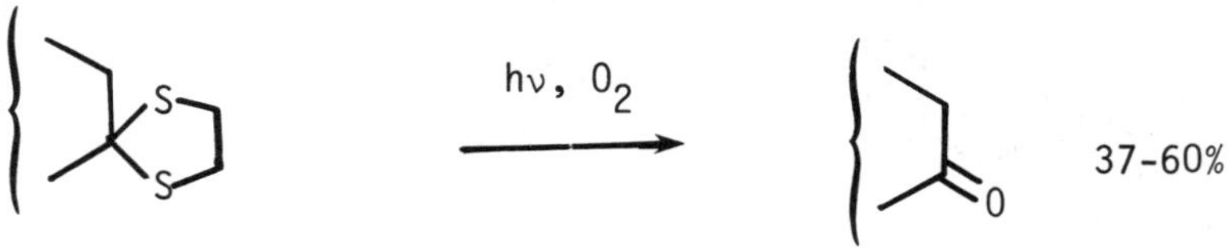

Chem Pharm Bull <u>27</u>, 538 (1979)

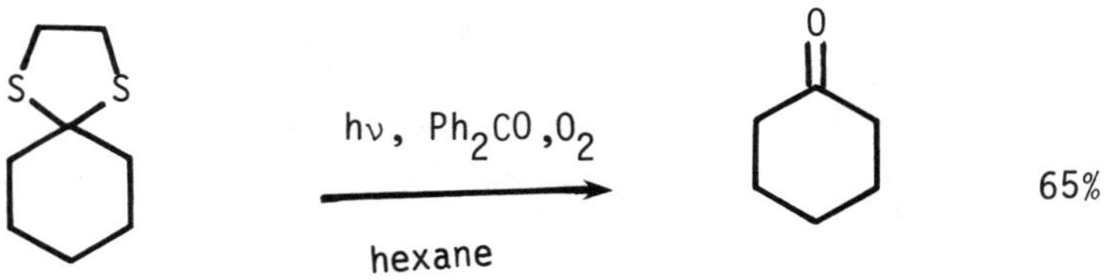

JCS Chem Comm, 680 (1977)

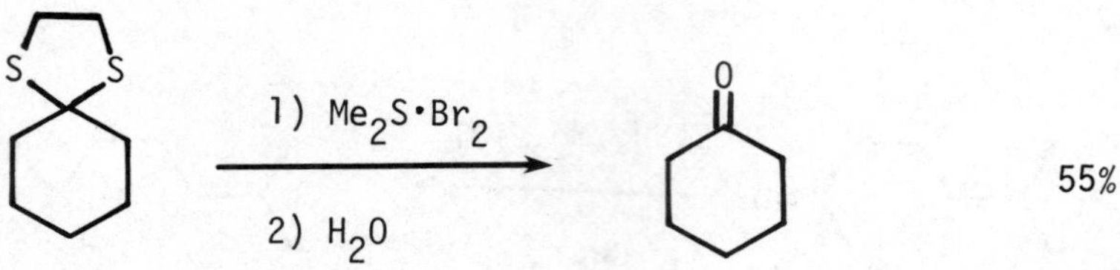

Synthesis, 720 (1979)

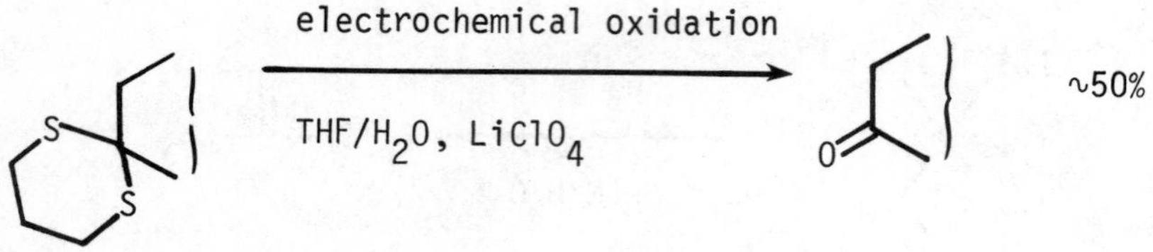

JCS Chem Comm, 255 (1978)

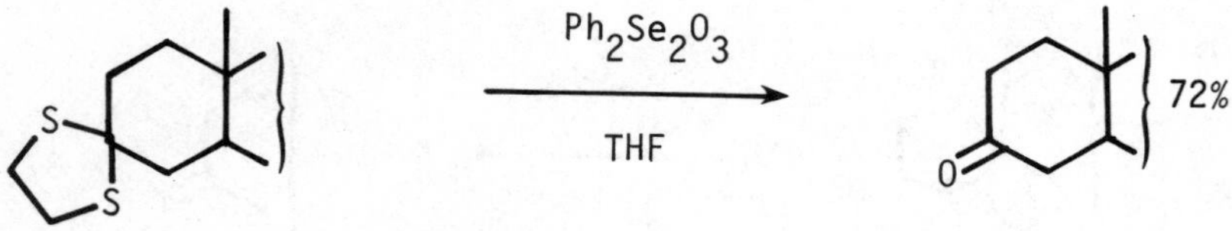

JCS Chem Comm, 751 (1977)

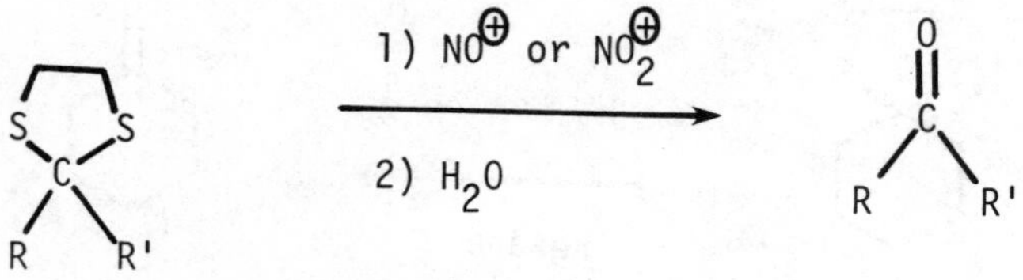

Synthesis, 273 (1979)

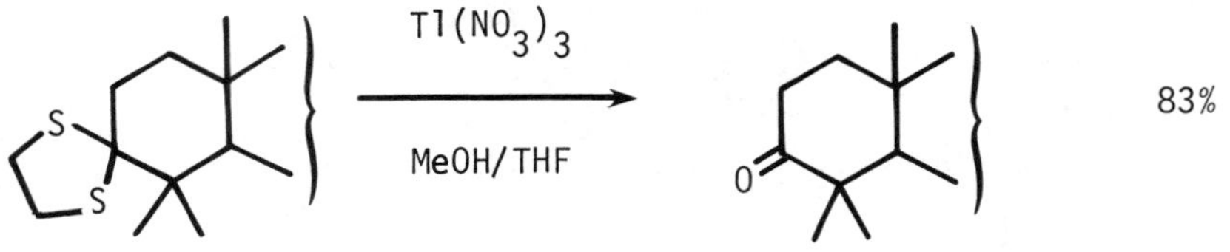

83%

Chem Pharm 26, 3743 (1978)

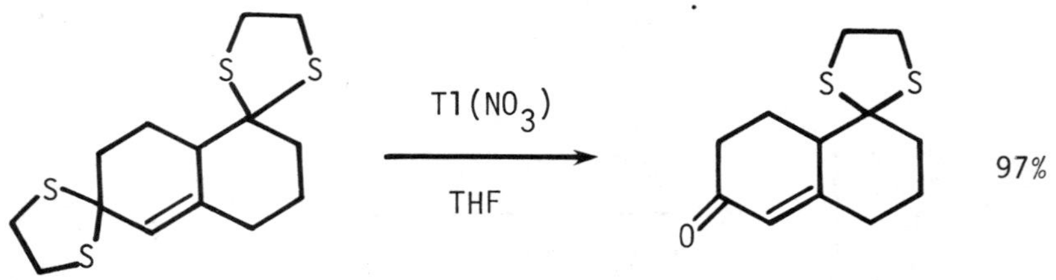

97%

Synth Comm 9, 301 (1979)

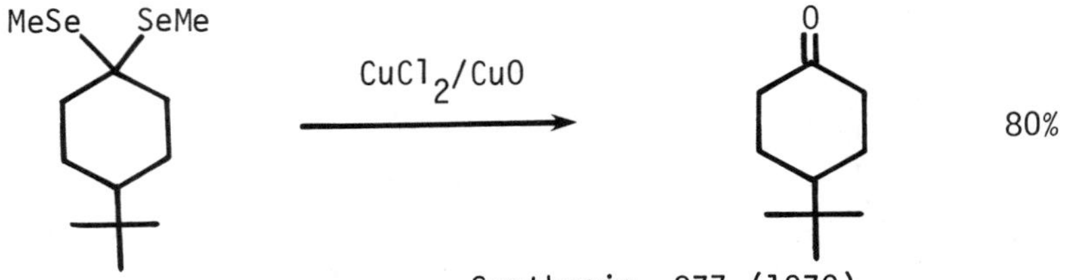

80%

Synthesis, 877 (1979)

59%

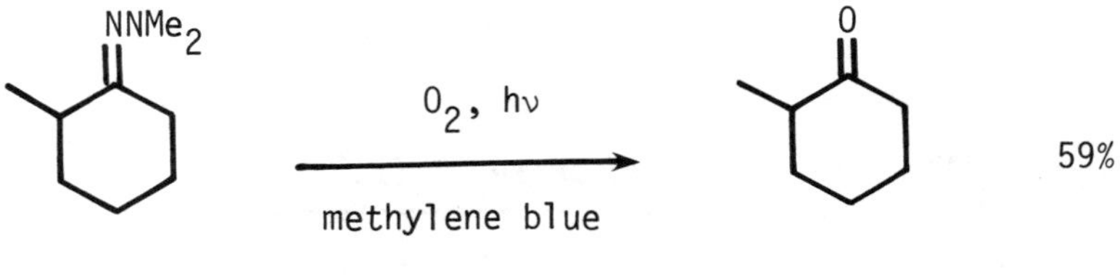

Synthesis, 893 (1977)

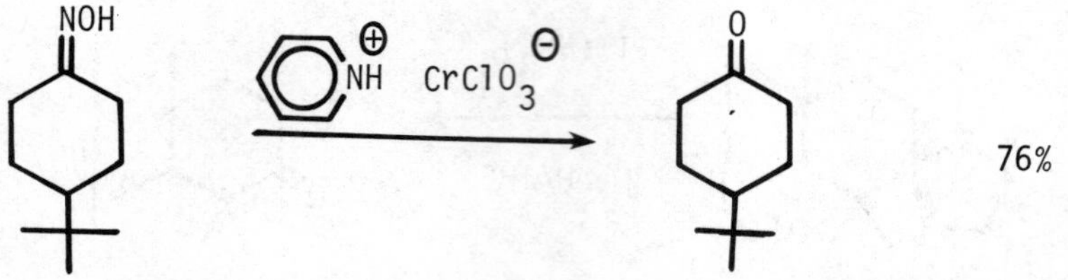

NOH
NH
$CrClO_3$
O
76%

Synthesis, 212 (1978)

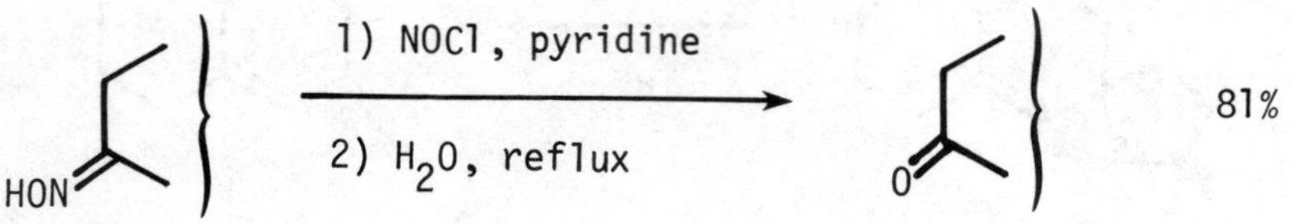

1) NOCl, pyridine
2) H_2O, reflux
HON
O
81%

Chem and Ind, 454 (1977)

NOH
C
R
R'
NOCl
O
~80%

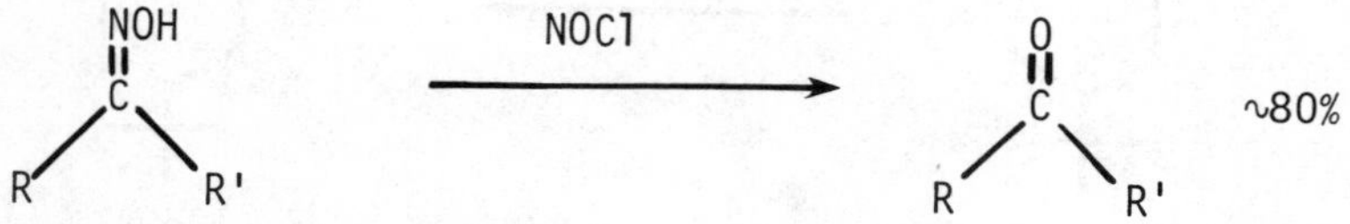

Indian J Chem 15B, 578 (1977)

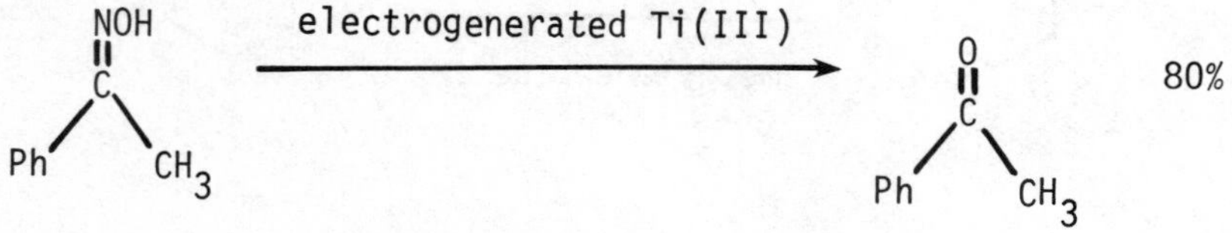

NOH
C
Ph
CH_3
electrogenerated Ti(III)
O
80%

Can J Chem 56, 2269 (1978)

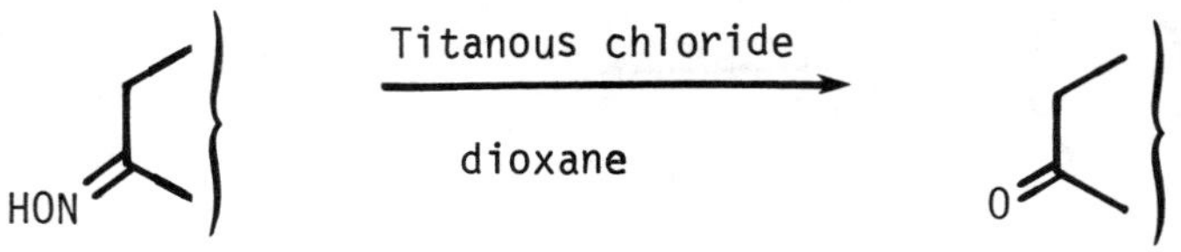

Chem and Ind, 742 (1977)

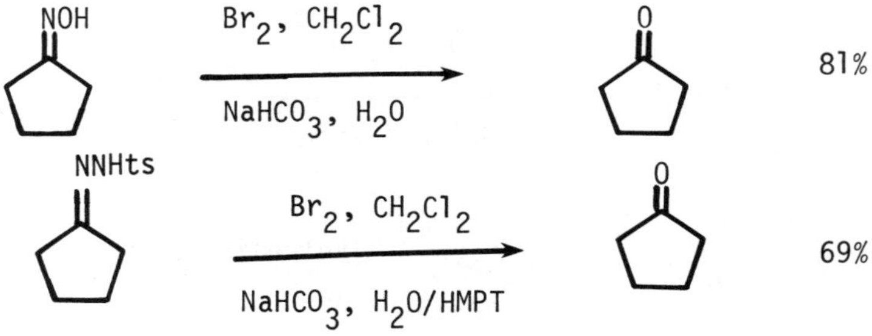

81%

69%

Synthesis, 113 (1979)

89%

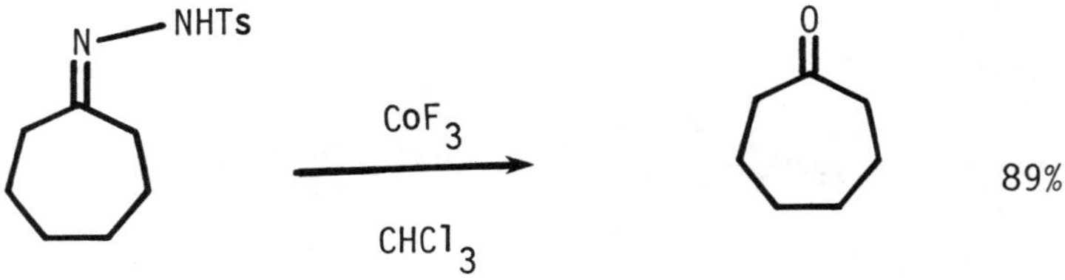

Synthesis, 308 (1979)

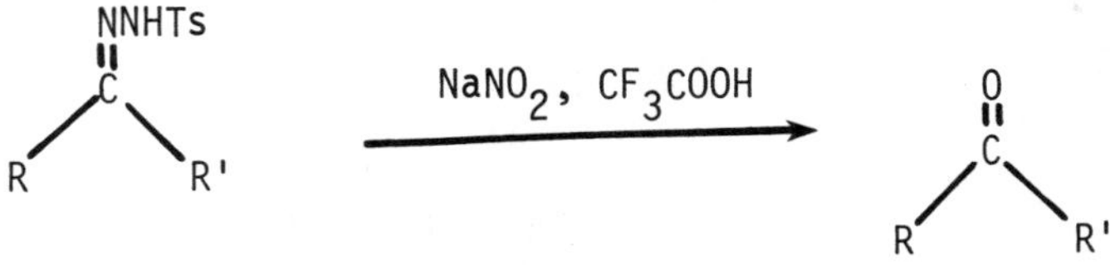

Synthesis, 207 (1979)

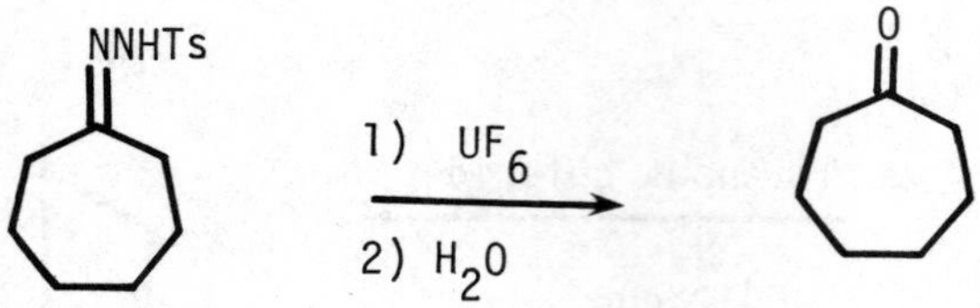

JACS <u>100</u>, 5396 (1978)

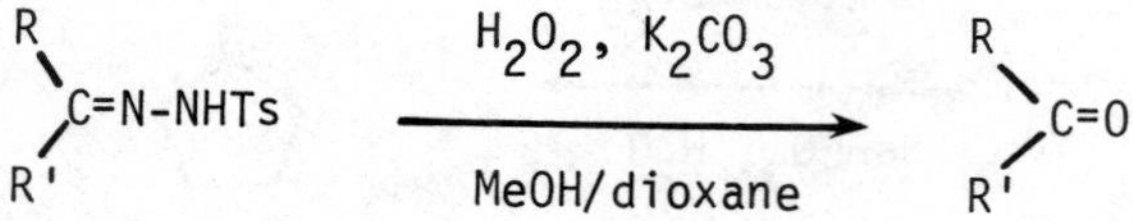

Synthesis, 919 (1978)

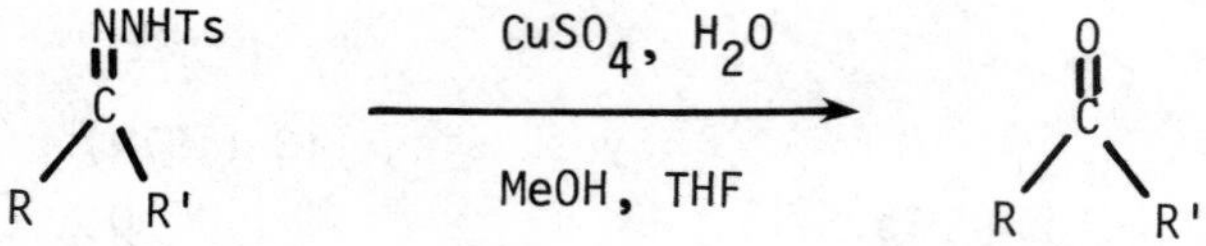

Gazz Chim Ital <u>108</u>, 137 (1978)

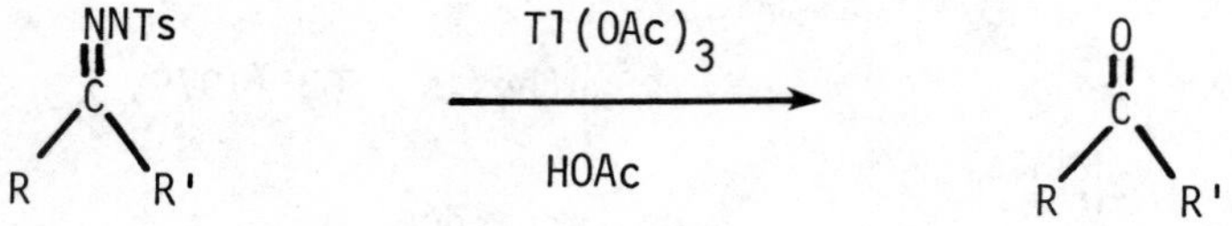

Tetr Lett, 4583 (1979)

Hydrazones, Oximes, Semicarbazones $\xrightarrow{(PhSeO)_2O}$ Ketones

JCS Chem Comm, 445 (1977)

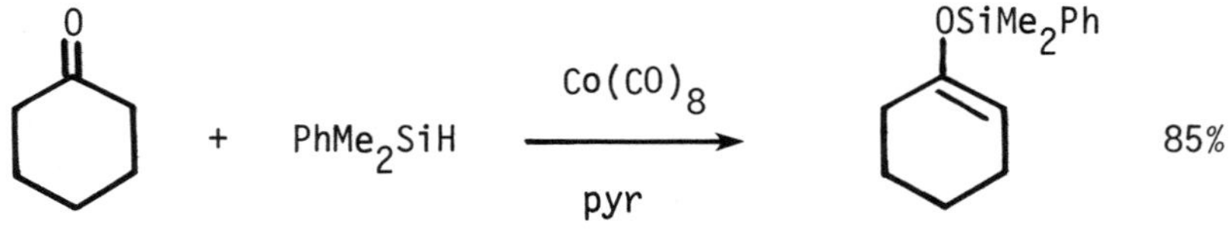

O
+
PhMe2SiH
Co(CO)8
pyr
OSiMe2Ph
85%

Tetr Lett, 2671 (1977)

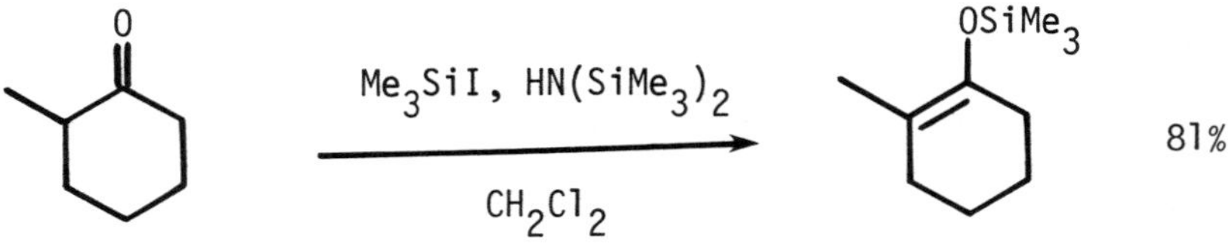

O
Me3SiI, HN(SiMe3)2
CH2Cl2
OSiMe3
81%

Synthesis, 730 (1979)

O
PhS-TMS
OTMS
SPh

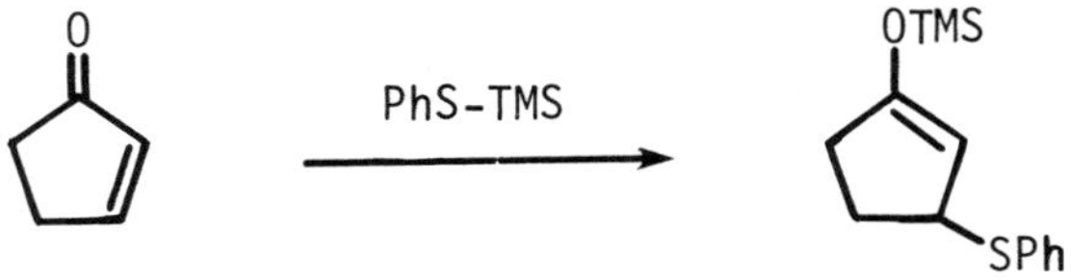

JACS 99, 5009 (1977)

Selective formation of enolate ions to protect carbonyl groups from reduction by LiAlH$_4$. Used to effect selective reductions of steroid diones and triones.

JCS Perkin I, 1075 (1977)

CHAPTER 13
PREPARATION OF NITRILES

Section 181   <u>Nitriles from Acetylenes</u>

$$PhCH_2CH_2C{\equiv}CH \quad \xrightarrow[\ominus\ CN,\ NaBH_4]{[Ni(CN)_4]^{2-}} \quad PhCH_2CH_2\underset{CN}{CHCH_3} \qquad 92\%$$

JCS Chem Comm, 1110 (1979)

Section 182   <u>Nitriles from Carboxylic Acids and Acid Halides</u>

No additional examples

Section 183   <u>Nitriles from Alcohols</u>

No additional examples

Section 184   <u>Nitriles from Aldehydes</u>

$$C_6H_{13}-CHO \xrightarrow[\text{2) } \nabla]{\text{1) } (\underline{i}\text{-Pr})_3PhSO_2NHNH_2} C_6H_{13}-C\equiv N \qquad 92\%$$

Synthesis, 112 (1979)

$$\underline{n}\text{-}C_6H_{13}\text{-CH=NOH} \xrightarrow{\text{SeO}_2, \text{ CHCl}_3} C_6H_{13}-C\equiv N \qquad 63\%$$

Synthesis, 703 (1978)

$$R\text{-CHO} \xrightarrow[\text{2) SeO}_2]{\substack{\text{1) NH}_2\text{OH} \cdot \text{HCl} \\ \text{Pyridine/CHCl}_3}} R\text{-CN} \qquad \sim80\%$$

R = alkyl, aryl, heterocyclic

Synthesis, 722 (1979)

$$Ph\text{-CH=N-OH} \xrightarrow[\text{pyridine}]{\text{TFAA}} Ph\text{-}C\equiv N \qquad 90\%$$

Synthesis, 56 (1979)

$$\underline{n}\text{-}C_6H_{13}\text{-}CH\text{=}NOH \xrightarrow[\text{Et}_3N]{\overset{\oplus\ominus}{Me_3NSO_2}} C_6H_{13}\text{-}C\equiv N \qquad 83\%$$

Synthesis, 702 (1978)

$$C_5H_{11}\text{-}CH\text{=}NOH \xrightarrow[\text{Et}_3N]{ClSO_2NCO} C_5H_{11}\text{-}CN \qquad 75\%$$

Synthesis, 227 (1979)

Synthesis, 338 (1977)

$$Ph\text{-}CH\text{=}NOH \xrightarrow{\textcircled{P}\text{-}PPh_2\cdot CCl_4} Ph\text{-}C\equiv N \qquad 76\%$$

Synthesis, 41 (1977)

$$\underline{n}\text{-}C_7H_{15}\text{-}CH{=}NOH \xrightarrow{\quad P_2I_4, \text{ pyridine} \quad} C_7H_{15}\text{-}C{\equiv}N \qquad 64\%$$

Synthesis, 905 (1978)

Ph-CHO $\xrightarrow{\quad 1)\ \text{(pyridinone reagent)} \quad \atop 2)\ \Delta}$ Ph-CN   99%

JCS Perkin I, 1957 (1979)

$$\text{Ph-CHO} \xrightarrow[\text{NH}_3,\ \text{benzene},\ \text{Na}_2\text{SO}_4]{\text{nickel peroxide}} \text{Ph-CN} \qquad 73\%$$

Synth Comm 9, 529 (1979)

Synthesis, 301 (1978)

$$C_6H_{13}-CHO \quad \xrightarrow[\text{2) KCN, MeOH}]{\text{1) }(\underline{i}\text{-Pr})_3PhSO_2NHNH_2} \quad C_6H_{13}CH_2CN \quad 72\%$$

JCS Chem Comm, 280 (1977)

Section 185   <u>Nitriles from Alkyls, Methylenes and Aryls</u>

No additional examples

Section 186   <u>Nitriles from Amides</u>

$$\underline{n}\text{-}C_{17}H_{35}\text{-}\overset{\overset{\text{O}}{\|}}{C}\text{-}NH_2 \quad \xrightarrow{\textcircled{P}\text{-}PPh_2 \cdot CCl_4} \quad \underline{n}\text{-}C_{17}H_{35}\text{-}C\equiv N \quad 100\%$$

Synthesis, 41 (1977)

$$CH_3\text{-}(CH_2)_{16}\text{-}CONH_2 \quad \xrightarrow[\text{pyridine}]{(CF_3CO)_2O} \quad CH_3\text{-}(CH_2)_{16}\text{-}CN \quad 94\%$$

Tetr Lett, 1813 (1977)

$$Ph-\overset{\overset{\displaystyle S}{\|}}{C}-NH_2 \xrightarrow[\displaystyle Ph_3P]{\overset{\displaystyle N-COOEt}{\underset{\displaystyle N-COOEt}{\|}}} Ph-C{\equiv}N \qquad 59\%$$

JCS Chem Comm, 220 (1977)

JOC <u>44</u>, 3436 (1979)

$$Bz-\overset{\overset{\displaystyle O}{\|}}{C}-NH_2 \xrightarrow[\displaystyle Et_3N]{\displaystyle ClSO_2NCO} Bz-CN \qquad 84\%$$

Synthesis, 227 (1979)

$$CH_3CH_2-NH-CHO \xrightarrow[\displaystyle O_2,\ 400-500^\circ]{\displaystyle bismuth\ phosphomolybdate} CH_3CH_2CN \qquad 90\%$$

Chem & Ind, 852 (1979)

$$Ph-\overset{\overset{\displaystyle S}{\|}}{C}-NH_2 \xrightarrow[\displaystyle \ominus OEt]{\displaystyle BrCH_2CN} Ph-C{\equiv}N \qquad 85\%$$

Synth Comm <u>9</u>, 569 (1979)

Section 187   Nitriles from Amines

$$Ph-CH_2NH_2 \xrightarrow[\text{pyridine}]{\text{CuCl, }O_2} Ph-C{\equiv}N \qquad 35\%$$

Synthesis, 245 (1977)

Section 188   Nitriles from Esters

$$\underset{R-\overset{\displaystyle O}{\overset{\|}{C}}-OR'}{} \xrightarrow[\text{xylene, }\Delta]{Me_2AlNH_2} R-CN \qquad {\sim}70\%$$

R = alkyl, aryl

Tetr Lett, 4907 (1979)

Section 189   Nitriles from Ethers and Epoxides

No additional examples

Section 190   Nitriles from Halides

87%

JOC 43, 1017 (1978)

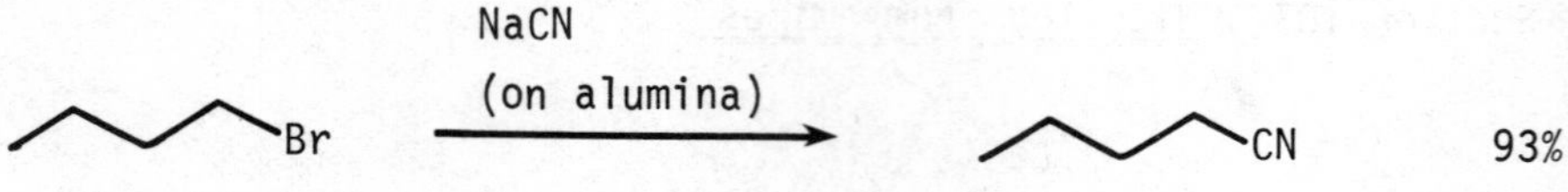

NaCN
(on alumina)

~~~~Br → ~~~~CN   93%

JOC 44, 2029 (1979)
JOC 44, 3436 (1979)

Et$_4$NCN

~~~~~Br → ~~~~~CN   80%

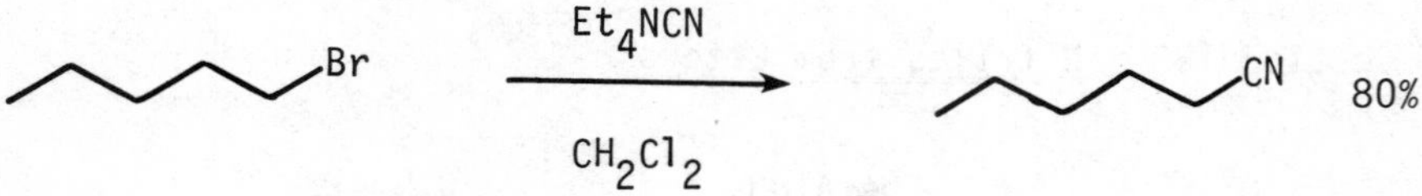

CH$_2$Cl$_2$

Liebigs Ann Chem, 1946 (1978)

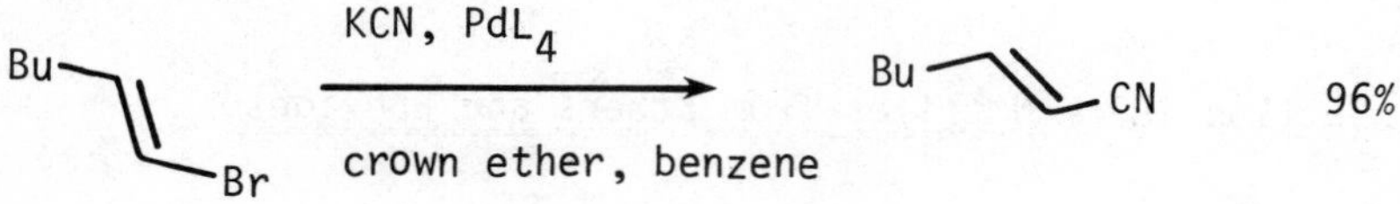

KCN, PdL$_4$

Bu~~~Br → Bu~~~CN   96%

crown ether, benzene

Tetr Lett, 4429 (1977)

NaCN, alumina

→   90%

Pd(II) catalyst

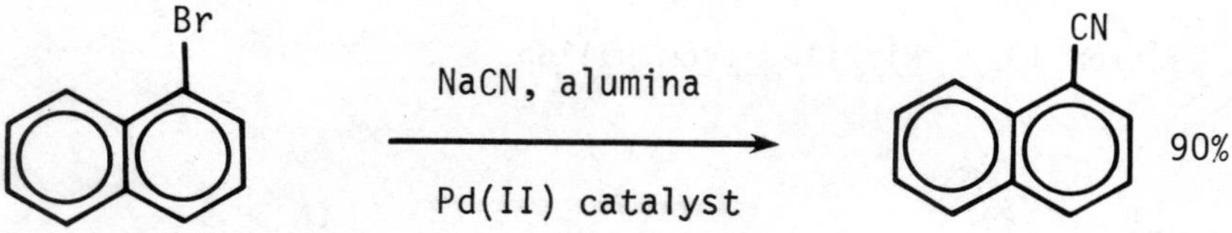

JOC 44, 4443 (1979)

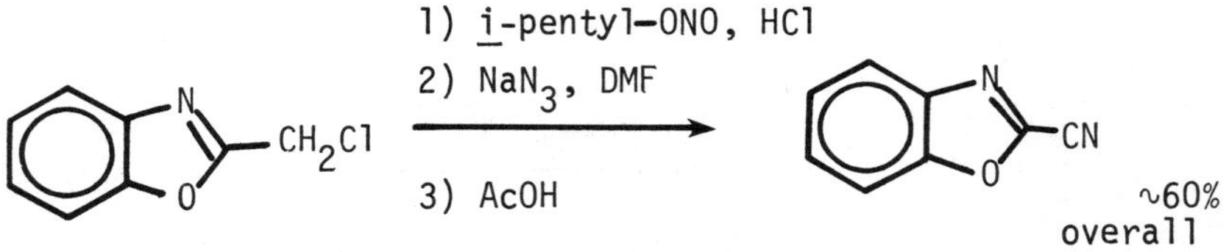

1) i-pentyl-ONO, HCl
2) NaN3, DMF
3) AcOH
CH2Cl
CN
~60% overall

Synthesis, 102 (1979)

Section 191 Nitriles from Hydrides

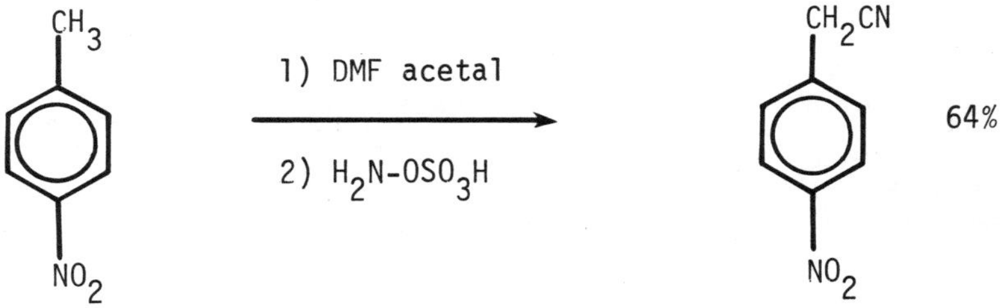

CH3
NO2
1) DMF acetal
2) H2N-OSO3H
CH2CN
NO2
64%

Tetr Lett, 1361 (1979)

BrCN
AlCl3, CS2
CN
92%

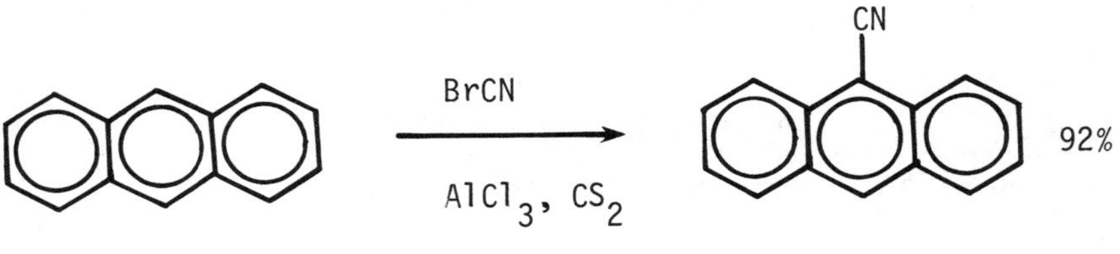

Tetrahedron 35, 2927 (1979)

Section 192   <u>Nitriles from Ketones</u>

1) $(\underline{i}\text{-Pr})_3PhSO_2NHNH_2$

2) KCN, MeOH

71%

JCS Chem Comm, 280 (1977)

$C=N-CH_2Ts$

$\underline{t}$-BuOK

80%

JOC <u>42</u>, 3114 (1977)

Section 193   <u>Nitriles from Nitriles</u>

Conjugate reductions and Michael alkylations of olefinic nitriles are found in Section 74 (Alkyls from Olefins).

Ph-C≡C-CN

1) Bu[CuBr]MgCl

2) H₃O⊕

96%

Synthesis, 454 (1978)

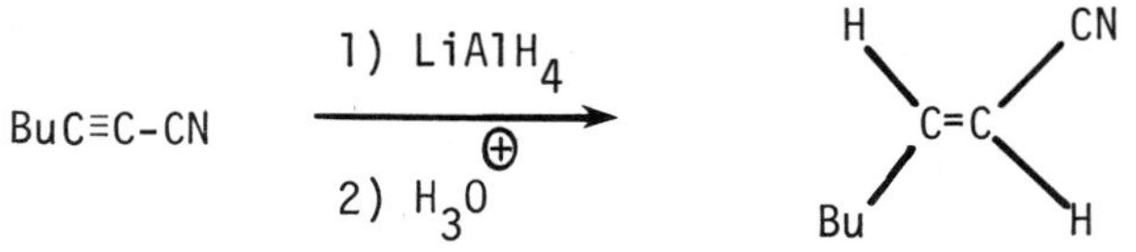

$$BuC{\equiv}C{-}CN \xrightarrow[\text{2) } H_3O^{\oplus}]{\text{1) } LiAlH_4}$$

90%

Synthesis, 430 (1979)

Section 194   Nitriles from Olefins

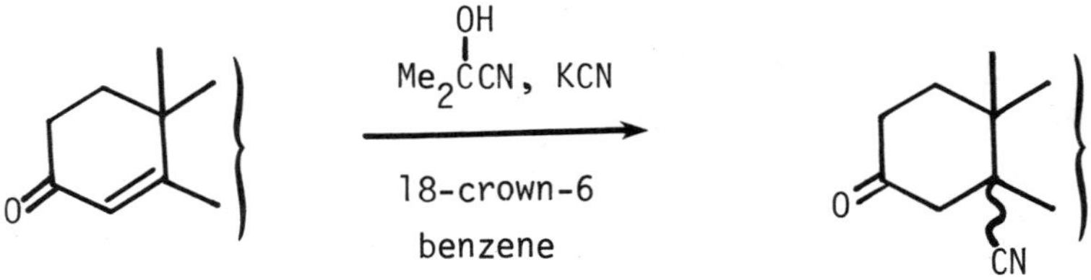

86%

Tetr Lett, 1117 (1977)

Section 195   Nitriles from Miscellaneous Compounds

$$C_6H_{13}{-}CH_2{-}NO_2 \xrightarrow[\text{CH}_2\text{Cl}_2]{\text{Et}_3\text{N/SO}_2} C_6H_{13}{-}C{\equiv}N$$

78%

Synthesis, 36 (1979)

$$C_9H_{19}\text{-}CH_2NO_2 \xrightarrow[\text{Et}_3N]{P_2I_4} C_9H_{19}\text{-}CN \qquad 78\%$$

Tetr Lett, 3995 (1979)

CHAPTER 14
PREPARATION OF OLEFINS

Section 196 Olefins from Acetylenes

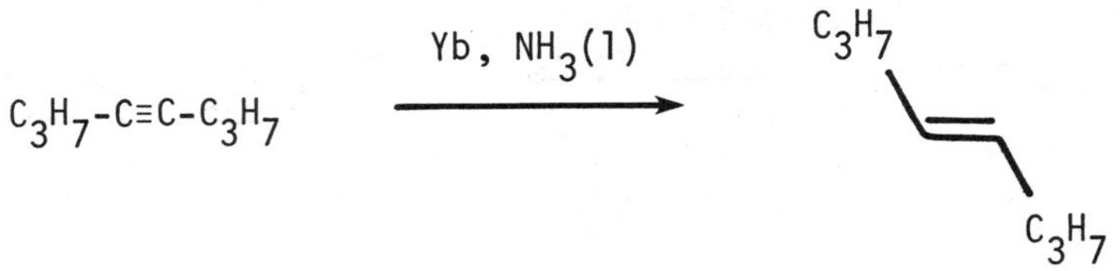

C3H7-C≡C-C3H7
Yb, NH3(l)
C3H7
C3H7

85%

JOC 43, 4555 (1978)

Bu-C≡C-(CH2)6-OH
1) LiAlH4
2) H2O
Bu
H
C=C
H
(CH2)6-OH

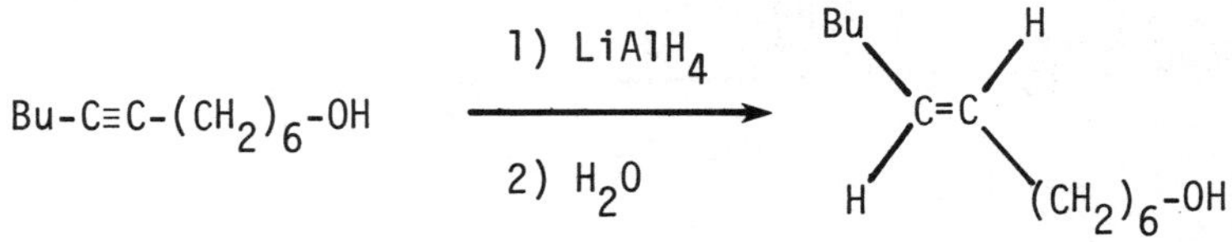

93%

Synthesis, 561 (1977)

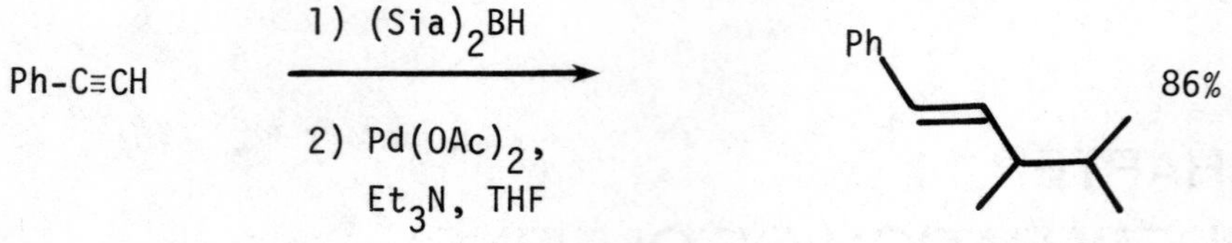

$Ph-C{\equiv}CH$ →
1) $(Sia)_2BH$
2) $Pd(OAc)_2$, Et_3N, THF

86%

JCS Chem Comm, 852 (1977)

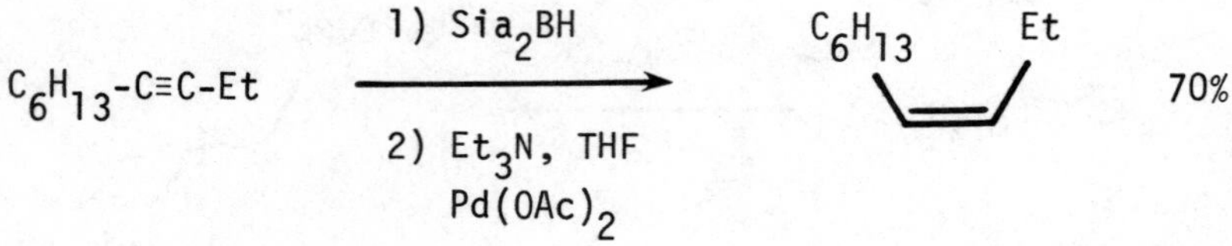

$C_6H_{13}-C{\equiv}C-Et$ →
1) Sia_2BH
2) Et_3N, THF $Pd(OAc)_2$

70%

JCS Chem Comm, 702 (1978)

$HC{\equiv}C-C_6H_{13}$ →
$LiAlH_4-NiCl_2$
$H_2C=CH-C_6H_{13}$

96%

Tetr Lett, 4481 (1977)

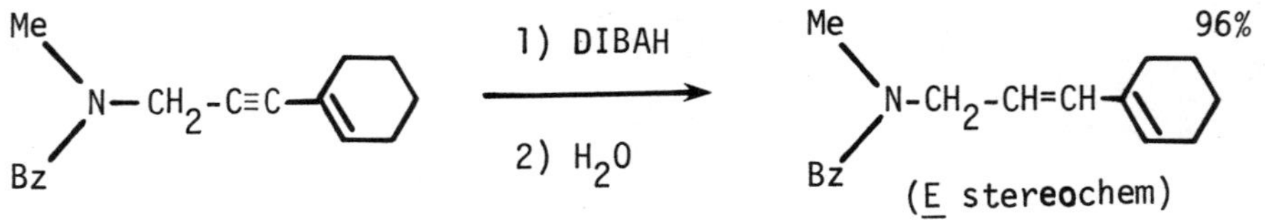

Me
N—CH2-C≡C—
Bz
1) DIBAH
2) H2O
Me
N-CH2-CH=CH—
Bz
(E stereochem)
96%

Tetr Lett, 3145 (1979)

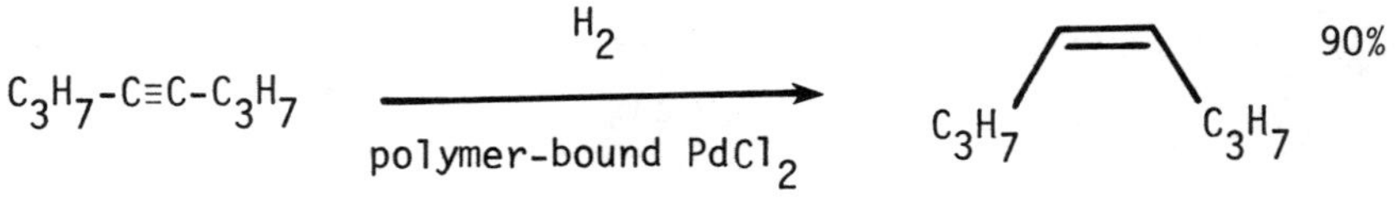

C3H7-C≡C-C3H7
H2
polymer-bound PdCl2
C3H7
C3H7
90%

JOC 43, 4686 (1978)

Et
C
C
Et
H2, NaH, Ni(OAc)2
THF/EtOH
Et
Et
98%

Tetr Lett, 3955 (1977)

CH3-C≡C-Pr
LiAlH4
NiCl2
H
H
C=C
CH3
Pr
92%

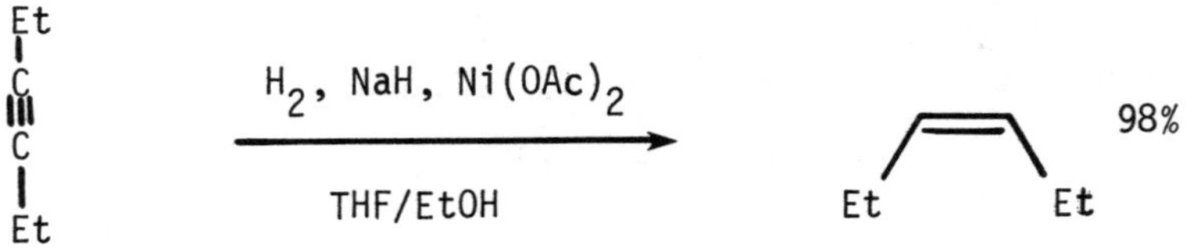

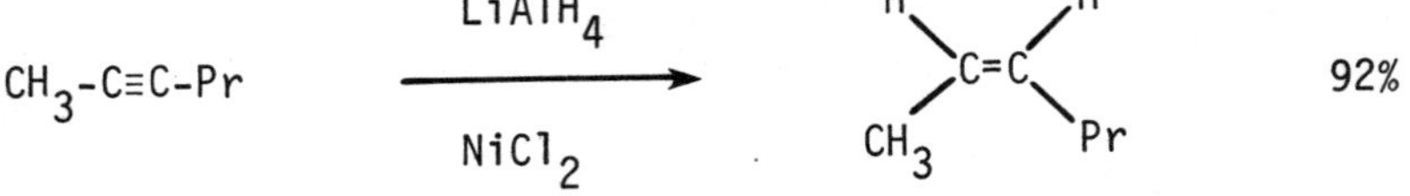

JOC 43, 2567 (1978)

$CH_3-C≡C-C_3H_7$ $\xrightarrow[\text{Cu-O-}\underline{\text{t}}\text{-Bu}]{MgH_2}$ [structure: $C=C$ with CH_3, C_3H_7, H, H]   81%

JOC 43, 757 (1978)

$(C_6H_{11})_2BH$ $\xrightarrow[\text{2. PhI(OAc)}_2]{\text{1. BuC≡CH}}$ [structure: $C=C$ with H, Bu, C_6H_{11}, H]   60%

Chem Lett, 665 (1978)

[structure: Bu, H on $C=C$ bonded to B with benzodioxaborole ring] $\xrightarrow[\text{benzene}]{\begin{array}{c}\text{PhI}\\\text{NaOEt, PdL}_4\end{array}}$ [structure: $C=C$ with Bu, H, H, Ph]   100%

JCS Chem Comm, 866 (1979)

$C_7H_{15}-C≡CH$ $\xrightarrow[\text{2) } H_3O^{\oplus}]{\text{1) } [\text{structure: } Me, CH_2ZnBr]}$ [structure: Me, C_7H_{15} diene]   72%

BSC France, 1173 (1976)

$C_6H_{13}-C{\equiv}CH$

$$\xrightarrow[\text{2)}\quad\raisebox{0pt}{\includegraphics{}}\text{Br}]{\substack{\text{Me}_2\text{S}\\\text{1) MeCu}\cdot\text{MgBr}}}$$

C=C with CH₃ group

Tetr Lett, 1363 (1978)

$PhC{\equiv}CD$

$$\xrightarrow[\text{2) H}_2\text{O}]{\text{1) Me}_3\text{Al-Cl}_2\text{ZrCp}_2}$$

Ph, D / C=C / Me, H   98%

JACS 100, 2252 (1978)

OH | —C≡CH

$$\xrightarrow[\underline{\text{i}}\text{-Pr}_2\text{NH, dioxane}]{\text{H}_2\text{C=O, CuBr}}$$

OH   97% | —CH=C=CH₂

JCS Chem Comm, 859 (1979)

Review: "The Selective Hydrogenation of Triple Bonds with Organometallic Transition Metal Compounds"

Pure and Appl Chem 50, 941 (1978)

Section 197   <u>Olefins from Carboxylic Acids and Acid Halides</u>

No additional examples

Section 198   <u>Olefins from Alcohols</u>

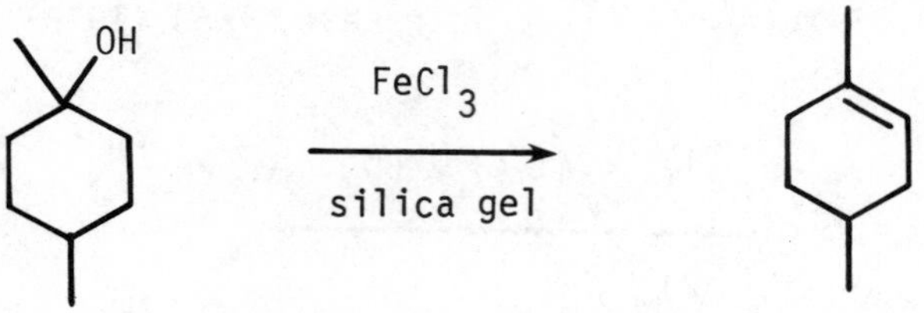

JOC <u>43</u>, 1020 (1978)

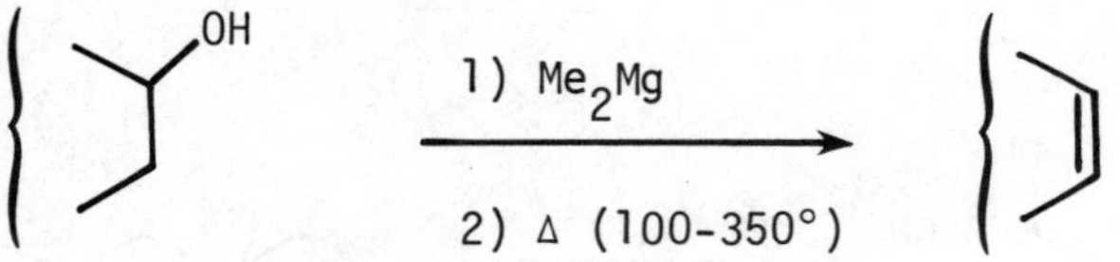

JOC <u>44</u>, 1221 (1979)

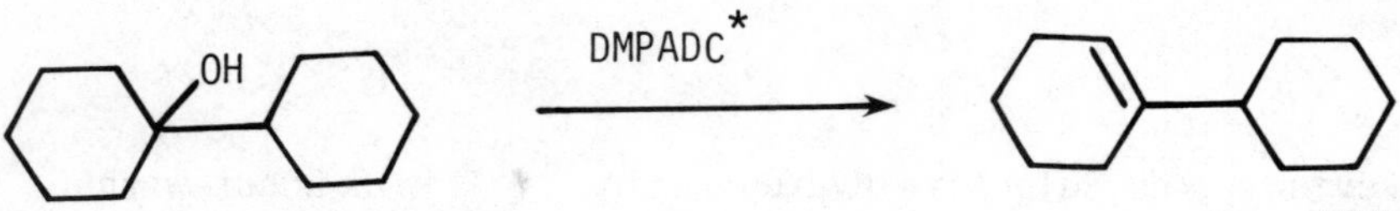

*<u>N</u>, <u>N</u> - Dimethylphosphoramidic dichloride

Chem Lett, 923 (1978)

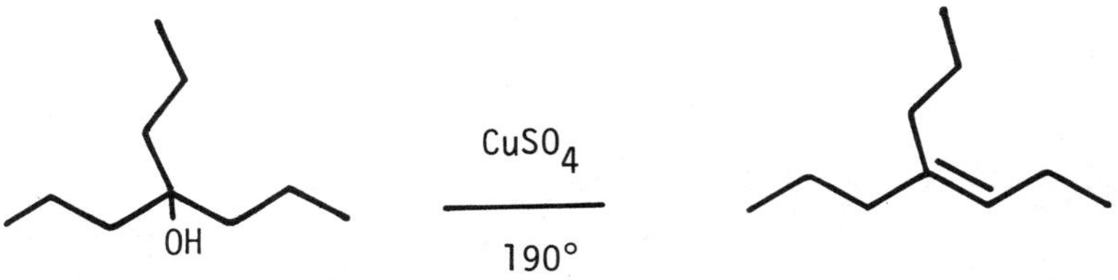

88%

Acta Chem Scand (B) 31, 721 (1977)

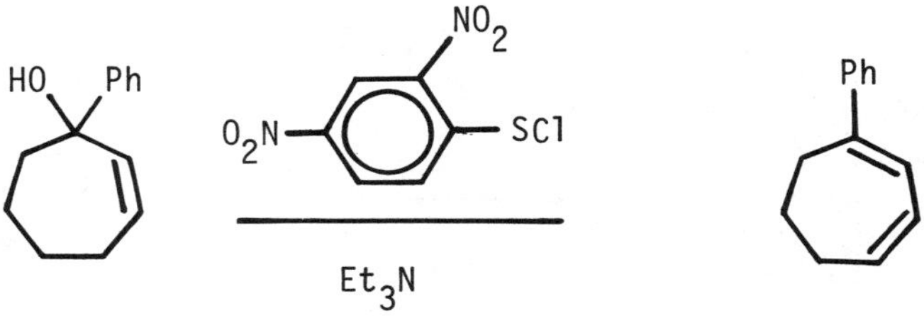

74%

JACS 100, 5981 (1978)

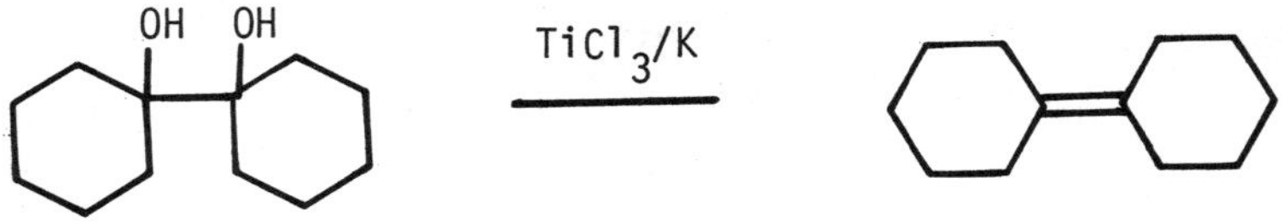

85%

JOC 43, 3255 (1978)

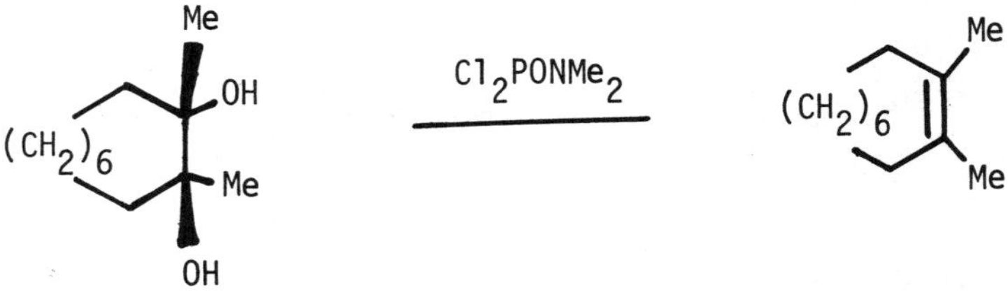

90%

JOC 42, 1311 (1977)

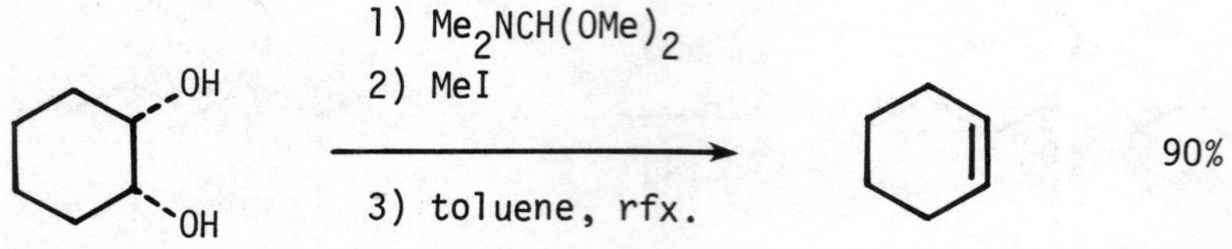

1) Me$_2$NCH(OMe)$_2$
2) MeI

3) toluene, rfx.

90%

Tetr Lett, 737 (1978)

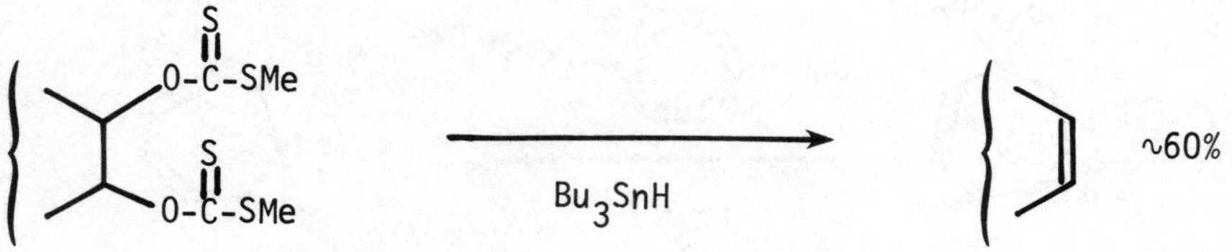

Bu$_3$SnH

~60%

Examples using carbohydrates

JCS Chem Comm, 866 (1977)
JCS Perkin I, 2378 (1979)

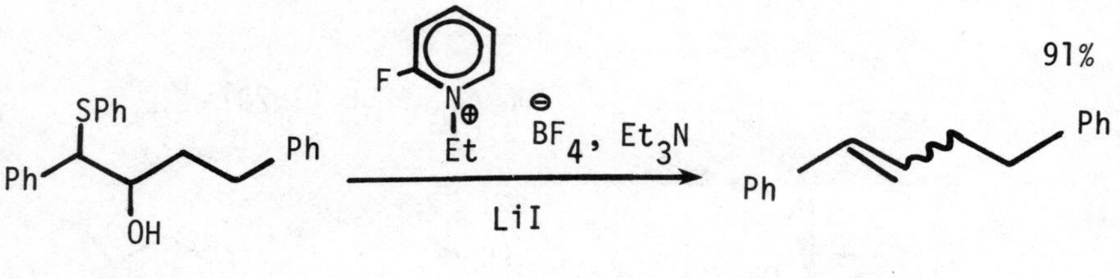

91%

BF$_4$, Et$_3$N

LiI

Chem Lett, 413 (1978)

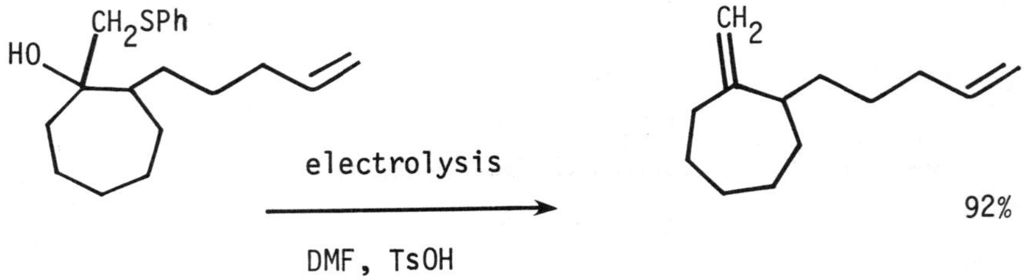

CH2SPh
HO
electrolysis
DMF, TsOH
CH2
92%

Tetr Lett, 2807 (1978)

Section 199 Olefins from Aldehydes

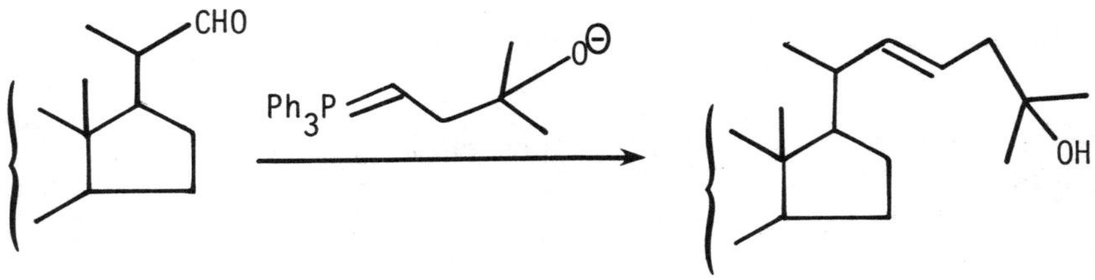

CHO
Ph3P
O
OH

JOC 43, 790 (1978)

Ph3P-CH2
Br
Br
1) NaNH2
2) PhCHO
Ph
Br

Indian J Chem 15B, 290 (1977)

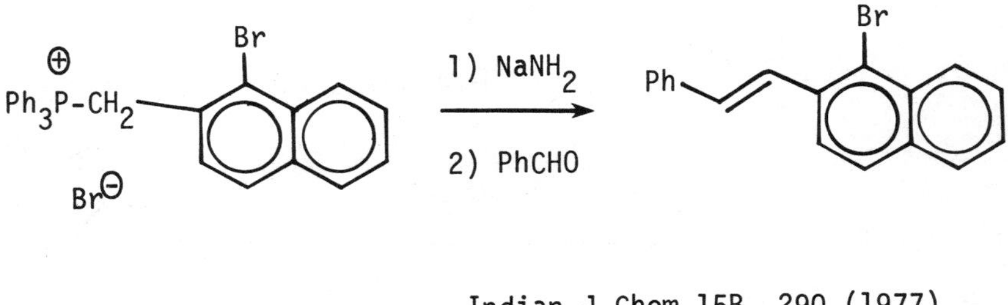

Ph-CHO $\xrightarrow[\text{KOH, THF}]{(EtO)_2\overset{\overset{O}{\|}}{P}CH_2X}$ PhCH=CHX   90-98%

X = CN, COOEt

Synthesis, 884 (1979)

Ph-CHO $\xrightarrow{Ph_3P=CHCl}$ $\underset{H}{\overset{Ph}{>}}C=CHCl$ 81%

JCS Chem Comm, 446 (1978)

$C_5H_{11}CHO$ $\xrightarrow{(Me_2N)_3P=CCl_2}$ $C_5H_{11}CH=CCl_2$ 94%

Tetr Lett, 1239 (1977)

$CH_3-(CH_2)_{10}-CHO$ $\xrightarrow{CH_2I_2-Zn-Me_3Al}$ $CH_3-(CH_2)_{10}\underset{H}{\overset{\overset{CH_2}{\overset{\|}{C}}}{<}}$ 81%

Tetr Lett, 2417 (1978)

PhCH$_2$CH$_2$CH=NNHTs

     + $\longrightarrow$ PhCH$_2$CH$_2$CH=CHEt   70%

EtCH(Li)CN

JACS <u>101</u>, 249 (1979)

PhCHO   + $\longrightarrow$

                                                                              84%

Chem Lett, 197 (1978)

2   $\xrightarrow{\text{TiCl}_3/\text{K}}$ C$_4$H$_9$CH=CHC$_4$H$_9$ 77%

JOC <u>43</u>, 3255 (1978)

2 Ph-CHO $\xrightarrow{\text{WCl}_6, \text{ LiAlH}_4}$ 73%

JOC <u>43</u>, 2477 (1978)

$$\begin{array}{c} \text{O NMe} \\ \backslash\backslash/ \\ \text{1) Ph-S-Me} \end{array}$$

THF, Ph_3CH, BuLi

$$CH_3(CH_2)_8CHO \xrightarrow{\hspace{3cm}} CH_3(CH_2)_8CH=CH_2 \qquad 56\%$$

2) Al/Hg

THF/H_2O/HOAc

JACS <u>101</u>, 3602 (1979)

Section 200   <u>Olefins from Alkyls, Methylenes and Aryls</u>

This section contains dehydrogenations to form olefins and
unsaturated ketones, esters, and amides. It also includes the
reduction of aromatic rings to olefins. Hydrogenation of aryls
to alkanes and dehydrogenations to from aryls are included in
Section 74 (Alkyls, Methylenes, and Aryls from Olefins).

$$\xrightarrow[\text{HCl, 80°}]{\text{t-BuOH, PdCl}_2}$$

80%

Synthesis, 773 (1977)

$$\xrightarrow{\text{(PhSe)}_2\text{O}}$$

39-92%

JCS Chem Comm, 130 (1978)

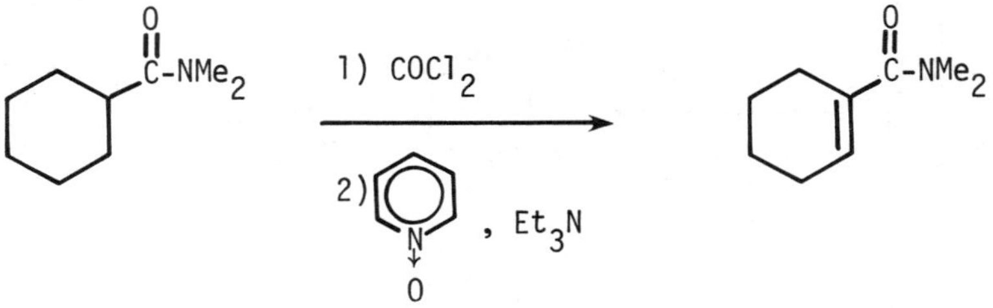

JACS 101, 4381 (1979)

70%

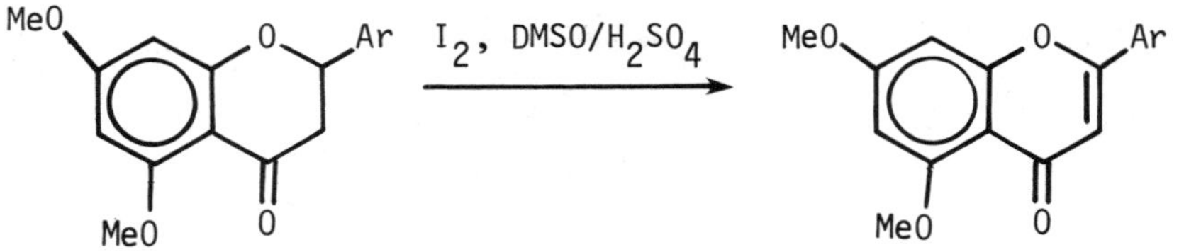

~100%

Chem & Ind, 315 (1979)

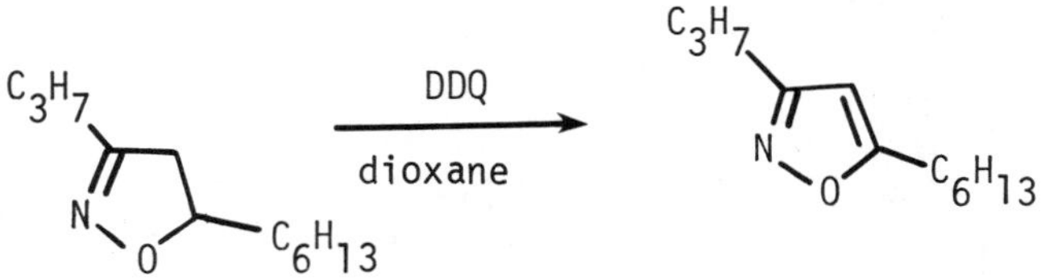

89%

J Chem Res (S), 311 (1979)

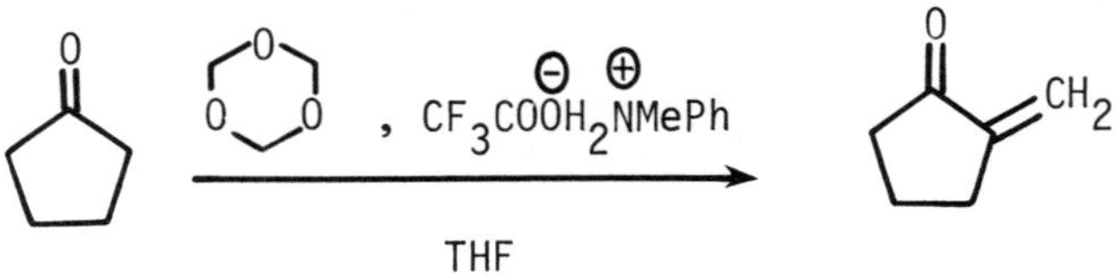

90%

Tetr Lett, 2111 (1978)

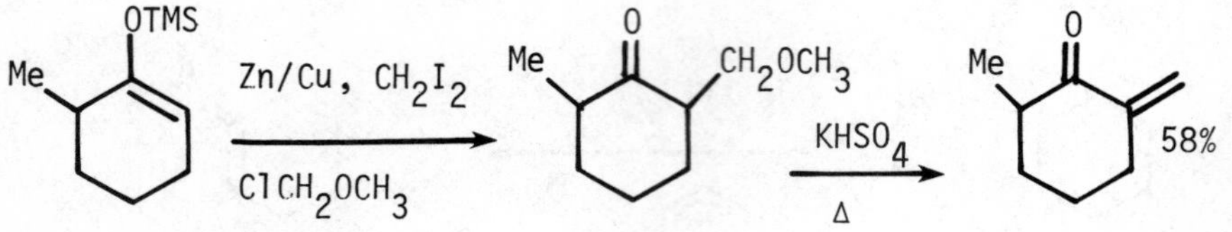

JACS <u>101</u>, 984 (1979)

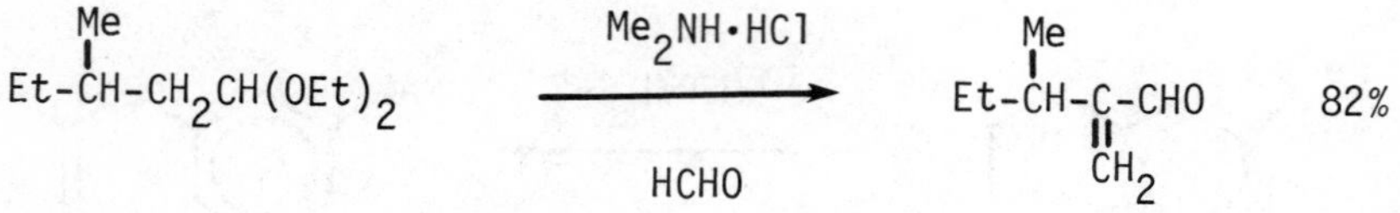

J Chem Research (S), 262 (1978)

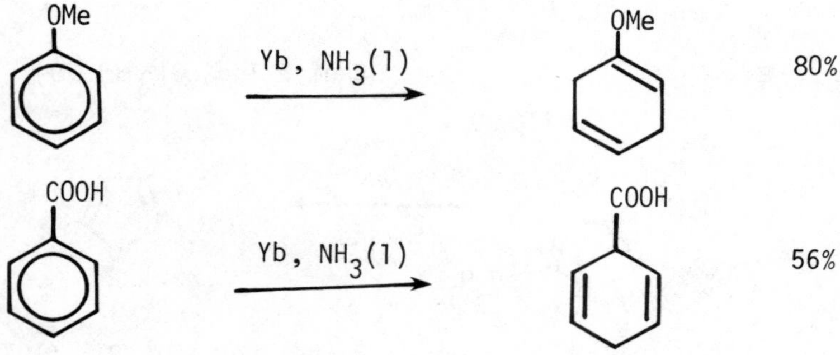

JOC <u>43</u>, 4555 (1978)

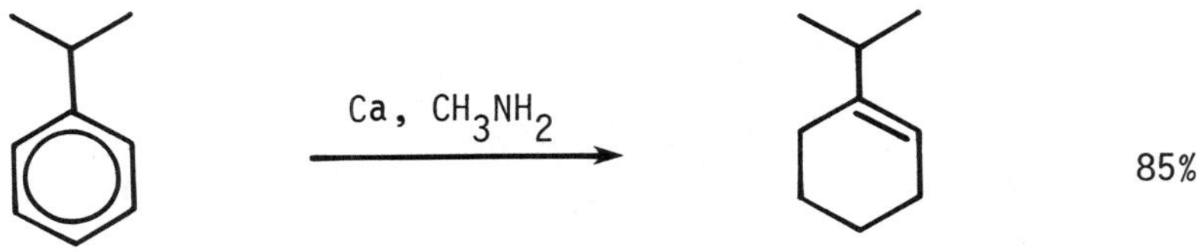

85%

JOC **44**, 3737 (1979)

Related methods: Alkyls and Aryls from Alkyls and Aryls (Section 65)   Alkyls and Aryls from Olefins (Section 74)

Section 201   Olefins from Amides

No additional examples

Section 202   Olefins from Amines

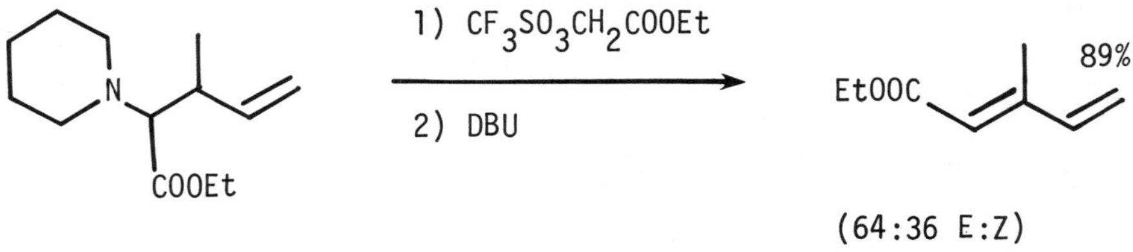

89%

(64:36 E:Z)

Tetr Lett 1241 (1977)

Section 203   Olefins from Esters

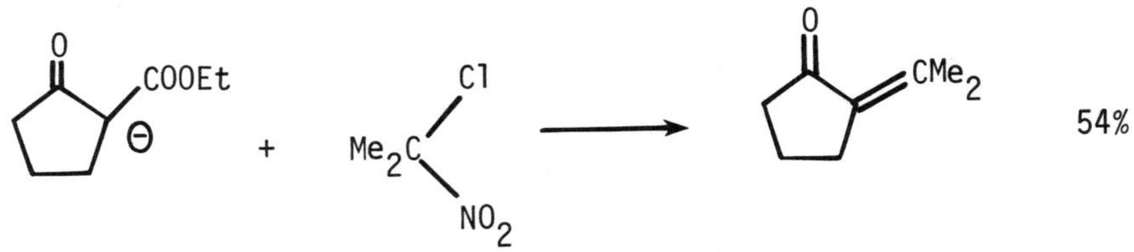

54%

Also works with lactones.   Chem Lett, 189 (1977)

1. TsCH$_2$MgI
2. NaBH$_4$

3. Electrochemical
   reduction

63%

Chem Lett, 69 (1978)

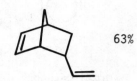

4 Ph$_3$P-CH$_2^{\oplus \ominus}$

$^{\ominus}$CH$_2$SOCH$_3$

DMSO

78%

JOC 44, 3157 (1979)

Section 204   Olefins from Epoxides

NaI, TFAA

CH$_3$CN, THF

95%

JOC 43, 1841 (1978)

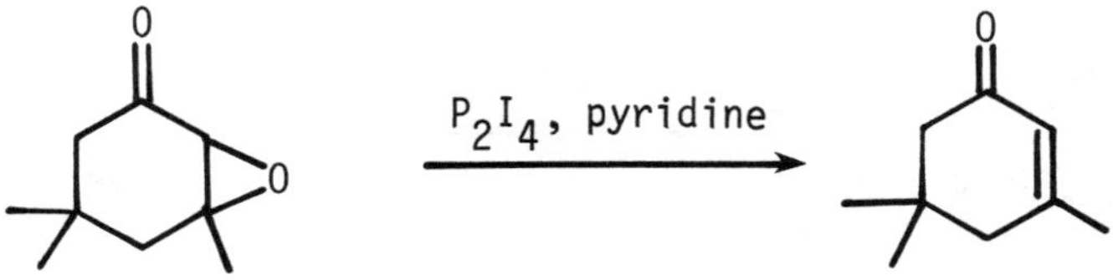

P_2I_4, pyridine

95%

Synthesis, 905 (1978)

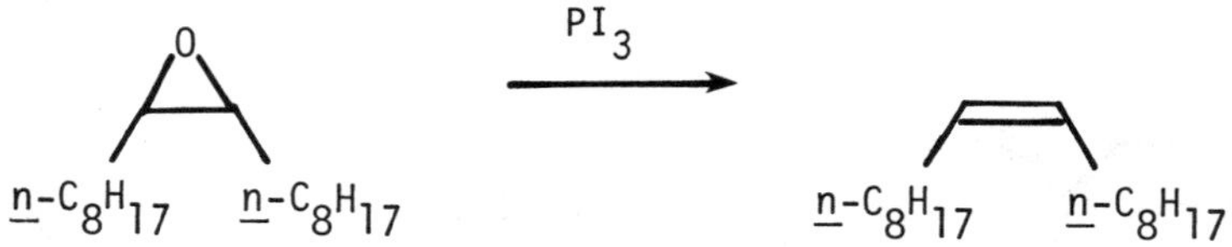

PI_3

83%

$\underline{n}$-C_8H_{17} $\underline{n}$-C_8H_{17} $\underline{n}$-C_8H_{17} $\underline{n}$-C_8H_{17}

Nouveau J Chem $\underline{3}$, 705 (1979)

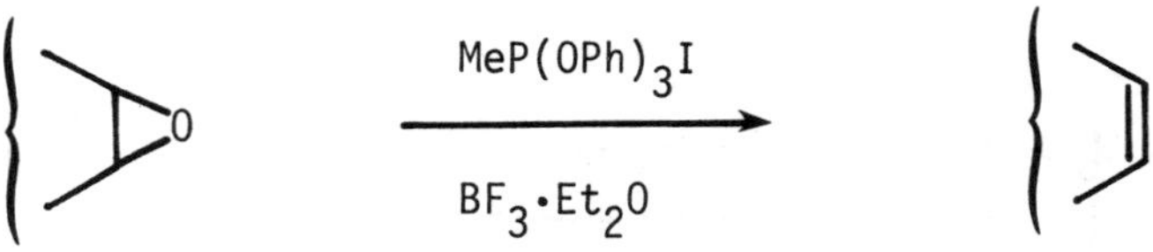

$MeP(OPh)_3I$

$BF_3 \cdot Et_2O$

~90%

JOC $\underline{43}$, 2076 (1978)

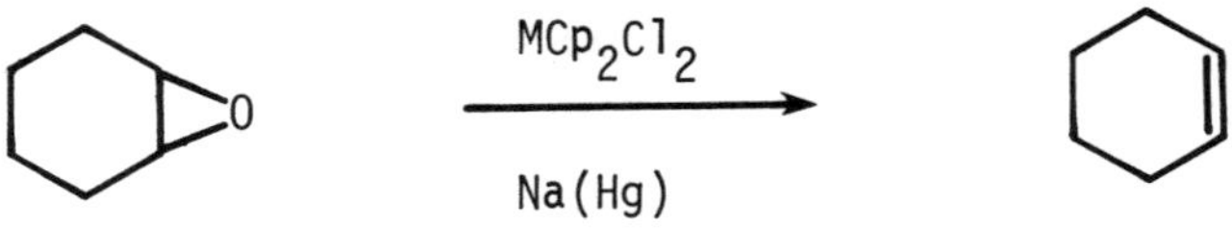

MCp_2Cl_2

Na(Hg)

M = No, W, Ti, Zr

JCS Chem Comm, 99 (1978)

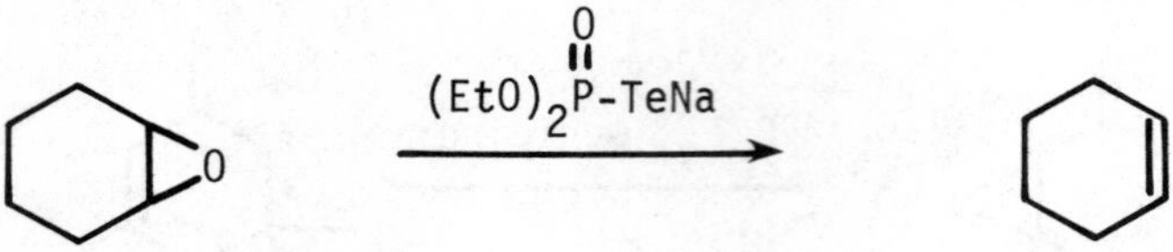

$$\text{(EtO)}_2\overset{\overset{\displaystyle O}{\|}}{P}\text{-TeNa}$$

88%

JCS Chem Comm, 658 (1977)

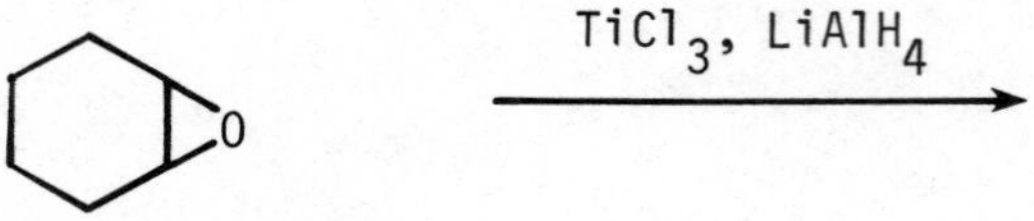

TiCl$_3$, LiAlH$_4$

69%

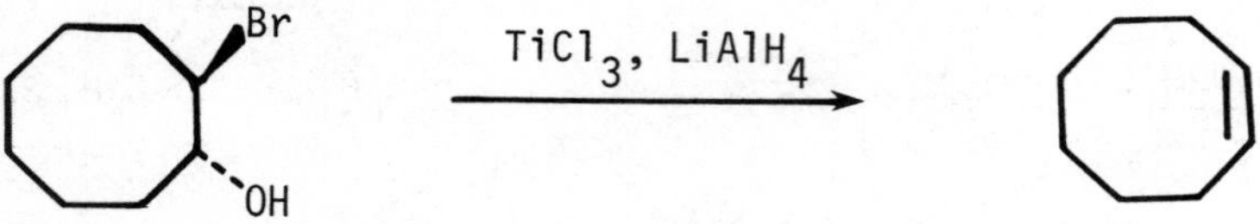

TiCl$_3$, LiAlH$_4$

96%

JOC 43, 3249 (1978)

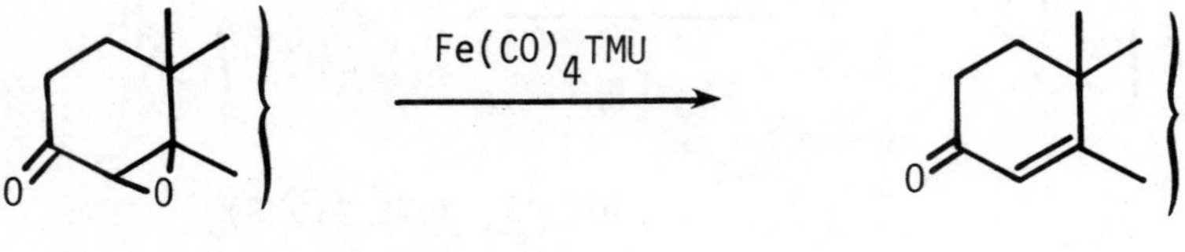

Fe(CO)$_4$TMU

95%

TMU = tetramethylurea

Tetr Lett, 4155 (1977)

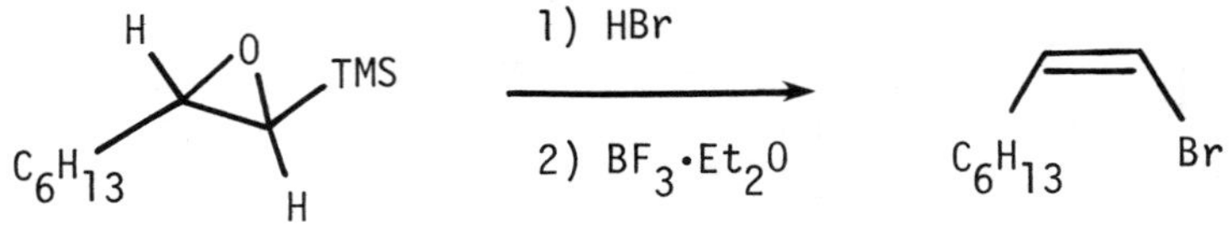
H
O
TMS
C_6H_{13}
H
1) HBr
2) $BF_3 \cdot Et_2O$
C_6H_{13}
Br

85%

JACS 99, 1993 (1977)

Section 205 Olefins from Halides and Sulfonates

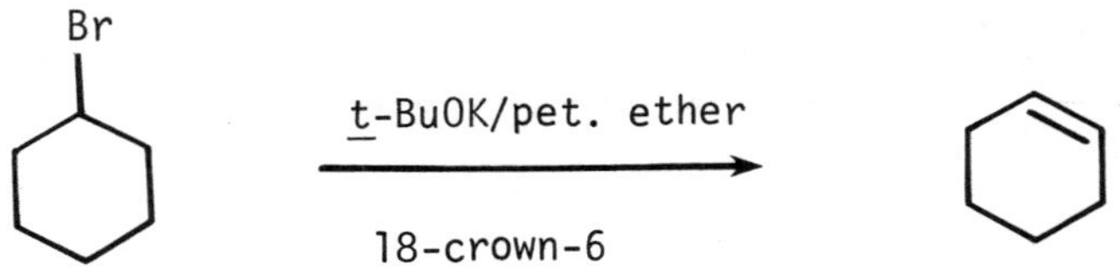
Br
t-BuOK/pet. ether
18-crown-6

83%

Synthesis, 372 (1979)

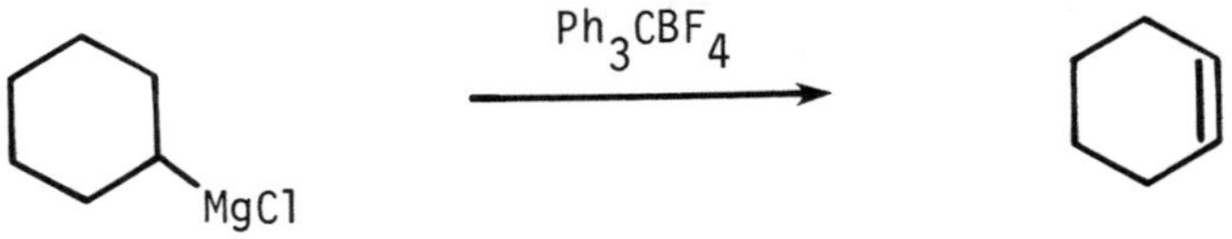
MgCl
Ph_3CBF_4

75%

Angew Int Ed 16, 44 (1977)

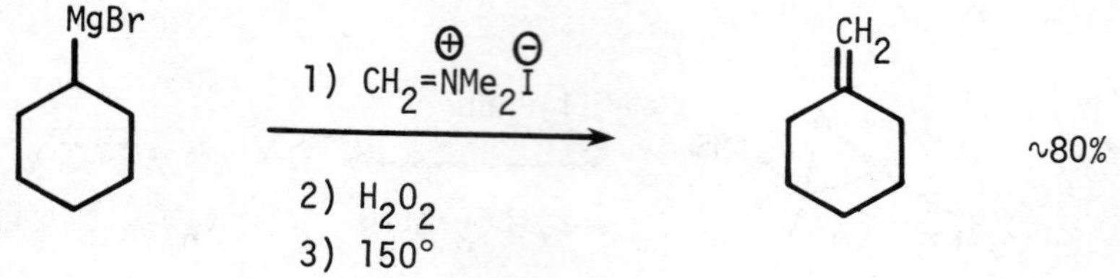

Tetr Lett, 1299 (1977)

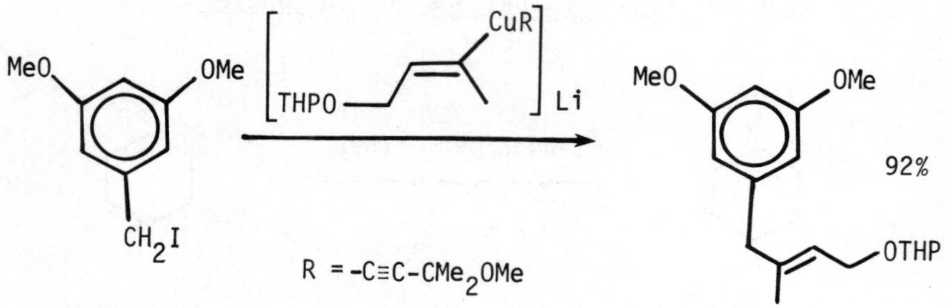

R = -C≡C-CMe$_2$OMe

JOC 43, 3418 (1978)

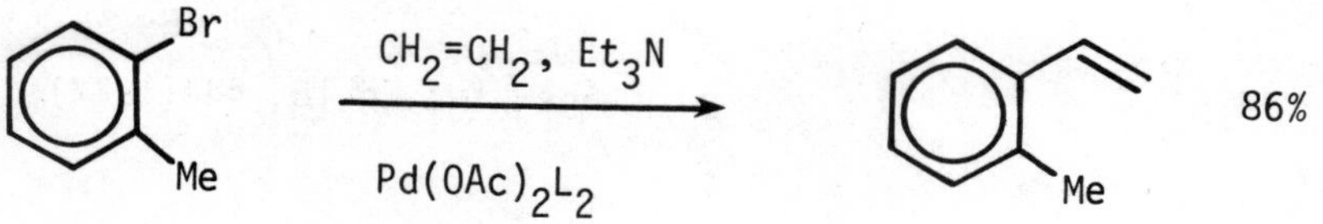

JOC 43, 2454 (1978)

$(C_6H_{13})_3B$ →   1) MeLi, CuI

2) ⟋⟍Br   → $C_7H_{14}-CH=CH_2$ 53%

BCS Japan 50, 2199 (1977)

$CH_2=CHSePh$ →   1) BuLi

2) $\underline{n}-C_{10}H_{11}Br$

3) O_3, $(\underline{i}-Pr)_2NH$

$\overset{H}{\underset{Bu}{}}C=C\overset{C_{10}H_{11}}{\underset{H}{}}$ 80%

JOC 43, 4252 (1978)

1) $KN(SiMe_3)_2$

2) $NaIO_4$

3) toluene, rfx

~75%

Tetr Lett, 4917 (1978)

$[(n-C_5H_5)Cr(NO)_2]_2$ 89%

Tetr Lett, 323 (1978)

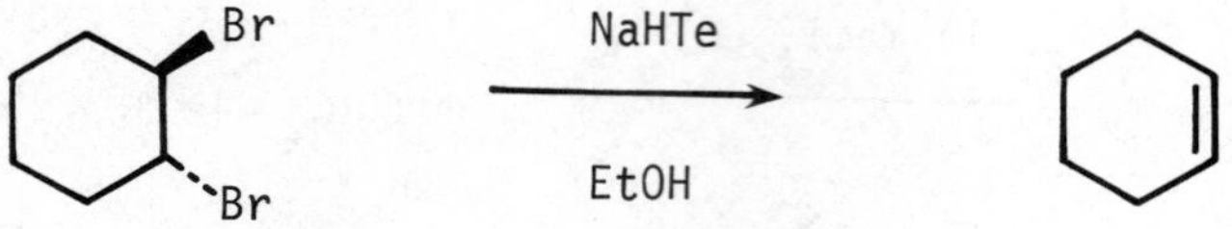

76%

Synthesis, 311 (1978)

$$\underset{\substack{| \quad |\\ Br \quad Br}}{Ph-CH-CH-Ph} \xrightarrow[\text{THF}]{VCl_3/LiAlH_4} PhCH=CHPh \qquad 97\%$$

Synthesis, 170 (1977)

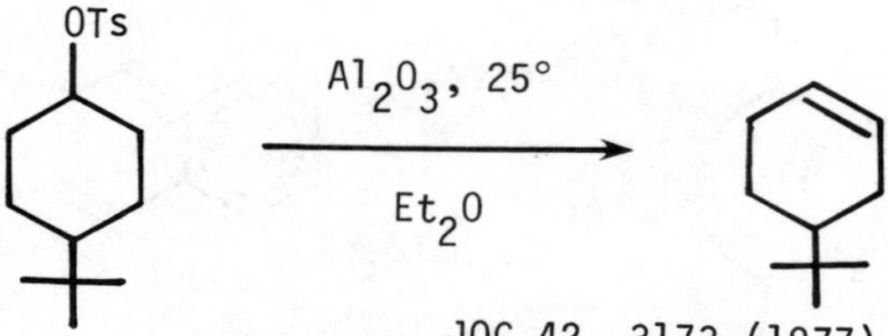

81%

JOC 42, 3173 (1977)

$$\underset{\substack{|\\ SO_2Ph}}{CH_3(CH_2)_7-CH-CH_2TMS} \xrightarrow{\overset{\ominus}{F}} CH_3(CH_2)_7-CH=CH_2 \qquad 80\%$$

Tetr Lett, 2649 (1979)

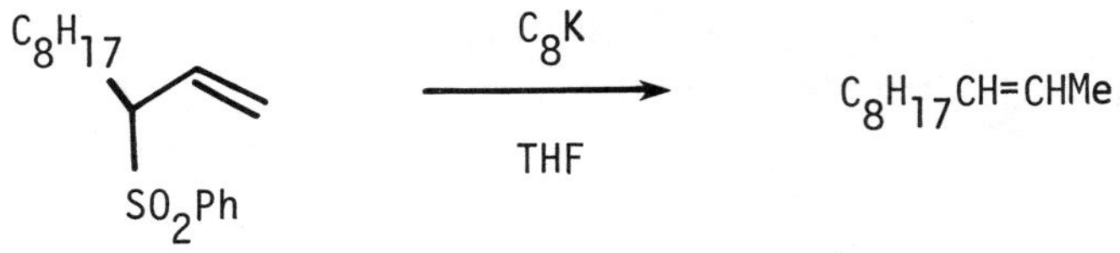

JCS Perkin I, 123 (1977)

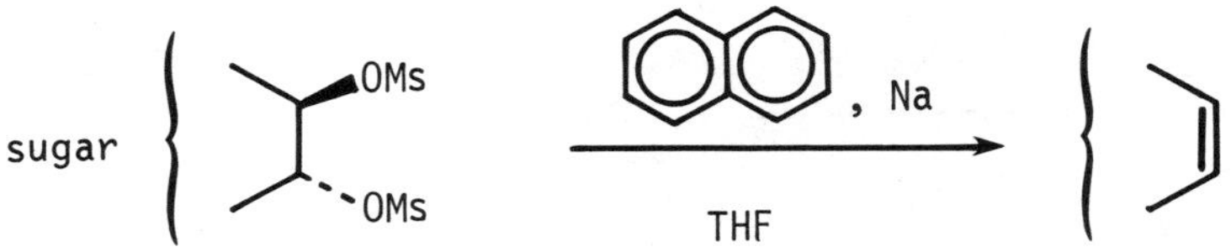

Chem Pharm 25, 2134 (1977)

Section 206   Olefins from Hydrides

No additional examples

Section 207   Olefins from Ketones

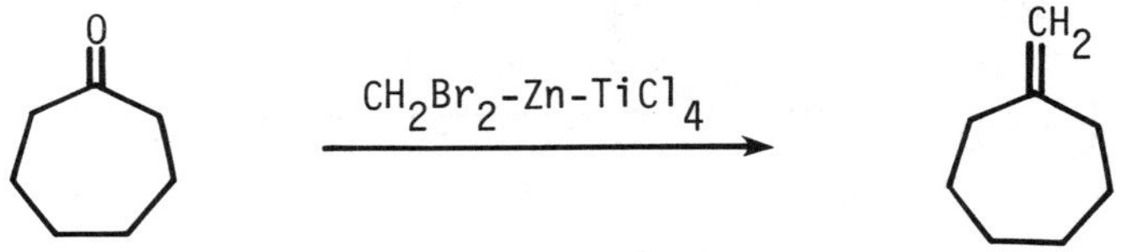

Tetr Lett, 2417 (1978)

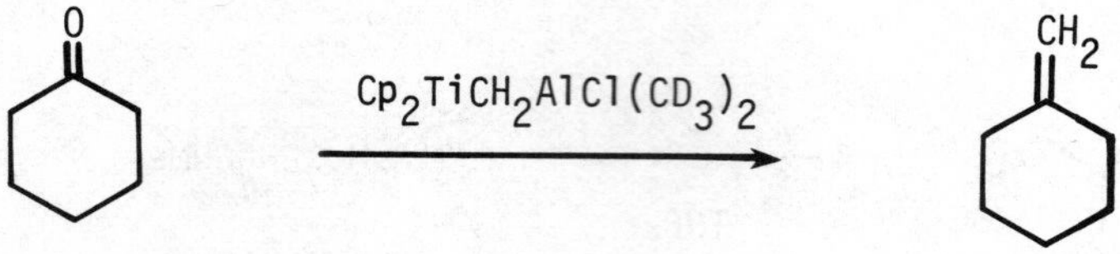
$Cp_2TiCH_2AlCl(CD_3)_2$
CH_2

65%

JACS 100, 3611 (1978)

$Ph_2C{=}O \xrightarrow[Ph_3P]{ClCH_2SiMe_3} Ph_2C{=}CH_2$ 78%

JOC 44, 413 (1979)

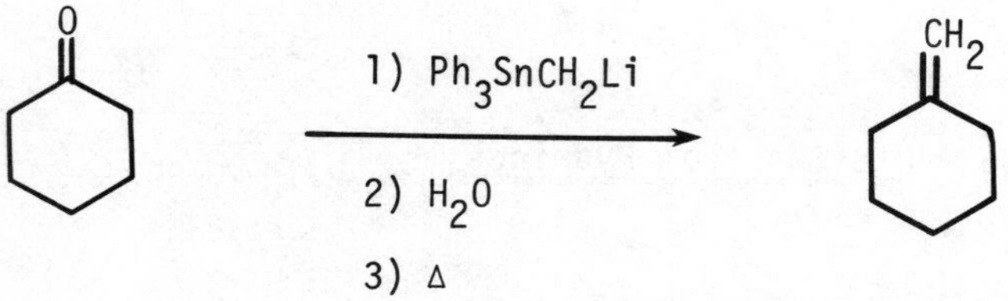
1) Ph_3SnCH_2Li
2) H_2O
3) Δ
CH_2

86%

Angew Int Ed 16, 862 (1977)

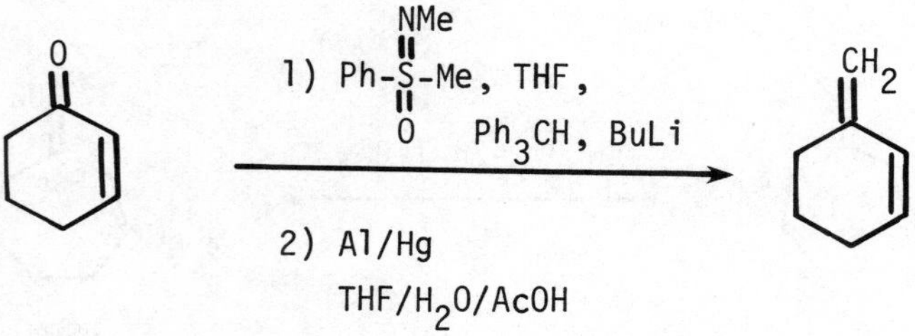
1) Ph-S(=NMe)(=O)-Me, THF,
Ph_3CH, BuLi
2) Al/Hg
$THF/H_2O/AcOH$
CH_2

45%

JACS 101, 3602 (1979)

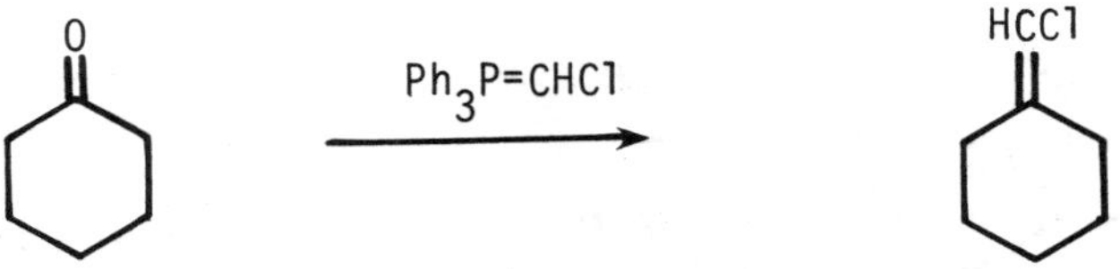

O
$Ph_3P=CHCl$
HCCl

94%

JCS Chem Comm, 446 (1978)

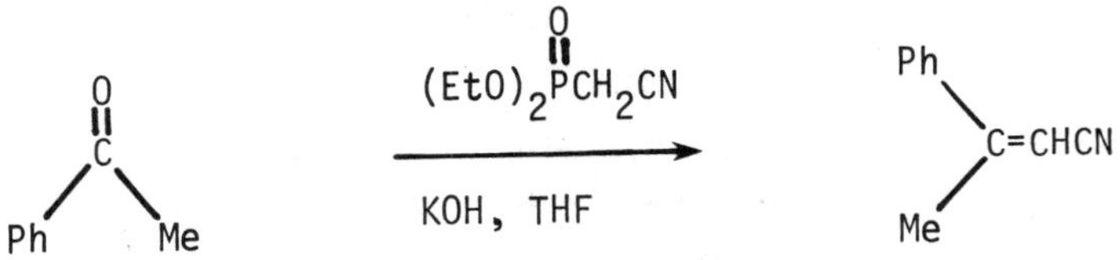

O
C
Ph
Me
O
$(EtO)_2PCH_2CN$
KOH, THF
Ph
C=CHCN
Me

88%

Synthesis, 884 (1979)

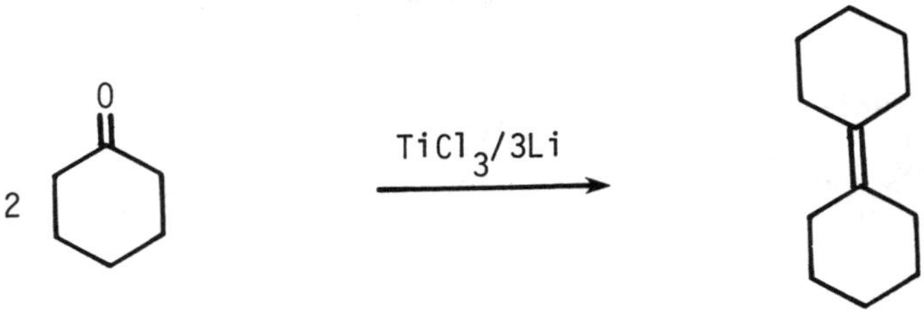

O
2
$TiCl_3/3Li$

79%

JOC 43, 3255 (1978)

Ph
C=O
C=O
Ph
$TiCl_3/LiAlH_4$
C
C
Ph
Ph

∿50%

ring size 3-12

JOC 43, 3609 (1978)

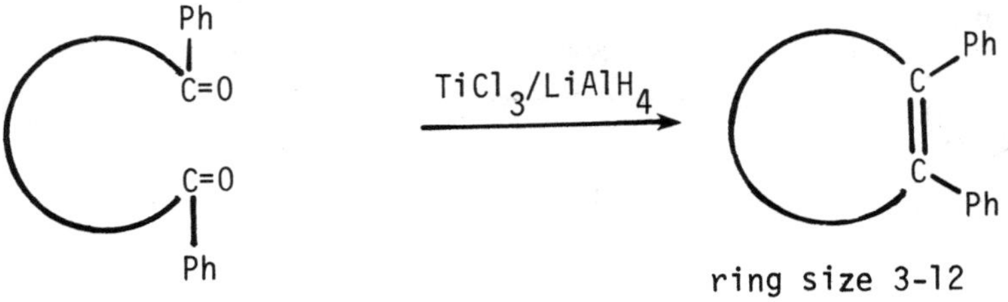

$R = H, Ph, Bu$
$n = 2-14$

~70-80%

JOC **42**, 2655 (1977)

$$2 \ C_3H_7-\overset{\overset{\displaystyle O}{\|}}{C}-Me \quad \xrightarrow[\text{THF/dioxane}]{TiCl_2, \text{ pyridine}} \quad C_3H_7-\overset{\overset{\displaystyle Me}{|}}{C}=\overset{\overset{\displaystyle Me}{|}}{C}-C_3H_7 \qquad 61\%$$

Synthesis, 553 (1977)

66%

JOC **43**, 1404 (1978)

1) EtMgBr
2) Li
3) $H_3O^{\oplus}$

87%

JCS Chem Comm, 847 (1978)

JOC $\underline{43}$, 2715 (1978)

Bull Chem Soc Japan $\underline{52}$, 1760 (1979)

Related methods: Olefins from Aldehydes, Section 199.

Section 208   Olefins from Nitriles

JOC $\underline{44}$, 2994 (1979)

PhCH$_2$CH$_2$CH=NNHTs

+   ⟶   PhCH$_2$CH$_2$CH=CHEt   70%

EtCH(Li)CN

JACS <u>101</u>, 249 (1979)

1) NaNH$_2$, benzene

2)   Br

3) H$_2$O

79%

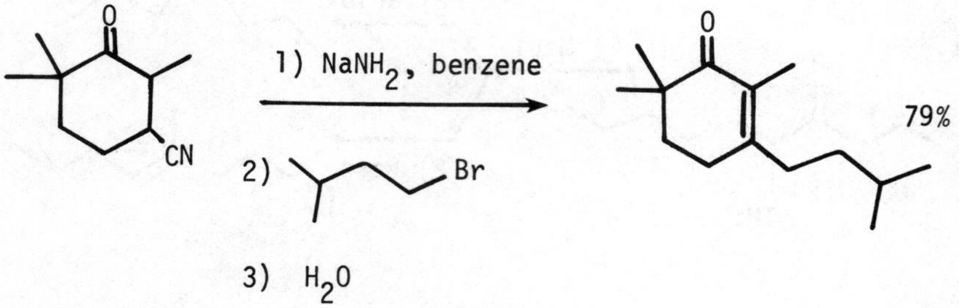

Tetr Lett, 3187 (1977)

Section 209   <u>Olefins from Olefins</u>

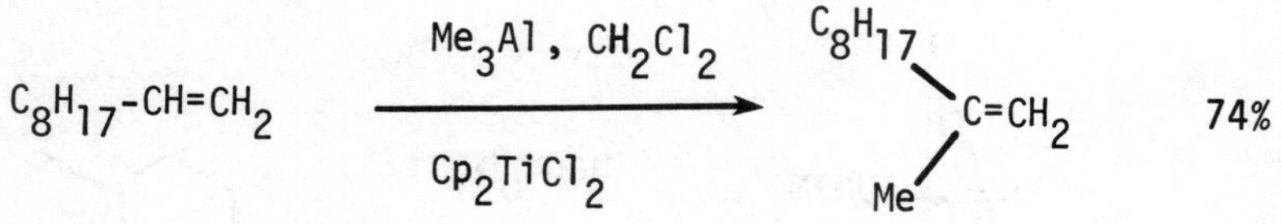

C$_8$H$_{17}$-CH=CH$_2$ $\xrightarrow[\text{Cp}_2\text{TiCl}_2]{\text{Me}_3\text{Al, CH}_2\text{Cl}_2}$ 74%

JOC <u>44</u>, 3603 (1979)

Section 210 Olefins from Miscellaneous Compounds

53%

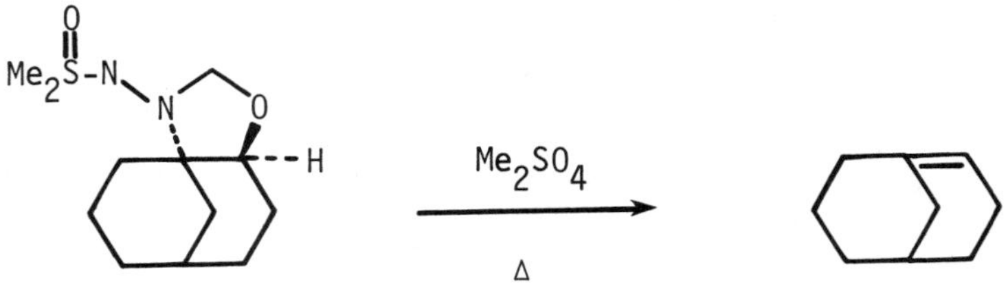

JACS 99, 1172 (1977)

$Ph_2C{=}CHNO_2$ $\xrightarrow[\text{DMF}]{Na_2S,\ PhSH}$ $Ph_2C{=}CH_2$ 94%

Tetr Lett, 1733 (1979)

84%

JOC 42, 2944 (1977)

$(C_5H_{11}\text{-}CH_2\text{-}CH(CH_3)\text{-}O)_3B$ $\xrightarrow[100°]{BF_3 \cdot Et_2O}$ $C_5H_{11}\text{-}CH{=}CHCH_3$ 88%

Synthesis, 717 (1977)

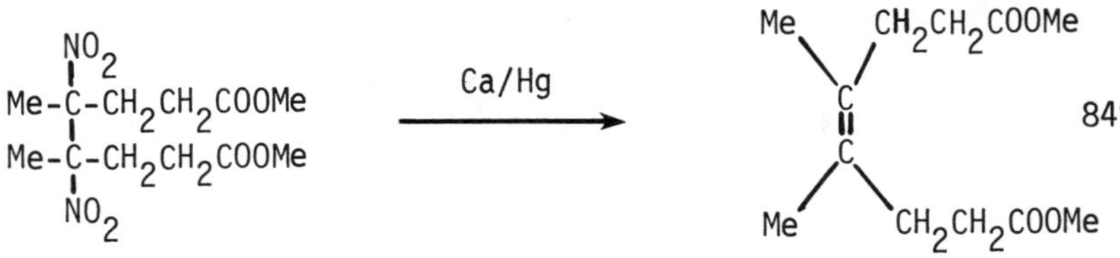

$$Ph_2C{=}S \xrightarrow{Co_2(CO)_8} Ph_2C{=}CPh_2 \qquad 71\%$$

JOC 42, 3522 (1977)

Section 210A Protection of Olefins

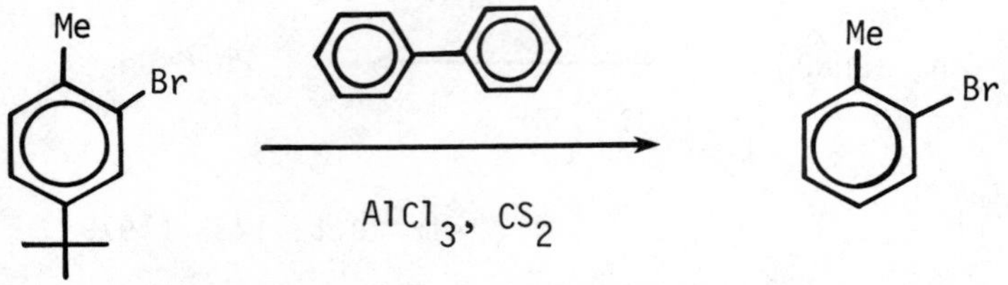
Me
Br
AlCl3, CS2
Me
Br

90%

Use of the t-butyl group to block positions on an aromatic ring.

JCS Perkin I, 176 (1979)

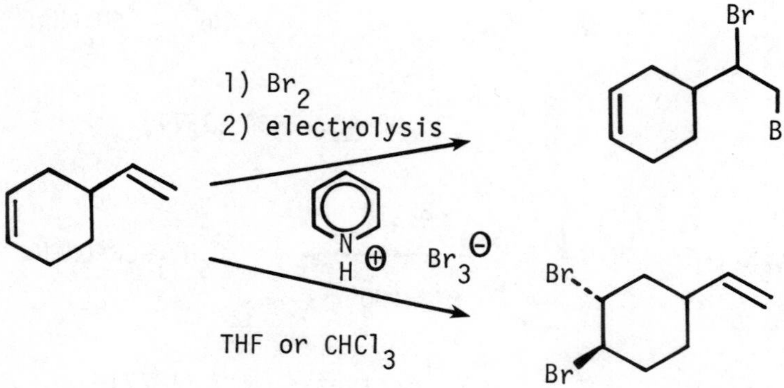
1) Br2
2) electrolysis
Br
Br
N
H
Br3
THF or CHCl3
Br
Br

62%

75%

Synthesis, 964 and 966 (1979)

Review: "Selective Synthesis of Aromatic Compounds Using
        Positional Protective Groups"

                              Synthesis, 921 (1979)

CHAPTER 15
PREPARATION OF
DIFUNCTIONAL COMPOUNDS

Section 300   <u>Acetylene - Acetylene</u>

$$Bu-C\equiv C-MgBr \xrightarrow[\text{CuBr/THF}]{HC\equiv C-CH_2OTs} Bu-C\equiv C-CH_2-C\equiv CH \qquad 75\%$$

Synthesis, 292 (1979)

Section 301   <u>Acetylene - Carboxylic Acid</u>

No additional examples

Section 302   <u>Acetylene - Alcohol</u>

$$Me-C\equiv C-\overset{\overset{O}{\|}}{C}-\underline{i}-Bu \xrightarrow[\text{Darvon}]{LiAlH_4} Me-C\equiv C-\overset{H}{\underset{\underline{i}-Bu}{\big\langle}}---OH \qquad 94\%$$

82%ee

JACS <u>99</u>, 8339 (1977)

$$t\text{-Bu-}\overset{\overset{\displaystyle O}{\|}}{C}\text{-C}\equiv\text{CH} \xrightarrow{\text{chiral LAH complex}^*} t\text{-Bu-}\overset{*}{\underset{\underset{\displaystyle OH}{|}}{CH}}\text{-C}\equiv\text{CH} \quad 90\%\text{ee}$$

*[LiAlH$_4$, N-methylephedrine, 3,5-dimethylphenol]

Tetr Lett, 2683 (1979)

$$C_6H_{13}C\equiv CCH_2\overset{\underset{\displaystyle CH_3}{|}}{C}=CH_2 \xrightarrow[\substack{2)\ H_2O_2,\ \ominus OH}]{1)\ 9\text{-BBN}} C_6H_{13}C\equiv CCH_2\overset{\underset{\displaystyle CH_3}{|}}{CH}CH_2OH \quad 98\%$$

JOC **44**, 2328 (1979)

$$H\text{-C}\equiv C\text{-CH}_2CH_2OH \xrightarrow[\substack{2)\ BuBr}]{1)\ LiNH_2,\ NH_3} Bu\text{-C}\equiv C\text{-CH}_2CH_2OH \quad 72\%$$

Helv Chim Acta **61**, 2275 (1978)

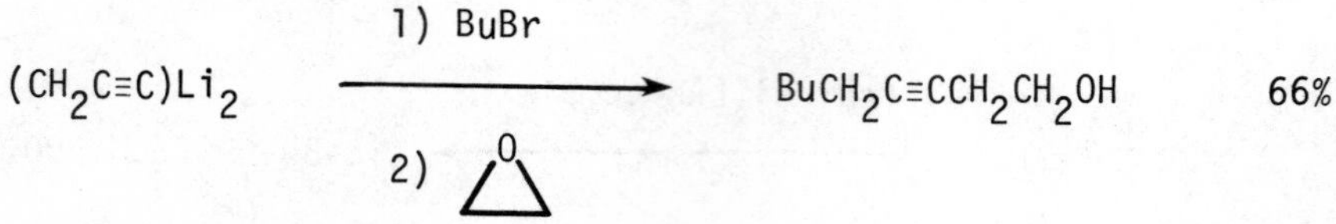
(CH2C≡C)Li2
1) BuBr
2)
O
BuCH2C≡CCH2CH2OH
66%

JCS Perkin I, 1218 (1979)

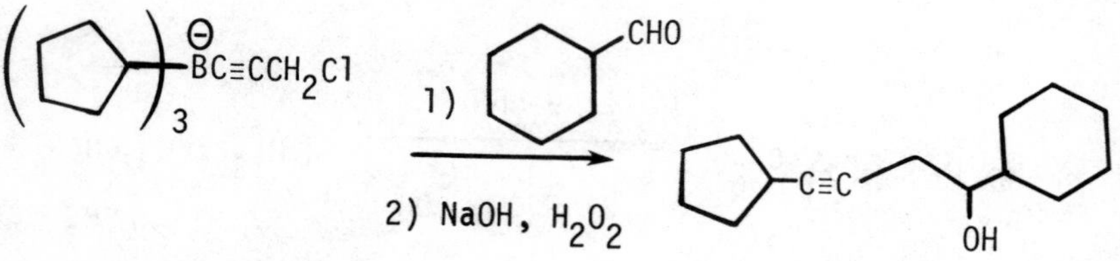
BC≡CCH2Cl
3
CHO
1)
2) NaOH, H2O2
C≡C
OH

JACS 100, 5561 (1978)

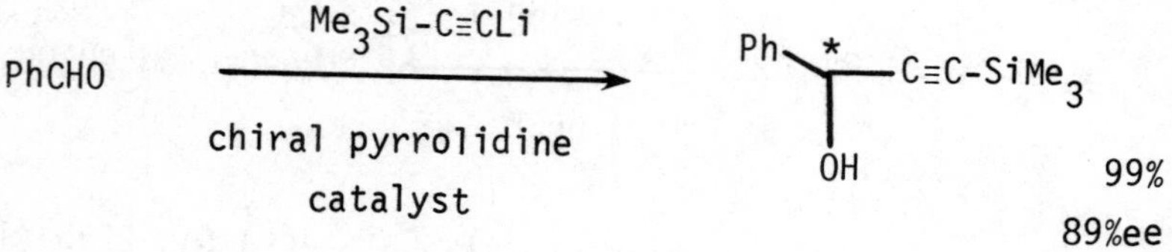
PhCHO
Me3Si-C≡CLi
chiral pyrrolidine
catalyst
Ph
*
C≡C-SiMe3
OH
99%
89%ee

Chem Lett, 447 (1979)

Section 303 Acetylene - Aldehyde

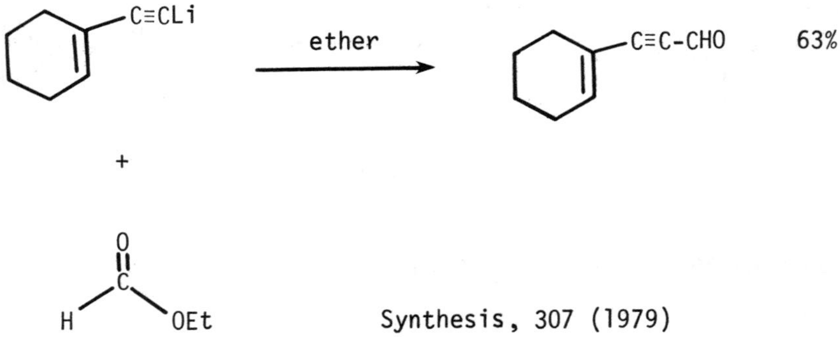
C≡CLi
ether
C≡C-CHO
63%
+
O
C
H
OEt

Synthesis, 307 (1979)

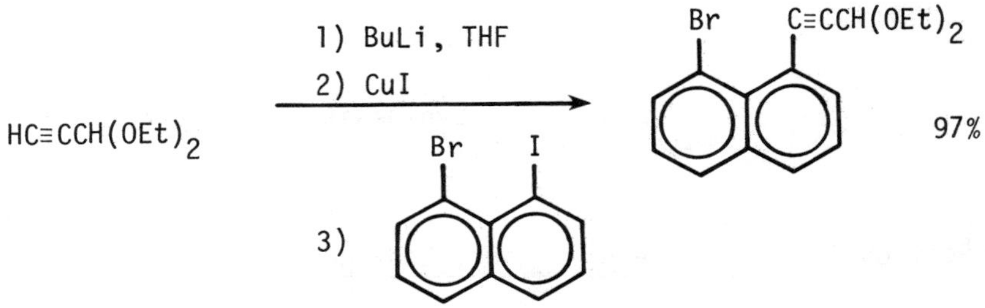
HC≡CCH(OEt)2
1) BuLi, THF
2) CuI
3)
Br
I
Br
C≡CCH(OEt)2
97%

JOC 42, 2626 (1977)

Section 304 Acetylene - Amide

No additional examples

Section 305   <u>Acetylene - Amine</u>

Ph-CH-C-Cl (O, OAc)

$$\begin{array}{c}Me \\ \diagdown \\ Ph \diagup \end{array} N-C\equiv C-SnPh_3$$

Ph-CH-C-C≡C-N (O, OAc, Me, Ph)   81%

Angew Int Ed <u>18</u>, 405 (1979)

Section 306   <u>Acetylene - Ester</u>

Li[Bu$_3$B-C≡C-COOEt] $\xrightarrow{\text{I}_2}$ Bu-C≡C-COOEt   79%

Synthesis, 679 (1977)

Section 307   <u>Acetylene - Ether, Epoxide</u>

Ph-CH-CH-C≡CH (OH OH) $\xrightarrow[\text{glyme}]{\text{TsCl, NaOH}}$ Ph-CH-CH-C≡CH (O epoxide)   80%

Synthesis, 706 (1977)

Section 308   <u>Acetylene - Halide</u>

$$Ph-C{\equiv}C-H \xrightarrow[\text{2) NCS}]{\text{1) BuLi}} Ph-C{\equiv}C-Cl \qquad 69\%$$

Synthesis, 296 (1979)

Section 309   <u>Acetylene - Ketone</u>

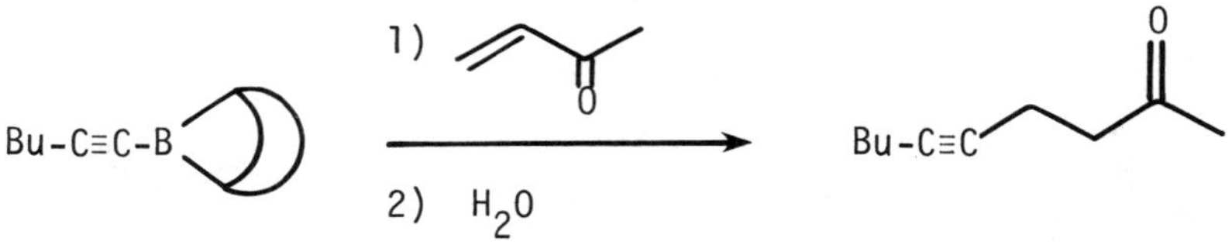

JACS <u>99</u>, 954 (1977)

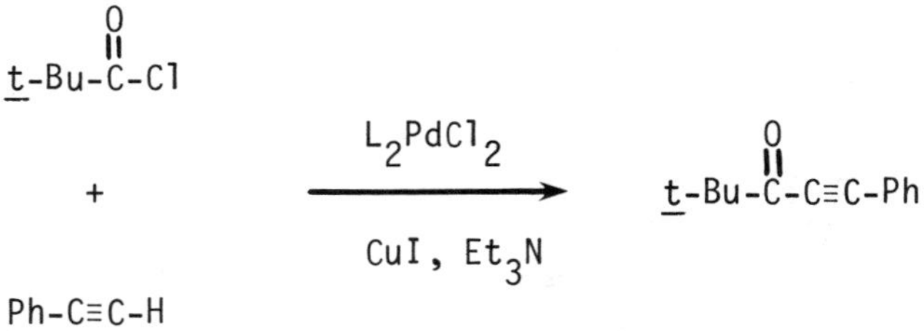

Synthesis, 777 (1977)

$Bu-C{\equiv}C-AlMe_2$

+

(cyclohex-2-enone structure)

DiBAH
——————→
$Ni(acac)_2$

(3-(1-hexynyl)cyclohexanone structure) $C{\equiv}C-Bu$

71%

JACS <u>100</u>, 2244(1978)

(pentane-2,4-dione structure)

1) $HC{\equiv}C-\overset{\overset{Ph}{|}}{C}HOH$
   $Co_2(CO)_6$
   $HBF_4{\cdot}Me_2O$
————————————→
2) $Fe(NO_3)_3$
   ethanol

(3-acetyl-4-phenyl product structure) 86%

Ph $C{\equiv}CH$

Tetr Lett, 4349 (1978)

$Ph_3C-OCH_2CH_2C{\equiv}\overset{\oplus}{C}\ \overset{\ominus}{Li}$

+

(tetrahydropyran-2-one structure)

THF
——————→

(product structure) OH 99%

Ph_3C-O $C{\equiv}C$ O

Tetr Lett, 937 (1978)

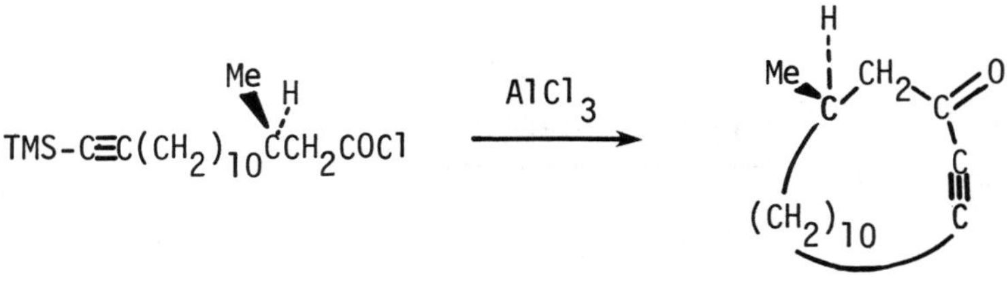

TMS-C≡C(CH$_2$)$_{10}$ĊCH$_2$COCl　　$\xrightarrow{\text{AlCl}_3}$

52%

Tetr Lett, 2301 (1978)

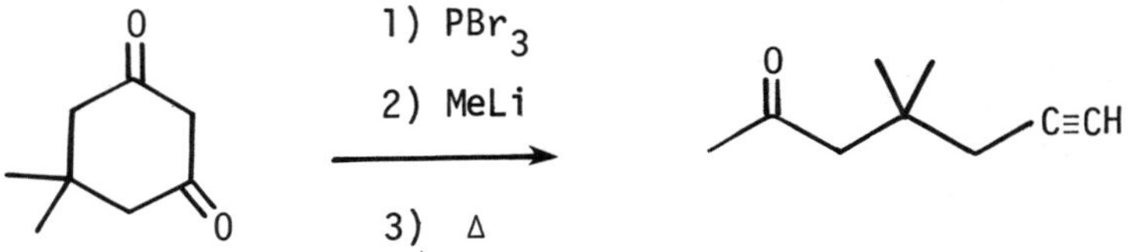

1) PBr$_3$

2) MeLi

3) Δ

—C≡CH

70%

JOC **42**, 2380 (1977)

Section 310　Acetylene - **Nitrile**

Review:　"α-Cyanoacetylenes"

Russ Chem Rev **46**, 374 (1977)

Section 311   <u>Acetylene - Olefin</u>

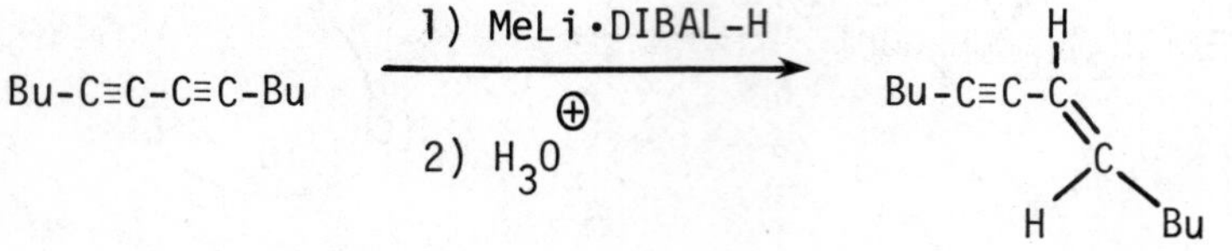

Bu-C≡C-C≡C-Bu $\xrightarrow[\text{2) H}_3\text{O}^{\oplus}]{\text{1) MeLi·DIBAL-H}}$ Bu-C≡C-C(H)=C(H)(Bu)   92%

Synthesis, 52 (1977)

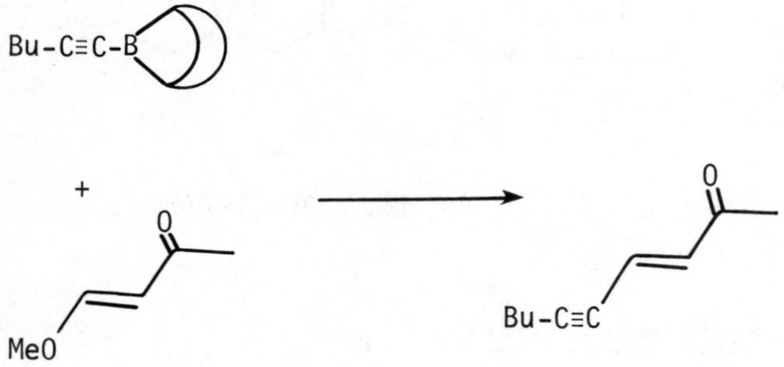

Bu-C≡C-B

+

(MeO-enone)   ⟶   Bu-C≡C- (enone)   85%

JOC 42, 3106 (1977)

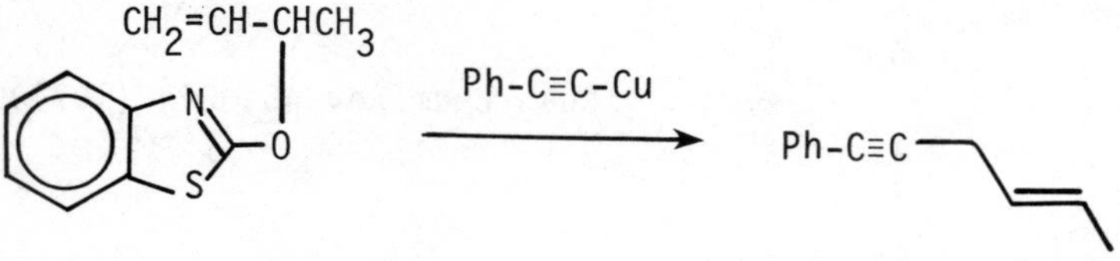

$\xrightarrow{\text{Ph-C≡C-Cu}}$ Ph-C≡C- (olefin)   70%

Tetr Lett, 3873 (1979)

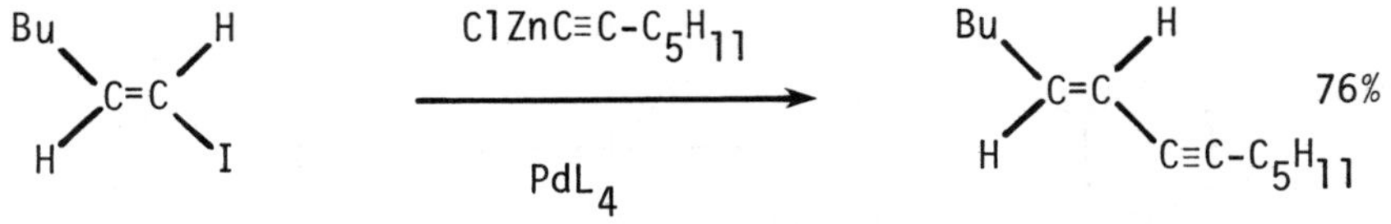

JCS Chem Comm, 683 (1977)

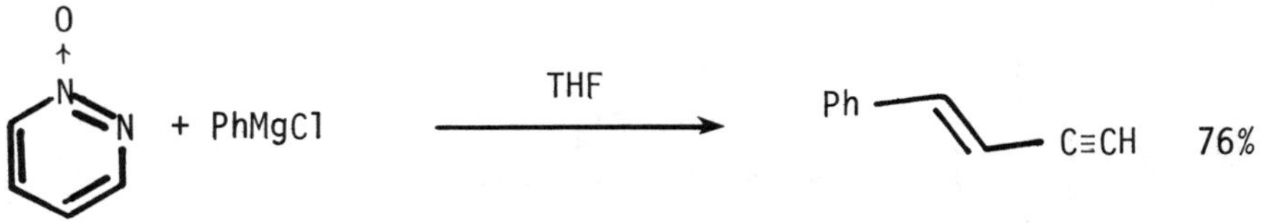

JCS Perkin I, 2136 (1979)

Section 312   Carboxylic Acid - Carboxylic Acid

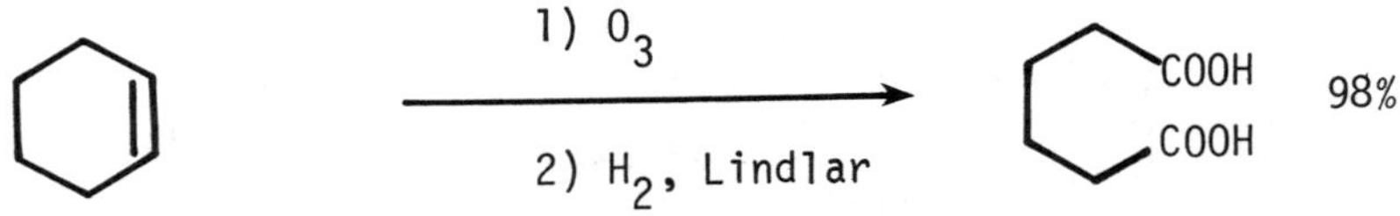

Bull Akad USSR Chem 25, 1790 (1977)
JOC (USSR) 14, 48 (1978)

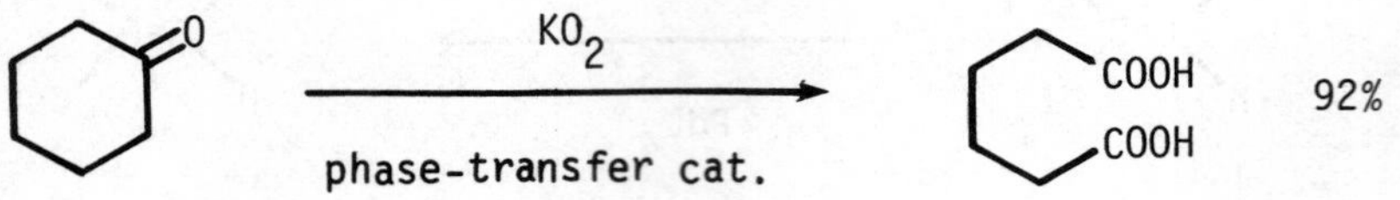

Tetr Lett, 3689 (1978)

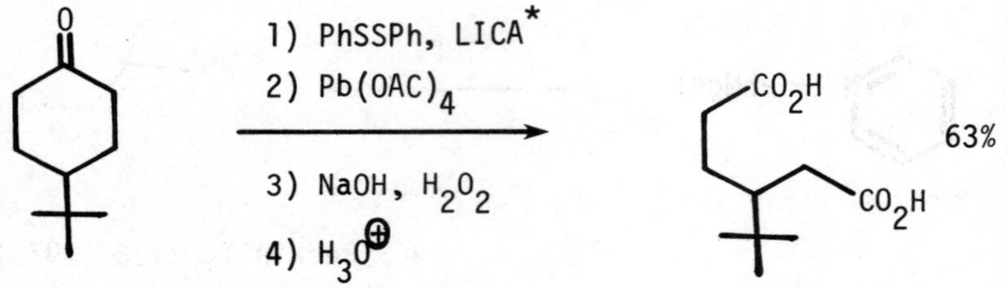

*LICA = lithium isopropylcyclohexylamide

JACS 99, 4405 (1977)

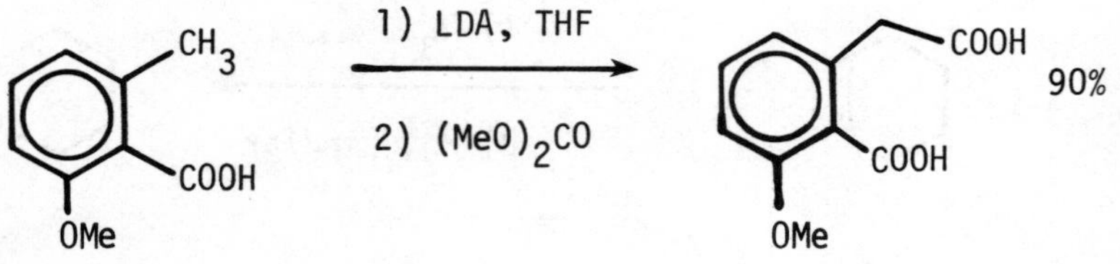

Synthesis, 245 (1977)

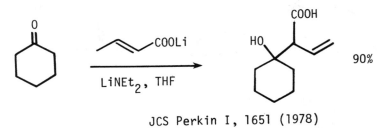

CH₂OTs structure with OTs

1) CH₂(COOEt)₂,

NaNH₂, dioxane

2) saponification

HOOC—COOH bicyclic structure

65%

Synth Comm <u>7</u>, 1 (1977)

Section 313   <u>Carboxylic Acid - Alcohol</u>

$$\underset{\text{Me-C-COOH}}{\overset{O}{\parallel}}$$

1) (EtO)₃P

2) NaOH

$$\underset{\text{Me-CH-COOH}}{\overset{OH}{|}}$$ 94%

JOC <u>42</u>, 2797 (1977)

cyclohexanone structure

⟍⟍⟍—COOLi

LiNEt₂, THF

HO⟍ cyclohexane with COOH and vinyl group

90%

JCS Perkin I, 1651 (1978)

n-C₅H₁₁-CHO

1) TMS-O-CH=C(OTMS)₂,

   SnCl₄

2) H₃O⁺

n-C₅H₁₁ structure with OH, COOH, OH

70%

Synthesis, 27 (1979)

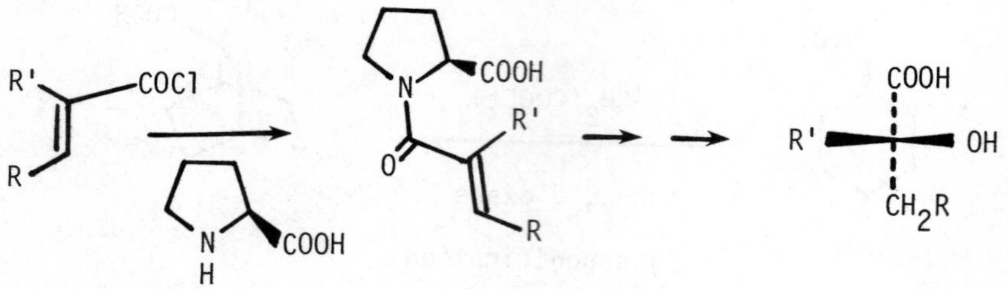

Tetr Lett, 1005 (1977)

Section 314   Carboxylic Acid - Aldehyde

No additional examples

Section 315   Carboxylic Acid - Amide

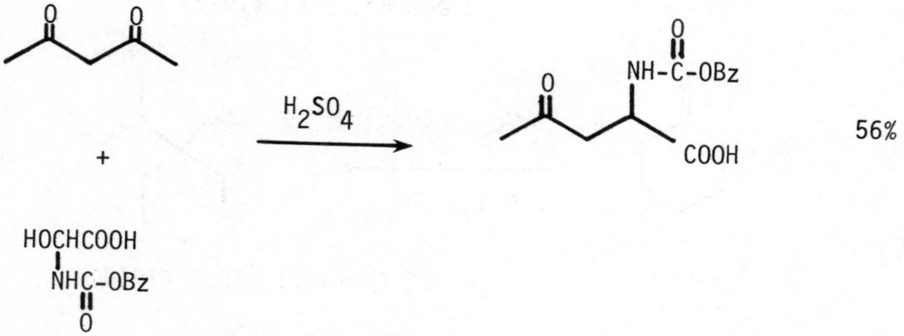

56%

Tetrahedron <u>34</u>, 467 (1978)

$$CH_2=\underset{\underset{\displaystyle NHCOOCH_2Ph}{|}}{C}-COOH \qquad \xrightarrow[\text{(S,S)-PPPM-Rh}]{H_2,} \qquad CH_3-\overset{*}{\underset{\underset{\displaystyle NHCOOCH_2Ph}{|}}{CH}}-COOH \qquad \sim90\%$$

$$(R)\ 59\%ee$$

Chem Lett, 777 (1977)

$$\underset{Ph}{\overset{\displaystyle COOH}{\diagdown}}C=CH \qquad \xrightarrow[\text{chiral Rh phosphines}]{H_2} \qquad Ph-CH_2-\overset{*}{CH}\overset{\displaystyle \diagup COOH}{\diagdown}_{NHAc} \quad 90\text{-}100\%$$

78-99%ee

JACS 99, 6262 (1977)
Tetr Lett, 3497 (1977)
JACS 100, 5491 (1978)
J Organometal 150, C14 (1978)

$$PhCH=C\overset{\displaystyle \diagup COOH}{\diagdown}_{NHAc} \qquad \xrightarrow[\substack{\text{polymer-bound chiral} \\ \text{Rh catalyst}}]{H_2} \qquad PhCH_2-\overset{*}{CH}\overset{\displaystyle \diagup COOH}{\diagdown}_{NHAc} \quad 100\%$$

78%ee

JACS 100, 264 and 268 (1978)

Related Methods: Section 316 (Acid-Amine)
                    Section 344 (Amide-Ester)
                    Section 351 (Amine-Ester)

Section 316   Carboxylic Acid - Amine

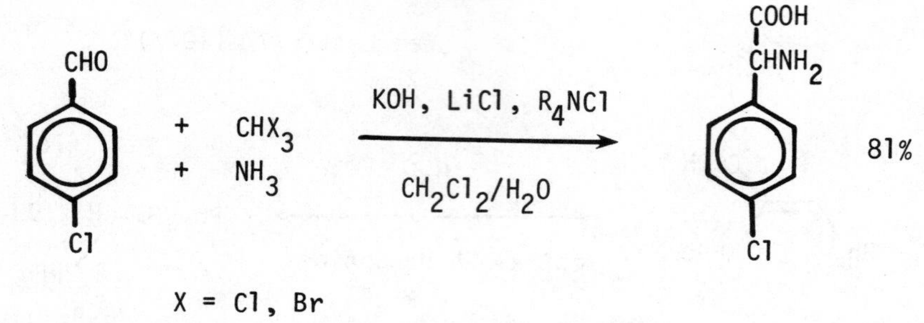

X = Cl, Br

Synthesis, 852 (1977)
Synthesis, 26 (1979)

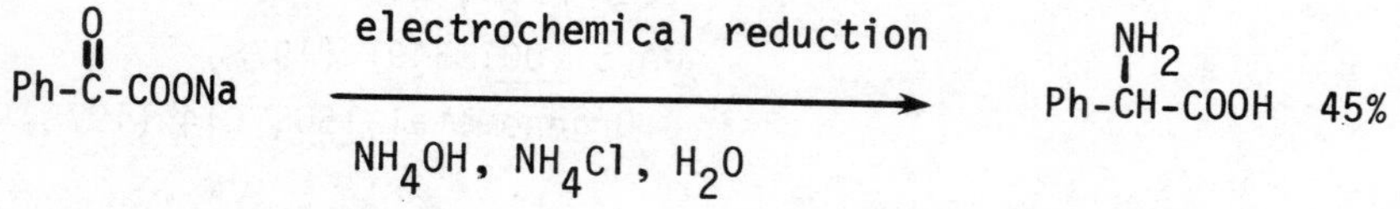

Aust J Chem 31, 73 (1978)

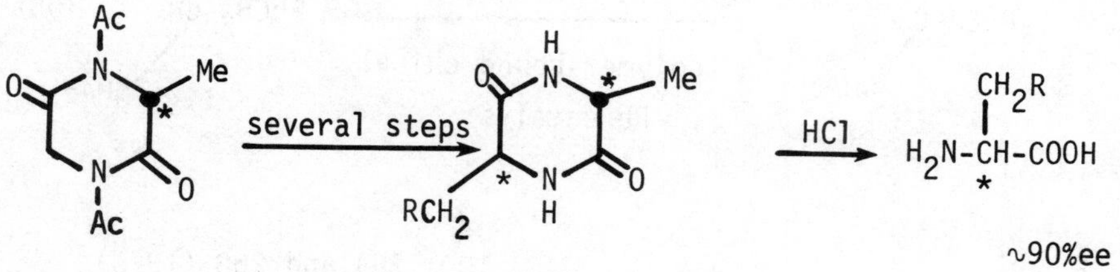

Tetr Lett, 4483 (1979)

R-CH-COOMe $\xrightarrow[\text{2) MeOH}]{\text{1) KOCN}}$ R, COOMe $\xrightarrow{\text{HCl}}$ R, COOH ~60%
| CH   CH
Br   NHCOOMe   NH$_2$

Angew Int Ed 18, 474 (1979)

1) BzBr

2) KOH

$\overset{\text{NH}_2}{\underset{\text{Me}}{\text{Bz---C---COOH}}}$ 86%

100% ee

Angew Int Ed 17, 117 (1978)

1) (i-Pr)$_2$NET

2) allyl Br

3) HCl/AcOH

$\overset{\text{Allyl}}{\underset{\text{NH}_2}{\text{Bz---C---COOH}}}$

93%

Chem Ber 112, 128 (1979)

R-CH=CH-COOH $\xrightarrow[\text{2) H}_2\text{, Pd/C}]{\text{1) NH}_2\text{OH·HCl, NaOAc}}$ ~50-80%

$\underset{\text{NH}_2}{\text{R-CH-CH}_2\text{COOH}}$

Synth Comm 9, 705 (1979)

NBz
‖
i-Bu — C — Me

1) BrCN
2) Et$_3$N
3) H$_3$O$^{\oplus}$

Me
|
i-Bu — C — NH$_3$ $^{\oplus}$ 74%
|
COOH

Indian J Chem 18B, 273 (1979)

Me
|
Ph-CH$_2$N=C-COOBz

1) BzCl, TBABr
2) H$_2$, Pd/C

Me
|
H$_2$N — C — COOH   86%
|
Bz

JCS Perkin I, 1730 (1977)

Ph$_2$C=N-CH$_2$CN

+

EtBr

1) BzEt$_3$NCl
50% NaOH, toluene

2) HCl

Et
|
H$_2$N-CH-COOH   90%

Tetr Lett, 4625 (1978)

(MeS)$_2$C=NCH$_2$CO$_2$Et

1) KOt-Bu
2) EtI

3) HCO$_3$H

Et
|
H$_2$N-CH-COOH   81%

Liebigs Annalen, 2066 (1979)

i-Pr-CHO

+

CN-CH$_2$-CONH$_2$

1) KOH
$\longrightarrow$
2) HCl, H$_2$O

$$\text{i-Pr-CH-CHCOOH}$$
with OH on the first carbon and NH$_2$ on the second

90%

Synthesis, 216 (1979)

(TMS)$_2$NCH$_2$COOTMS

+

(isopropyl)—CHO

1) LDA
$\longrightarrow$
2) HCl, MeOH

Me$_2$CH-CH-CHCOOH
with OH and NH$_2$·HCl

90%

JOC 44, 3967 (1979)

CHO
|
CHO

+

CH$_2$COOH
|
C=O
|
COOH

1) -CO$_2$
$\longrightarrow$
2) NH$_4$OH

3) NaBH$_4$

HO—(pyrrolidine ring)—N(H)—COOH

41%

JOC 42, 3440 (1977)

1) CH$_2$=CHCN
   Et$_3$N, pyr
───────────────
2) Dowex (H$^{\oplus}$)

84%

Synthesis, 150 (1979)

H$_2$, Pd/C
───────────────

74%

Indian J Chem 15B, 573 (1977)

Several methods for synthesis of 6-substituted tryptophans are presented.

JOC 44, 3741 and 3748 (1979)

Asymmetric synthesis of amino acids by catalytic reduction of azalactones by substituted α-acylaminoacrylic acids.

Bull Acad USSR Chem 27, 957, 1186, and 1190 (1978)

Review: "Production and Utilization of Amino Acids"

Angew Int Ed <u>17</u>, 176 (1978)

Related methods: Section 315 (Acid-Amide)
Section 344 (Amide-Ester)
Section 351 (Amine-Ester)

Section 317   <u>Carboxylic Acid - Ester</u>

$$HOOC-(CH_2)_4-COOH \xrightarrow[\text{continuous extraction}]{EtOH/H_2SO_4} HOOC-(CH_2)_4-COOEt \quad \sim 90\%$$

Synth Comm <u>9</u>, 669 (1979)

1) H_2NNMe_2

2) $H_3O^{\oplus}$

80%

Tetr Lett, 1403 (1978)

Section 318   Carboxylic Acid - Ether, Epoxide

1) CHCl$_3$, NaOH
   THF

2) NaOH, MeOH

54%

JOC 43, 2702 (1978)

Section 319   Carboxylic Acid - Halide

$$CH_3(CH_2)_{13}CH_2COOH \xrightarrow[\substack{\text{chlorosulfuric} \\ \text{acid}}]{Cl_2, O_2} CH_3(CH_2)_{13}-\overset{Cl}{\underset{|}{C}H}-COOH$$

92%

Bull Chem Soc Japan 52, 255 (1979)

$$BuMgBr \xrightarrow[\text{2) } H_2O]{\text{1) } CF_2=C(SEt)_2} Bu-\overset{F}{\underset{|}{C}H}-COOH$$

81%

Chem Lett, 175 (1979)

$$Ph\diagdown\diagup\diagdown CHO \xrightarrow[\text{2) } H_3O^{\oplus}]{Li_2[(EtO)_2\overset{O}{\overset{\|}{P}}-\overset{\ominus}{\underset{\text{Cl}}{C}}-COO]} Ph\diagdown\diagup\diagdown\diagup\underset{\underset{Cl}{}}{\overset{COOH}{}}$$

96%

Synth Comm **8**, 19 (1978)

Section 320   Carboxylic Acid - Ketone

$$\underset{\sim}{n}-C_5H_{11}-\overset{O}{\overset{\|}{C}}-Cl \xrightarrow[\text{2) } H_2O]{\text{1) } LiCH(COOTMS)_2} \underset{\sim}{n}-C_5H_{11}-\overset{O}{\overset{\|}{C}}-CH_2COOH$$

91%

Synthesis, 787 (1979)

1) $CF_2=\overset{Li}{\overset{|}{C}}-OTs$

2) $H_3O^{\oplus}$

3) $^{\ominus}OH$

4) $H_3O^{\oplus}$

~68%

Tetr Lett, 4809 (1978)

NaOH

>90%

J Prakt Chem **319**, 213 (1977)

CN
|
R-C=NPh $\xrightarrow[\text{HCl}]{\text{acetone}}$ R-C-COOH   ~80%

(where the product is $R-\underset{\underset{\text{O}}{\|}}{C}-COOH$)

R = Subst. Ph, styryl

Indian J Chem <u>17B</u>, 169 (1979)

Also via: Ketoesters (Section 360)

Section 321   <u>Carboxylic Acid - Nitrile</u>

No additional examples

See also: Section 361 (Cyanoesters)

Section 322   <u>Carboxylic Acid - Olefin</u>

$$\underset{\text{cyclohexanone}}{} \xrightarrow[\text{2) } H_3O^{\oplus}]{\text{1) } (EtO)_2\overset{\underset{\text{O}}{\|}}{P}-\overset{\ominus}{C}H\overset{\ominus}{COO}} \underset{}{\overset{H}{\diagdown}\underset{C}{}\diagup^{COOH}}$$

70%

Synth Comm <u>8</u>, 463 (1978)

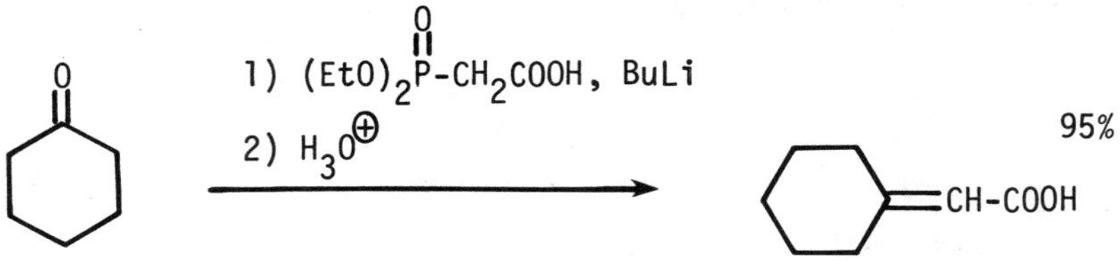

Synthesis, 133 (1978)

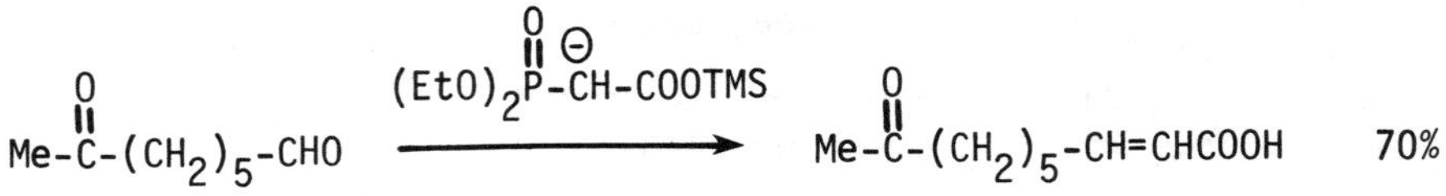

Also works with ketones.

Synthesis, 131 (1978)

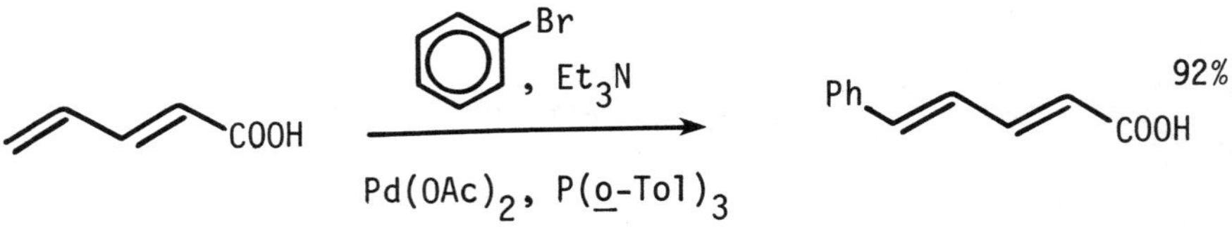

JOC 43, 5018 (1978)

Also via: Hydroxy acids (Section 313)
            Olefinic amides (Section 349)
            Olefinic esters (Section 362)
            Olefinic nitriles (Section 376)

Section 323   <u>Alcohol - Alcohol</u>

$$\xrightarrow{\overset{\oplus}{Ph_3P}\ \overset{\ominus}{MnO_4}}$$

80%

Tetrahedron <u>35</u>, 1109 (1979)

$$\xrightarrow[\underline{t}\text{-BuOOH, OsO}_4]{\text{acetone, Et}_4\text{NOAc}}$$

52%

JOC <u>43</u>, 2063 (1978)

$$\xrightarrow[\text{benzene}]{\text{Me}_8\text{Sn}_3,\ h\nu}$$

59%

Tetr Lett, 2847 (1978)

$Li[Bu_3BCH=CH_2]$

+

Me_2CHCHO

$$\xrightarrow{\text{2) H}_2\text{O}_2,\ \ominus\text{OH}}$$

74%

Tetrahedron <u>33</u>, 1949 (1977)

$$\begin{array}{c} OLi \\ | \\ Ph-CH-CH_2Li \end{array}$$

$+$

$EtCHO$

$\xrightarrow{\qquad 2)\ H_3O^{\oplus} \qquad}$

$$\begin{array}{c} OH \qquad\quad OH \\ | \qquad\qquad | \\ Ph-CH-CH_2-CH \\ | \\ Et \end{array}$$

62%

JOC **44**, 4798 (1979)

2 PhCH₂OTMS $\xrightarrow[\text{2) NaOH, EtOH}]{\text{1) t-BuOOH}}$ $\begin{array}{c} Ph-CH-CH-Ph \\ | \quad | \\ OH \quad OH \end{array}$ 69%

JOC **44**, 295 (1979)

93%

1) Bu₃B, H₂C=CHLi

2) H₂O₂, $^{\ominus}$OH

Tetrahedron **33**, 1945 (1977)

$(CH_3)_2C=O$ $\xrightarrow[\text{2) LiAlH}_4]{\text{1) } ArC\overset{O\cdots Li}{\underset{O-CHCH_3}{\diagup}}}$ $\begin{array}{c} OH \quad OH \\ | \quad\ \ | \\ Me_2C-CHCH_3 \end{array}$ 38%

$Ar = \left(\begin{array}{c} \text{(aryl structure)} \end{array} \right)$

JOC **43**, 4255 (1978)

Also via: Hydroxyesters (Section 327)
           Diesters (Section 357)

Section 324   <u>Alcohol - Aldehyde</u>

JACS <u>101</u>, 5848 (1979)

1) RMgX
2) NH_4Cl
3) $H_3O^{\oplus}$

$$\underset{OH}{\overset{CHO}{\underset{|}{Ph-\!\!\!-\!\!\!-R}}}$$

67-82%

>94%ee

(R = Me, Et, <u>i</u>-Pr, vinyl)

Chem Lett, 1253 (1978)

$$\underset{Me-C-CHO}{\overset{O}{\overset{\|}{}}}$$

1)

2) PhMgBr
3) H_2O, NH_4Cl

$$\underset{Me\quad OH}{\overset{CHO}{\underset{|}{-\!\!-\!\!-Ph}}}$$

76%

99%ee

Chem Lett, 705 (1979)

Review: "Aldol Condensations"

Fortschritte der Chem Forsch <u>67</u>, 1 (1976)

Related methods: Alcohol - Ketone (Section 330)

Section 325   <u>Alcohol - Amide</u>

$$HO-CH_2CH_2CH_2NH_2 \xrightarrow[\text{DMF}]{C_6F_5OAc} HO-CH_2CH_2CH_2-\underset{\underset{O}{\overset{|}{C}CH_3}}{N}H \qquad 91\%$$

JOC <u>44</u>, 654 (1979)

Section 326   <u>Alcohol - Amine</u>

79%

75%

JOC <u>43</u>, 2544 (1978)
JOC <u>43</u>, 2628 (1978)

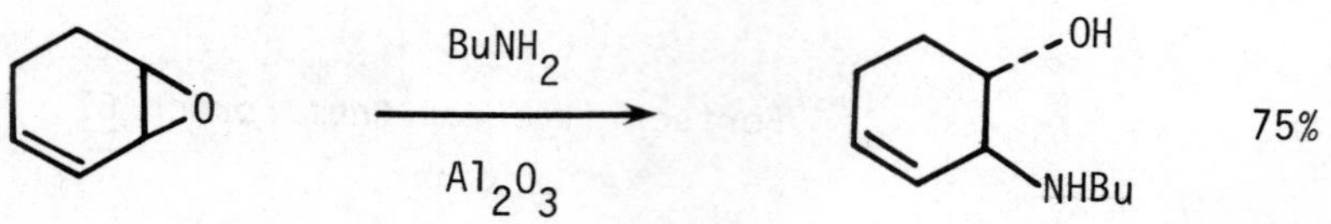

BuNH$_2$

Al$_2$O$_3$

75%

JACS <u>99</u>, 8208 and 8214 (1977)

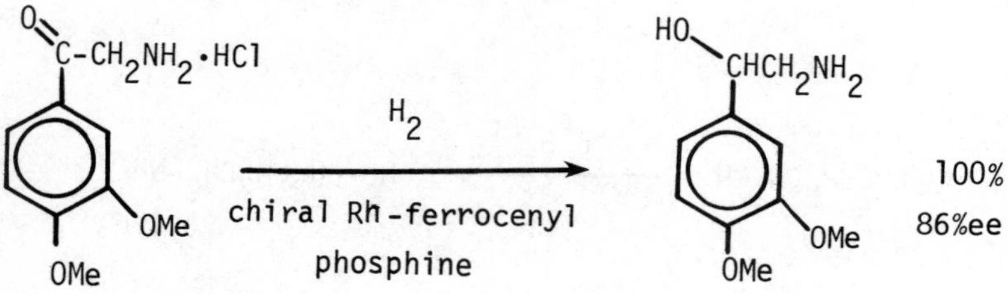

H$_2$

chiral Rh-ferrocenyl
phosphine

100%
86%ee

Tetr Lett, 425 (1979)

1) <u>t</u>-BuN-CH$_2$Li

2) LiAlH$_4$

3) Raney Ni

70%

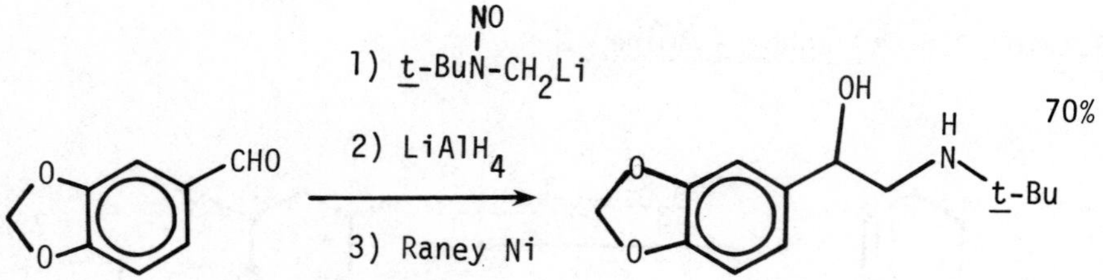

Synthesis, 423 (1979)

PhN-Li
|
Ph-CH-CH$_2$Li

$\xrightarrow{}$
2) H$_3$O$^{\oplus}$

NHPh OH
| |
Ph-CH-CH$_2$-CH   81%
            |
            CHMe$_2$

+

Me$_2$CH-CHO   JOC **44**, 4798 (1979)

PhCH$_2$CH$_2$CH=NOH $\xrightarrow{\text{BH}_3\cdot\text{ pyridine}}$ PhCH$_2$CH$_2$CH$_2$NHOH   92%

Also works with oximes of ketones.

JCS Perkin I, 643 (1979)

$\xrightarrow{\text{H}_2\text{NNH}_2,\text{ Rh}}$

77%

Tetrahedron **34**, 213 (1978)

Ph-CH=NOH $\xrightarrow[\underline{\text{i}}\text{-Pr-COOH}]{\text{NaBH}_4}$

OH
|
Ph-CH$_2$-N-CH$_2$-$\underline{\text{i}}$-Pr   65%

Synthesis, 856 (1977)

Section 327   Alcohol - Ester

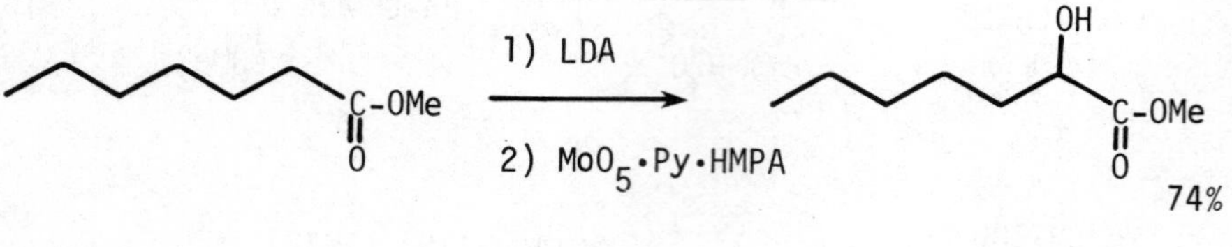

JOC 43, 188 (1978)

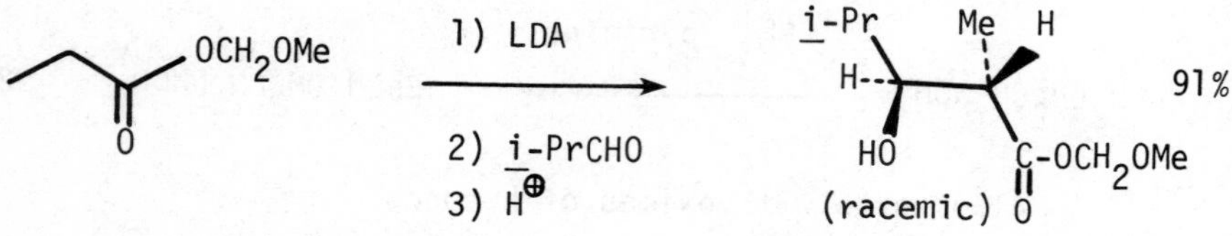

JACS 101, 2501 (1979)

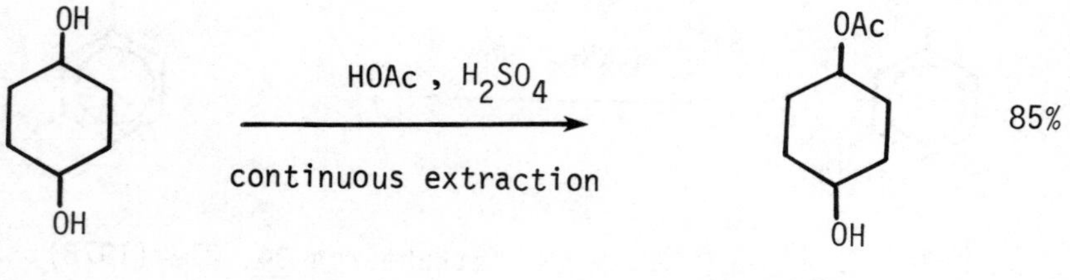

Tetr Lett, 1971 (1979)

Ph-C(=O)-COOEt

structure with SO$_2$Na and CONH$_2$ on pyridine ring with N-Bz

MgCl$_2$

Ph-CH(OH)-COOEt   70%

Bull Chem Soc Japan **52**, 1237 (1979)

$CH_3CCH_2COCH_3$ (diketone)

H$_2$, EtCOOMe

NiO·tartaric acid

$CH_3CHCH_2COCH_3$ (with OH and *)

up to 85%ee

Chem Lett, 1131 (1977)

(P)—C$_6$H$_4$—CH$_2$-O-C(=O)-(CH$_2$)$_8$-C(=O)-Cl

1) NaBH$_4$

2) cleave, esterify

MeO$_2$C(CH$_2$)$_8$CH$_2$OH   67%

Can J Chem **56**, 1562 (1978)

JCS Chem Comm, 162 (1977)

Also via: Hydroxyacids (Section 313)

Section 328   <u>Alcohol - Ether, Epoxide</u>

Can J Chem <u>55</u>, 3351 (1977)

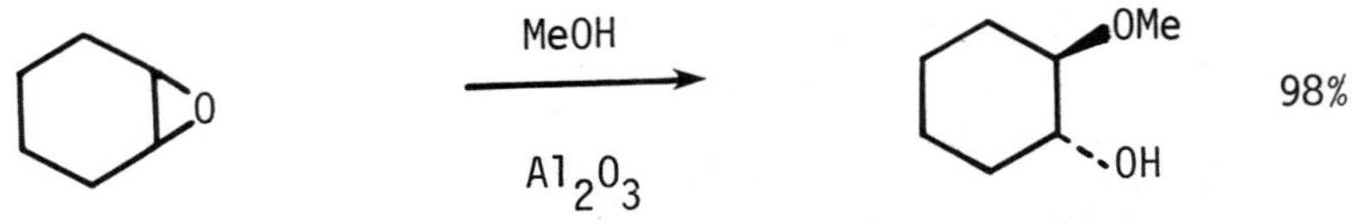

MeOH
Al_2O_3
OMe
OH
O
98%

JACS 99, 8208 and 8214 (1977)

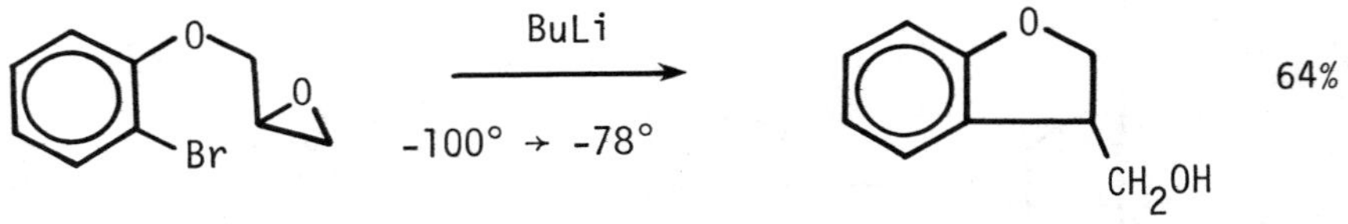

BuLi
-100° → -78°
O
O
Br
CH_2OH
64%

JOC 43, 3800 (1978)

Li
1) TMS-CHOMe
2) CsF
O
OH
OMe

JCS Chem Comm, 822 (1979)

OH
Ph_3SiOOH
OH
O
77%

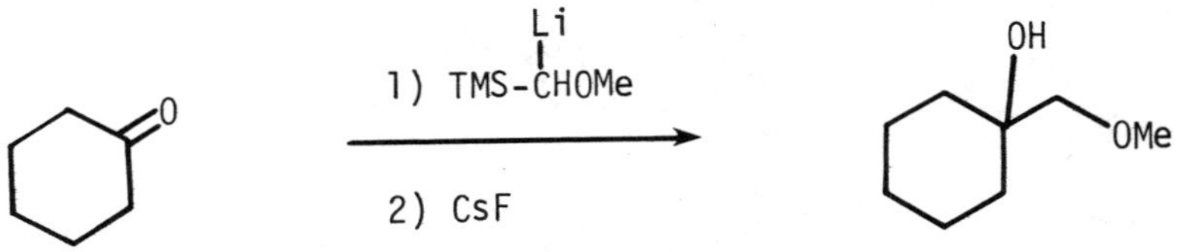

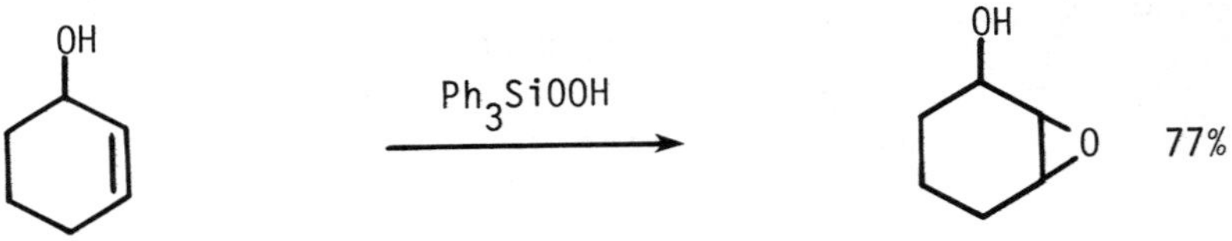

Tetr Lett, 4337 (1979)

Section 329   Alcohol - Halide

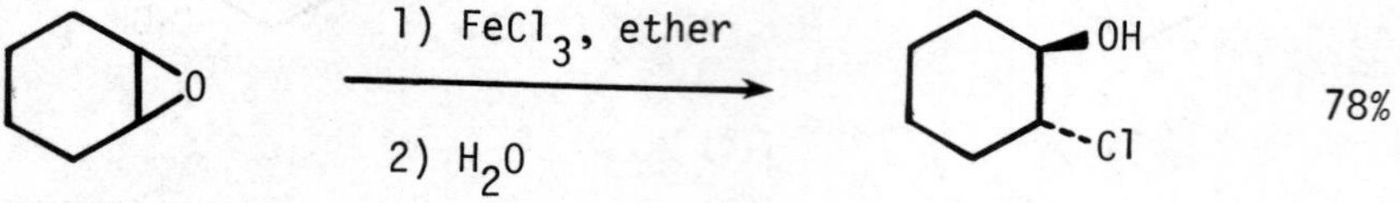

1) $FeCl_3$, ether

2) H_2O

78%

JOC 42, 343 (1977)

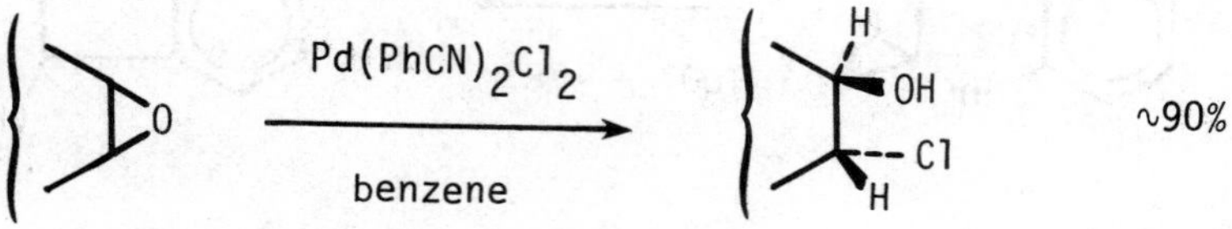

$Pd(PhCN)_2Cl_2$

benzene

~90%

Several examples using steroids.

JOC 44, 1569 (1979)

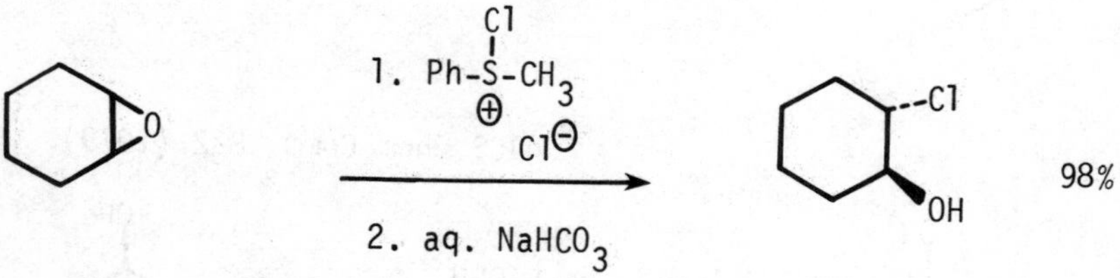

1. Ph-S-CH$_3$
   Cl
   ⊕
   Cl⊖

2. aq. $NaHCO_3$

98%

Chem Lett, 995 (1977)

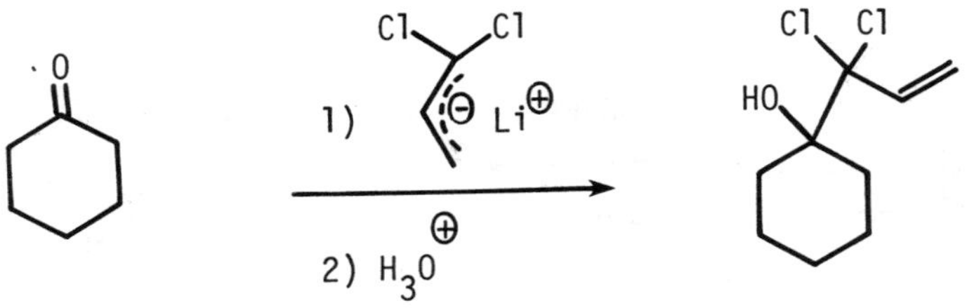

JACS <u>99</u>, 5317 (1977)

Section 330   <u>Alcohol - Ketone</u>

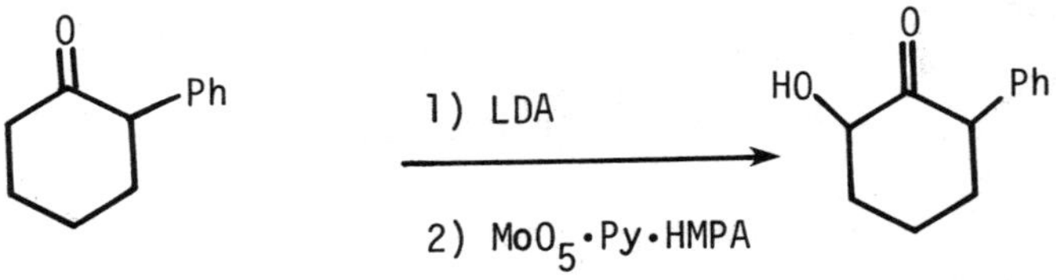

JOC <u>43</u>, 188 (1978)

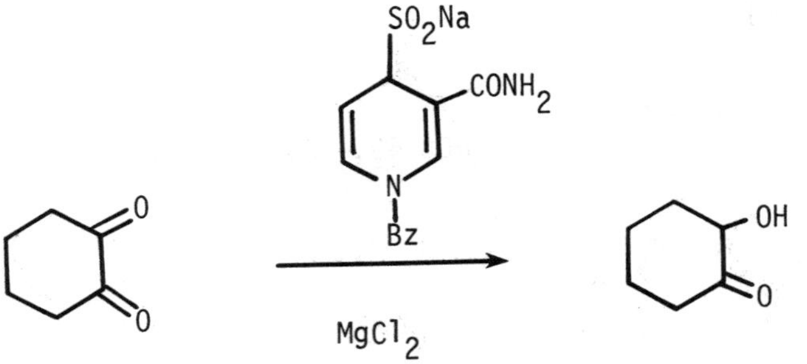

Bull Chem Soc Japan <u>52</u>, 1237 (1979)

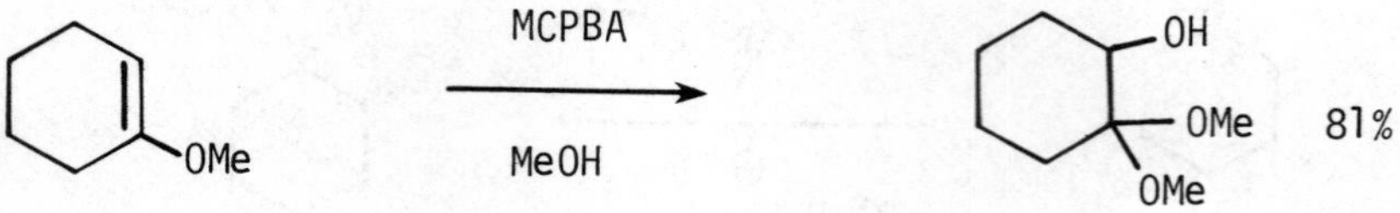

MCPBA
MeOH
OMe
OH
OMe
OMe
81%

Synthesis, 578 (1977)

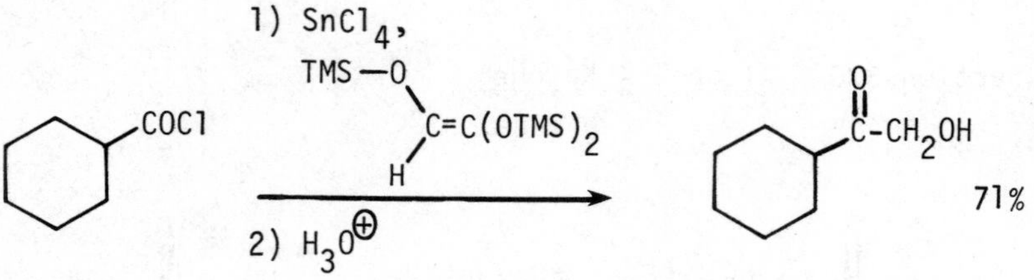

1) SnCl4,
TMS—O
C=C(OTMS)2
H
2) H3O⊕
COCl
C-CH2OH
O
71%

Tetr Lett, 2749 (1978)

OTMS
MCPBA
O
OH
70%

JOC 43, 1599 (1978)

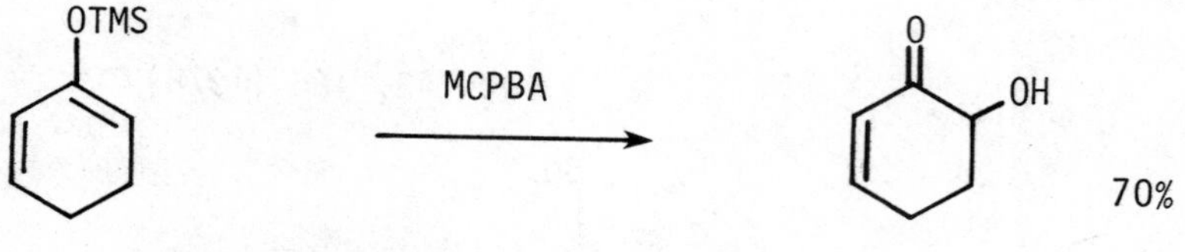

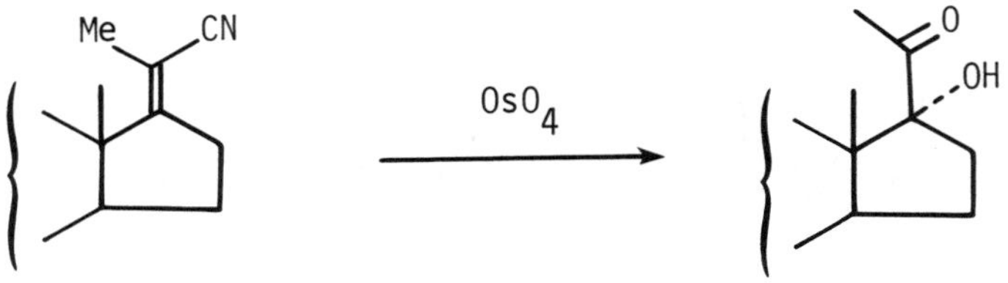

JOC 44, 702 (1979)

69%

1) LDA, DME

2) [cyclohexanone]

3) $H_3O^{\oplus}$

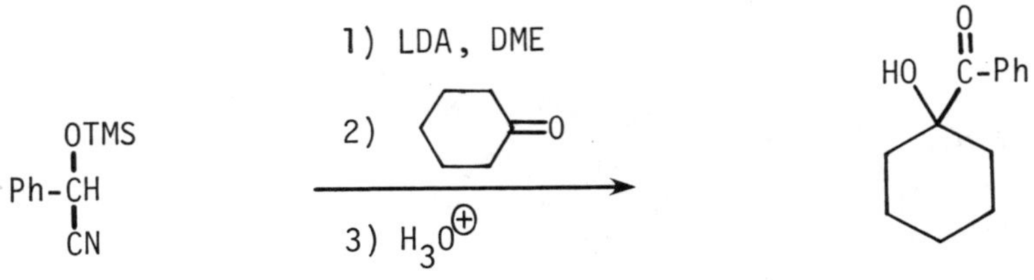

89%

Chem Ber 112, 2062 (1979)

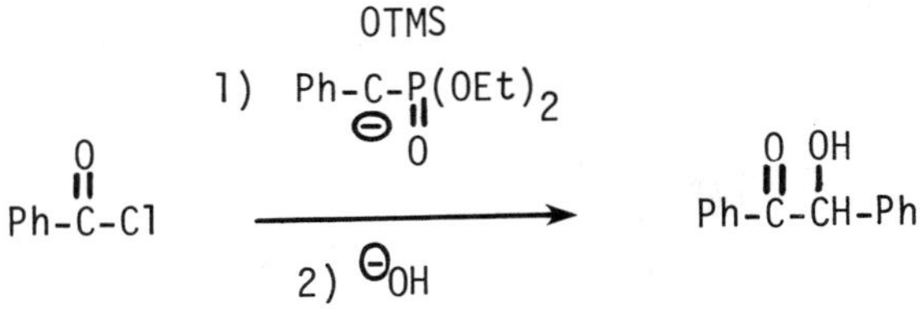

OTMS

1) $Ph-\overset{\ominus}{\underset{\underset{O}{\parallel}}{C}}-\overset{}{P}(OEt)_2$

$Ph-\overset{O}{\overset{\parallel}{C}}-Cl$

2) $\overset{\ominus}{O}H$

$Ph-\overset{O}{\overset{\parallel}{C}}-\overset{OH}{\overset{|}{C}}H-Ph$

77%

Chem Lett, 519 (1979)

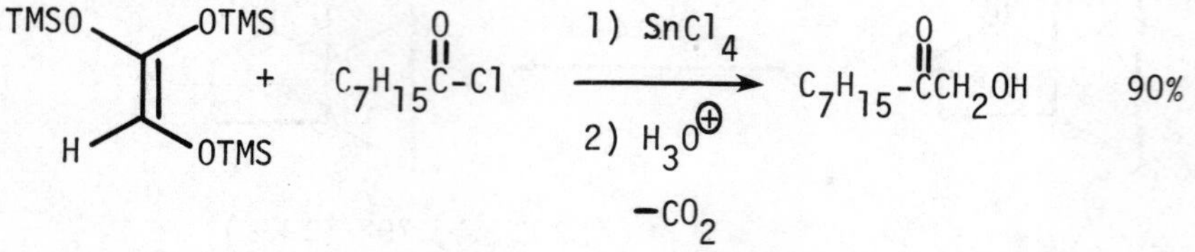

TMSO、 ＿OTMS
     C=C   +   C_7H_{15}C-Cl
H／ ＼OTMS

$$\xrightarrow[\substack{2) \ H_3O^{\oplus} \\ -CO_2}]{1) \ SnCl_4}$$

$C_7H_{15}-\overset{O}{\overset{\|}{C}}CH_2OH$ 90%

JOC **44**, 4617 (1979)

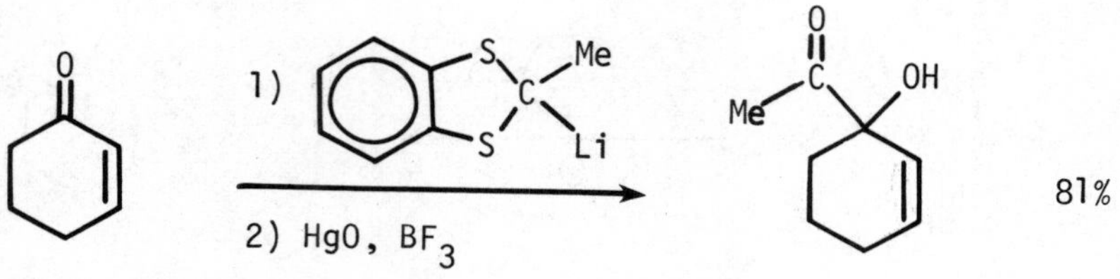

81%

Tetr Lett, 2345 (1978)

$PhCH_2CH_2-\overset{S}{\overset{\|}{C}}-SEt$

1) EtMgI

2) (cyclopentanone)

3) HgCl_2, CaCO_3, H_2O

58%

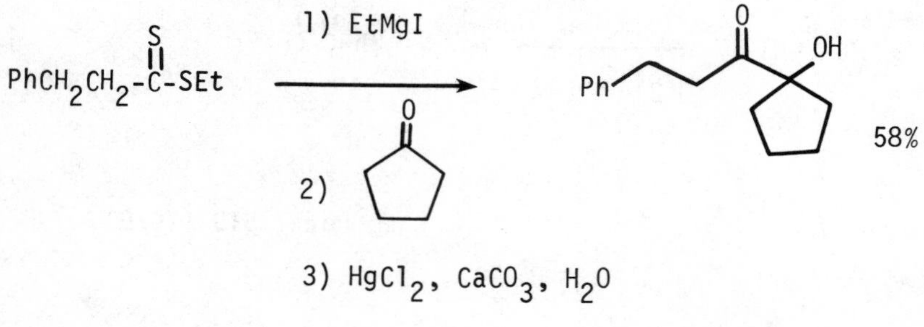

Tetr Lett, 4657 (1978)

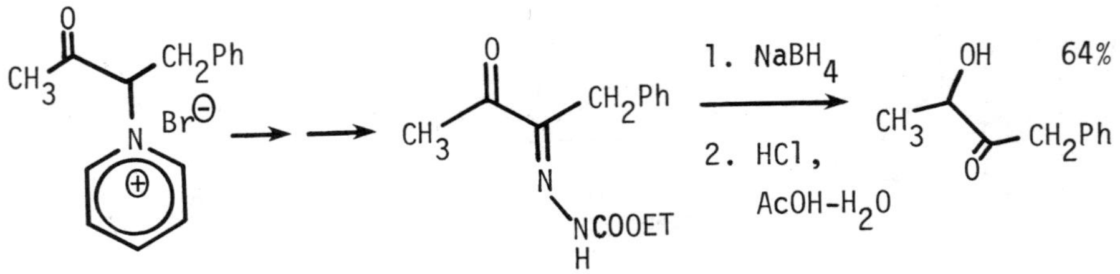

Chem Lett, 245 (1977)

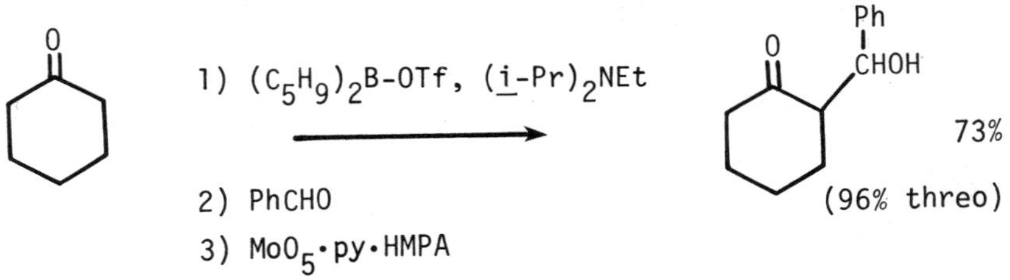

1) $(C_5H_9)_2$B-OTf, $(\underline{i}\text{-Pr})_2$NEt

2) PhCHO

3) $MoO_5 \cdot py \cdot HMPA$

73%

(96% threo)

JACS <u>101</u>, 6120 (1979)

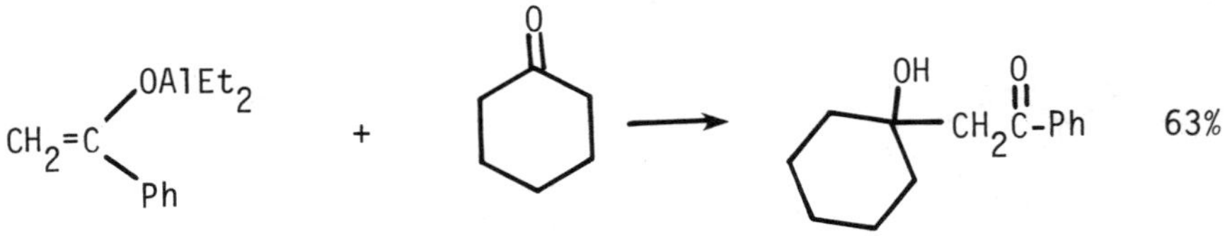

$CH_2=C$ $\overset{OAlEt_2}{\underset{Ph}{}}$ +

63%

Chem Lett, 379 (1979)

1) BuLi

2) $Ph_2C=O$

3) $H_3O^{\oplus}$

∿90%

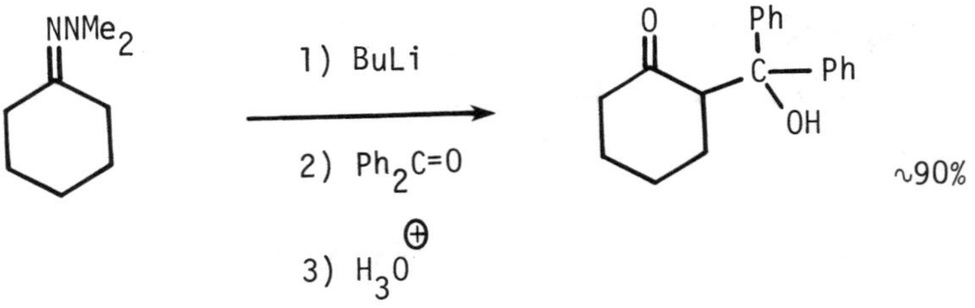

Chem Ber <u>111</u> 1362 (1978)

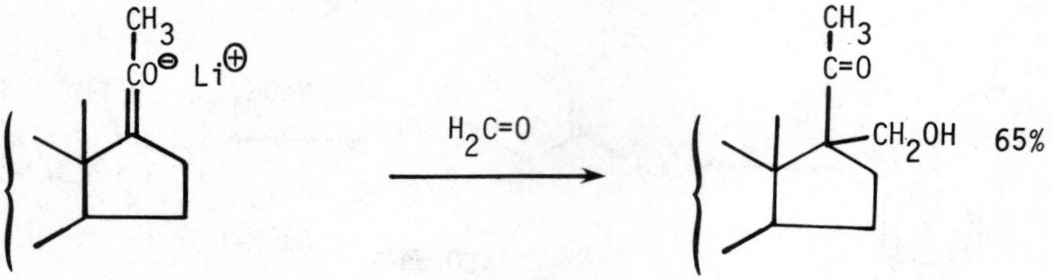
CH3
CO⊖ Li⊕
H2C=O
CH3
C=O
CH2OH
65%

Steroids 34, 597 (1979)

Review: "Aldol Condensations"

Fortschritte der Chem Forsch 67, 1 (1976)

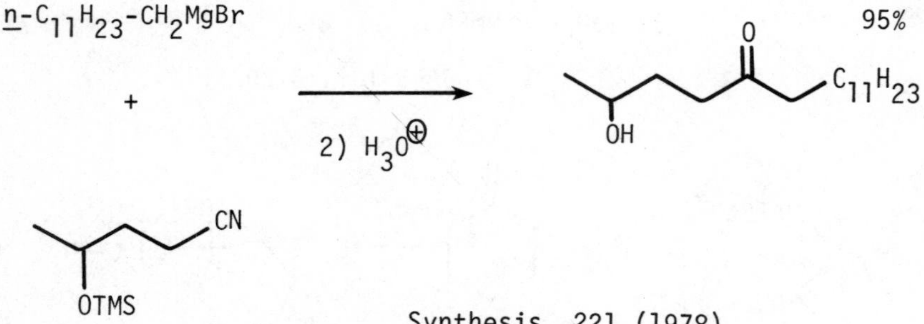
n-C11H23-CH2MgBr
+
2) H3O⊕
CN
OTMS
O
OH
C11H23
95%

Synthesis, 221 (1978)

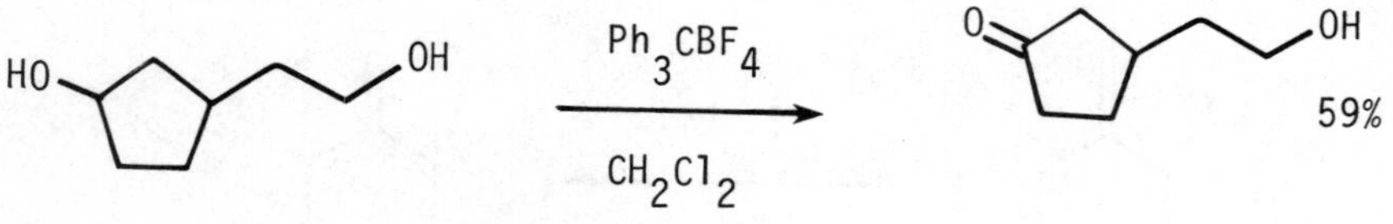
HO
OH
Ph3CBF4
CH2Cl2
O
OH
59%

Tetr Lett, 2771 (1978)

1) $H_2NOCH_2CH_2NMe_2$

2) $NaBH_4$

3) $H_2C=O$, $H_3O^{\oplus}$

R-C(=O)----C(=O)-R'   →   R-C(=O)----CH-R'   ~80%
                                        OH

Synthesis, 466 (1977)

Section 331   <u>Alcohol - Nitrile</u>

1) $PhSO_2C \equiv N \rightarrow O$

2) Na/Hg

→   (cyclohexane ring with OH and CN)   89%

JACS <u>101</u>, 1319 (1979)

1) Me_3SiCN

2) HCl, H_2O

→   (cyclohexane ring with HO and CN)   90%

Tetr Lett, 3773 (1978)

Section 332   Alcohol - Olefin

Allylic and benzylic hydroxylation (C=C-CH → C=C-C-OH, etc.) is
listed in Section 41 (Alcohols and Phenols from Hydrides).

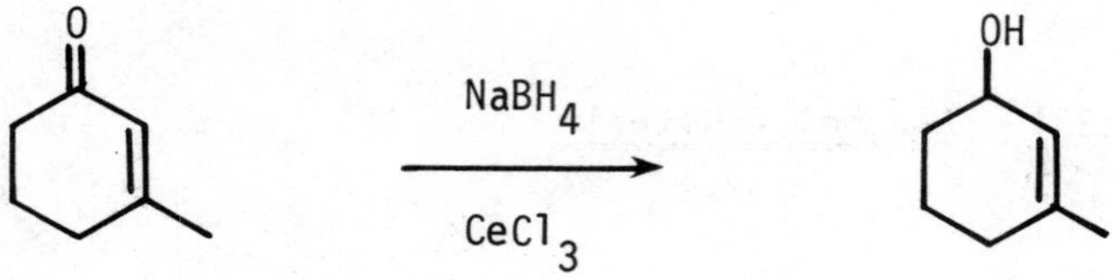

JACS 100, 2226 (1978)

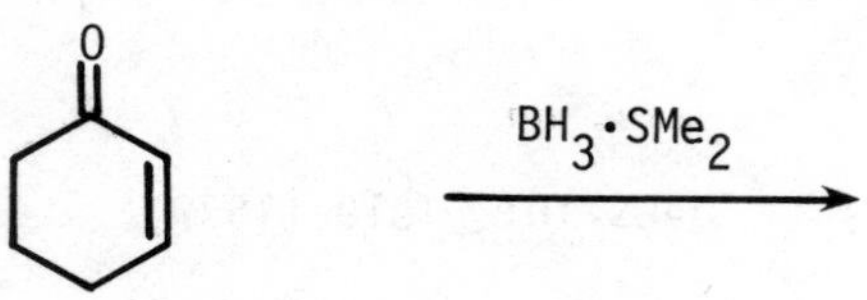

JOC 43, 1829 (1978)

PhCH=CHCHO $\xrightarrow[\text{RhCl}_3]{\text{H}_2, \text{ CO}}$ PhCH=CHCH$_2$OH   83%

BCS Japan 50, 2148 (1977)

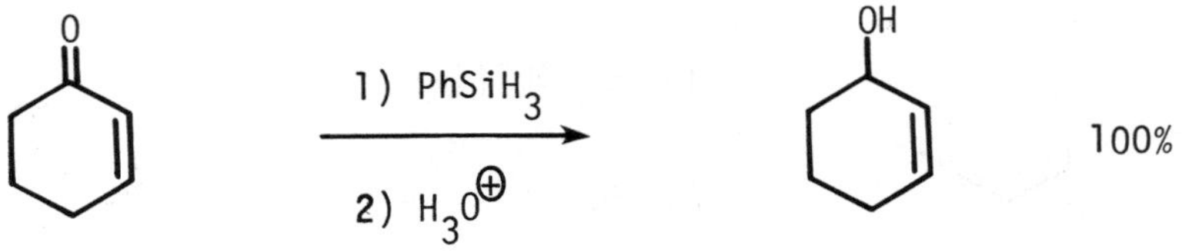

1) PhSiH$_3$

2) H$_3$O$^{\oplus}$

100%

Bull Akad USSR Chem <u>26</u>, 995 (1977)

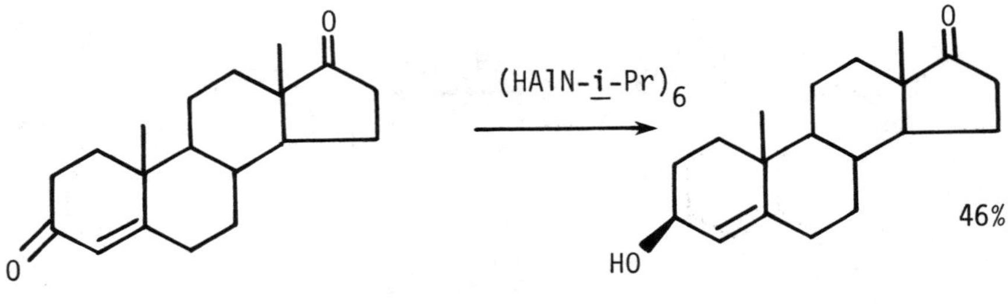

(HAlN-<u>i</u>-Pr)$_6$

46%

Tetr Lett, 2369 (1977)

1) O$_2$, hν

2) NaBH$_4$

3) Bu$_4$NF, MeCN

54%

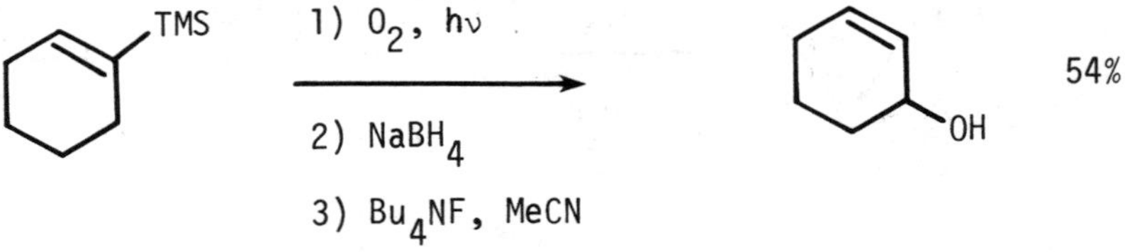

JACS <u>101</u>, 4420 (1979)

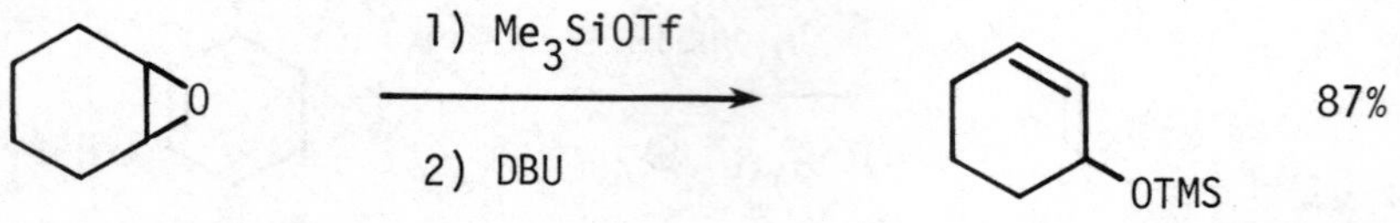

87%

JACS 101, 2738 (1979)

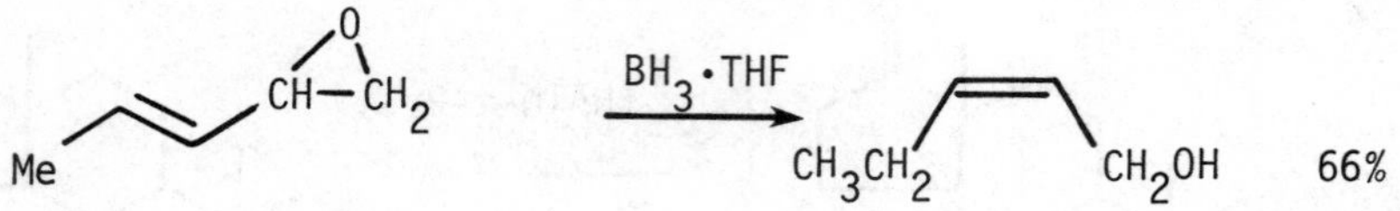

66%

Synthesis, 62 (1979)

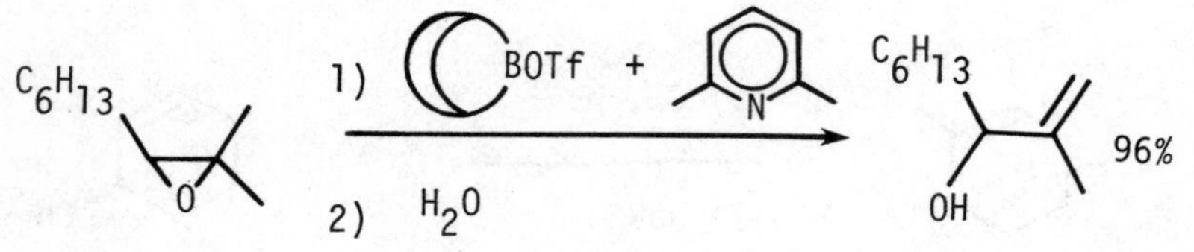

96%

Chem Lett, 1215 (1977)

99%

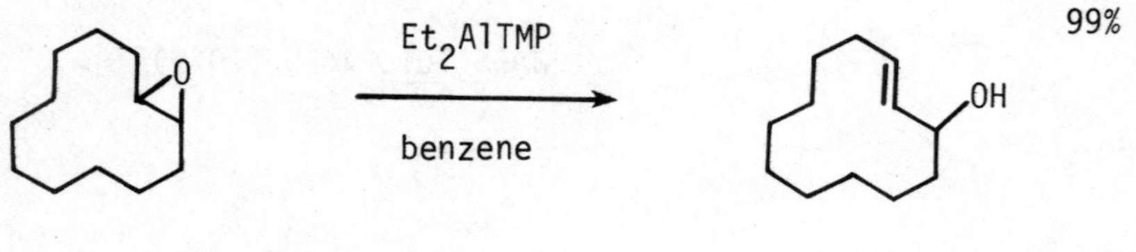

Bull Chem Soc Japan 52, 1705 (1979)

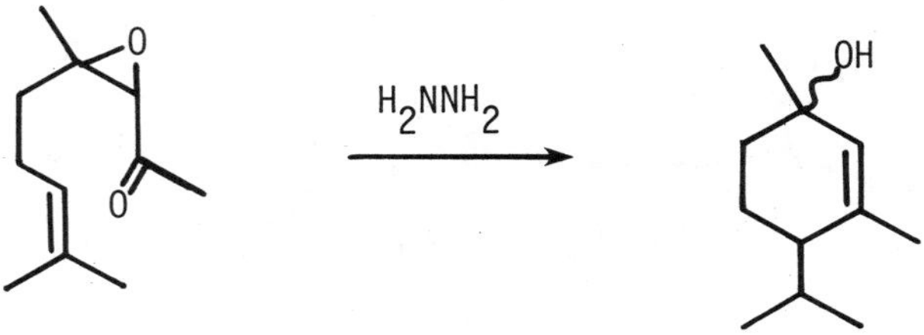

O
O
H_2NNH_2
OH

60%

JACS 99, 7067 (1977)

160°
$CH_3(CH_2)_7$-CHCH$_2$OH
S
O
Ph
$CH_3(CH_2)_6CH=CHCH_2OH$

94%

Tetr Lett, 4093 (1978)

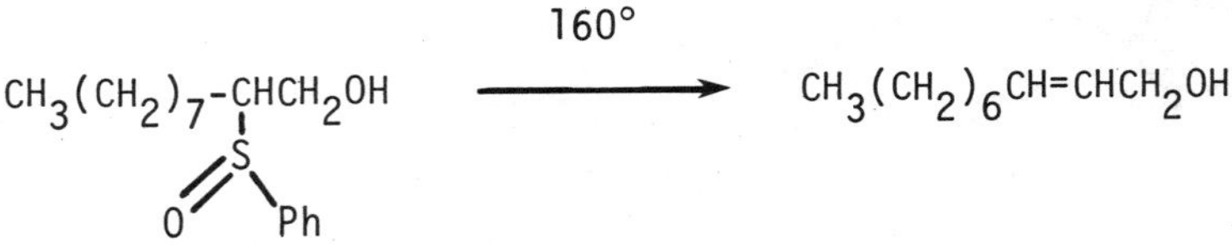

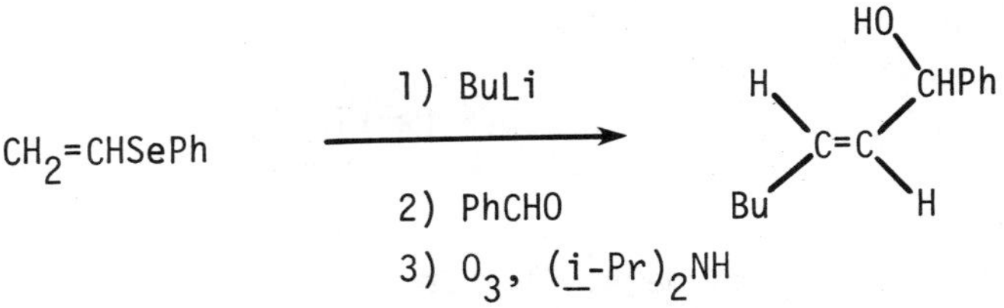

CH_2=CHSePh
1) BuLi
2) PhCHO
3) O_3, (i-Pr)$_2$NH
HO
CHPh
H
C=C
Bu
H

71%

JOC 43, 4252 (1978)

1) PhSeOH
2) Me_3COOH
OH

83%

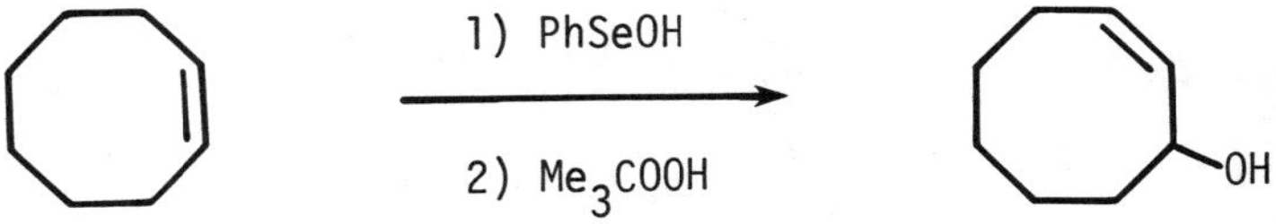

JOC 43, 1689 (1978)

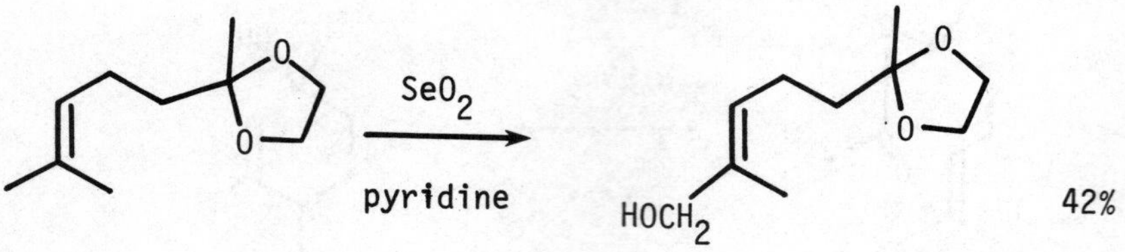

Synthesis, 215 (1978)

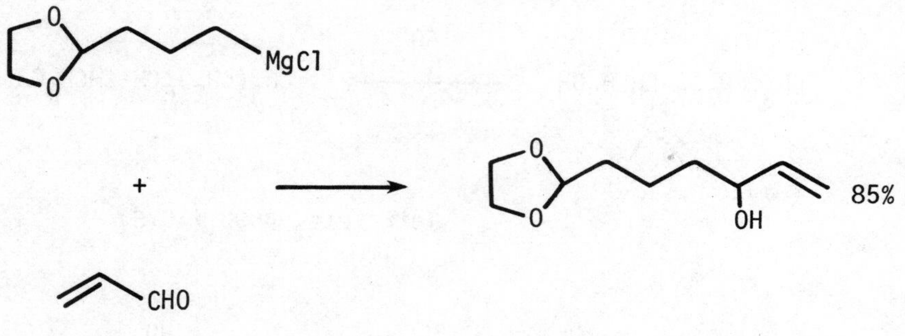

JCS Perkin I, 2353 (1977)

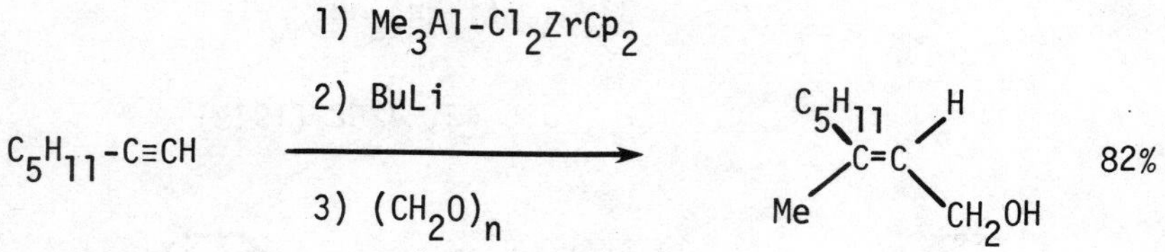

Tetr Lett, 2357 (1978)

$$C_7H_{13}CH_2-\underset{\underset{SeMe}{|}}{\overset{\overset{SeMe}{|}}{C}}H \quad \xrightarrow[\text{2) } CH_2O]{\text{1) BuLi}} \quad C_7H_{13}CH_2-\underset{\underset{SeMe}{|}}{C}HCH_2OH$$

$$\downarrow \underline{t}\text{-BuOOH, } Al_2O_3$$

$$C_7H_{13}CH=CH-CH_2OH \qquad 65\%$$

Tetr Lett, 1145 (1978)

$$\xrightarrow{\text{LiAlH}_4} \qquad 78\%$$

Tetr Lett, 4089 (1978)

1) VO(**acac**)$_2$, $\underline{t}$-BuOOH

2) MsCl, Et$_3$N

3) Na(Ca), NH$_3$(l)

$\sim$30%

Bull Chem Soc Japan <u>52</u>, 1757 (1979)

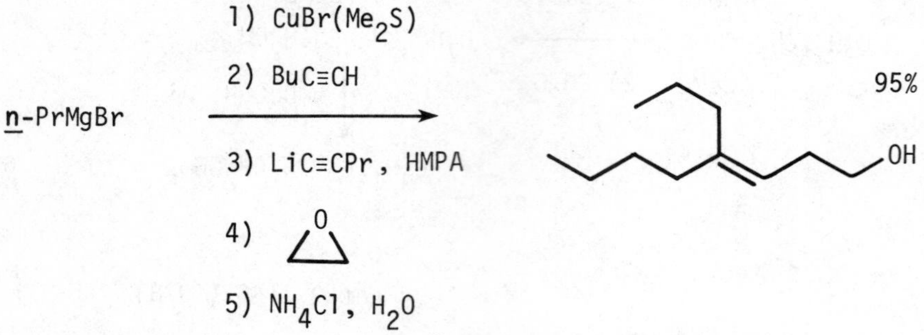
n-PrMgBr
1) CuBr(Me2S)
2) BuC≡CH
3) LiC≡CPr, HMPA
4) O
5) NH4Cl, H2O
OH
95%

Tetr Lett, 2465 (1978)

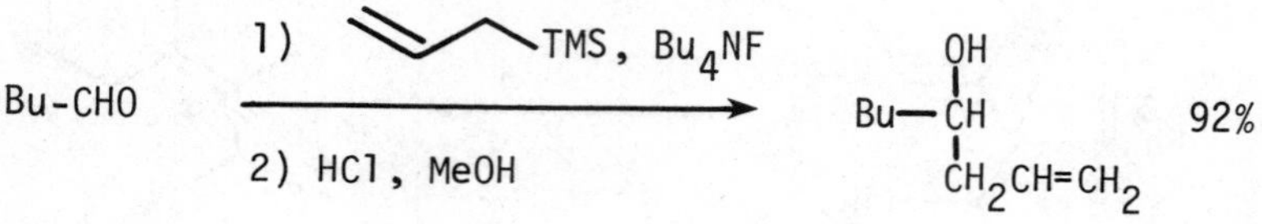
Bu-CHO
1) TMS, Bu4NF
2) HCl, MeOH
OH
Bu—CH
CH2CH=CH2
92%

Tetr Lett, 3043 (1978)

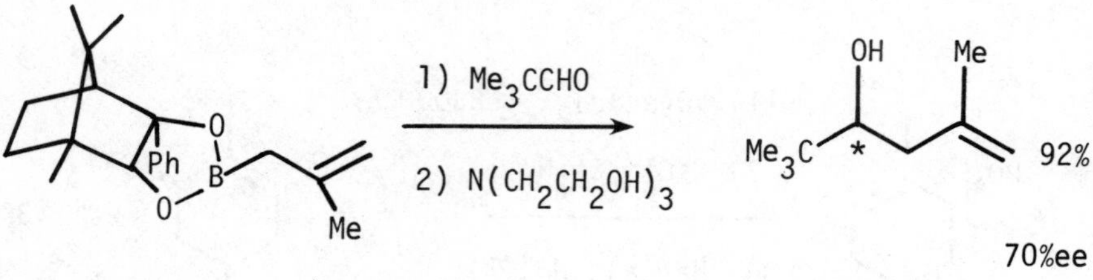
O
Ph
B
O
Me
1) Me3CCHO
2) N(CH2CH2OH)3
OH
Me
Me3C
*
92%
70%ee

Angew Int Ed 17, 768 (1978)

Review: "Simple Enols"

Chem Rev 79, 515 (1979)

Also via: Acetylenes - Alcohols (Section 302)

Section 333   Aldehyde - Aldehyde

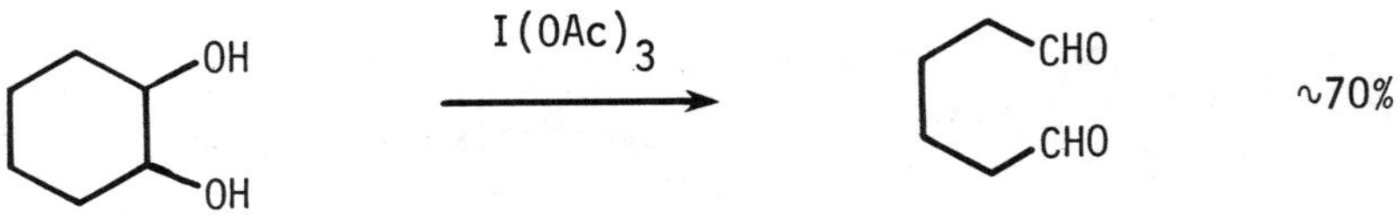

JCS Perkin I, 1483 (1978)

Section 334   Aldehyde - Amide

No additional examples

Section 335   Aldehyde - Amine

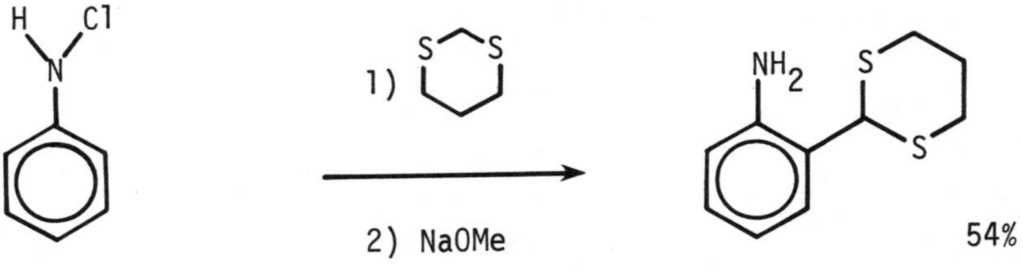

JACS 100, 7600 (1978)

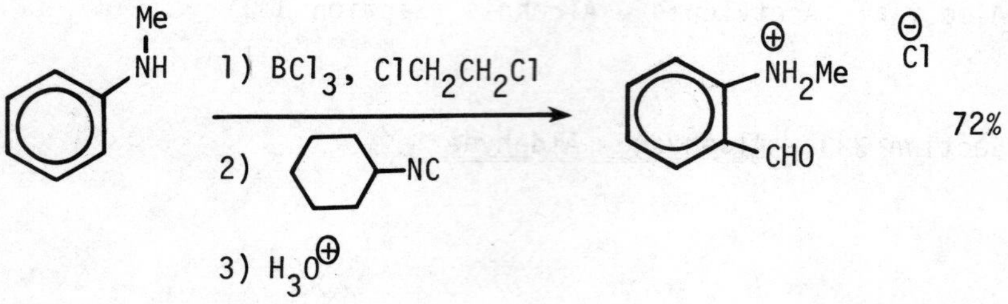

Synthesis, 99 (1979)

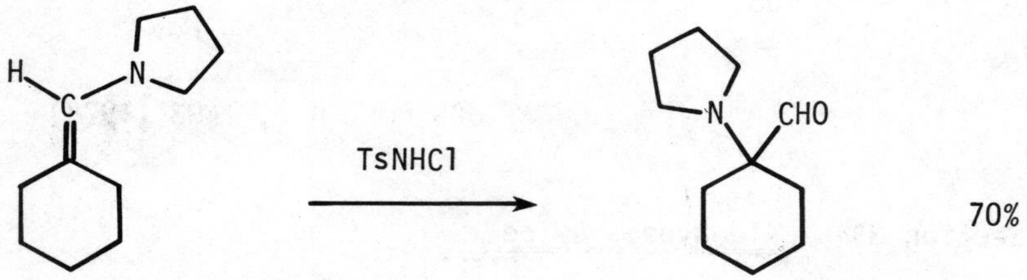

Angew Int Ed 18, 933 (1979)

Section 336   Aldehyde - Ester

No additional examples

Section 337 Aldehyde - Ether, Epoxide

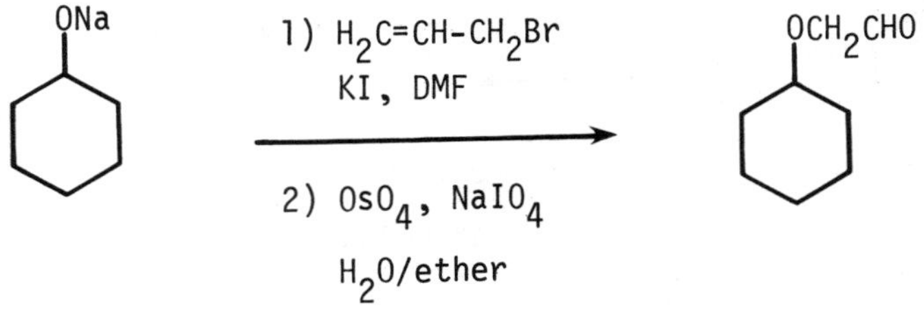
ONa
1) $H_2C=CH-CH_2Br$
KI, DMF
2) OsO_4, $NaIO_4$
H_2O/ether
OCH_2CHO

72%

Synthesis, 202 (1979)

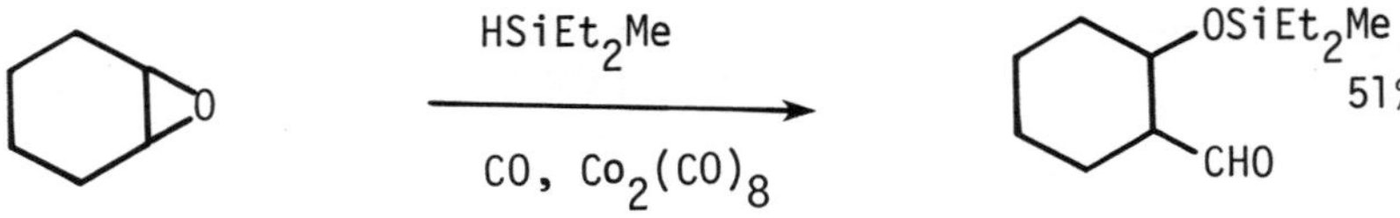
O
$HSiEt_2Me$
CO, $Co_2(CO)_8$
$OSiEt_2Me$
CHO

51%

Angew Int Ed 16, 789 (1977)

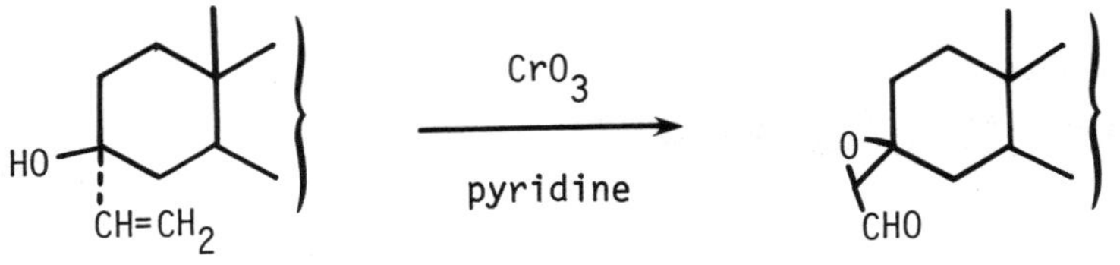
HO
$CH=CH_2$
CrO_3
pyridine
O
CHO

70%

JOC 42, 813 (1977)

Section 338   Aldehyde - Halide

1) KH

2) I_2

88%

Tetr Lett, 2817 (1979)

1) LiClC=CF_2

2) LiAlH$_4$

3) H_2SO_4

~40%

Synthesis, 458 (1978)

t-BuOCl

EtOH

76%

JOC 42, 1057 (1977)

Section 339   Aldehyde - Ketone

$$\underset{\text{Ph-C-CH}_2\text{Br}}{\overset{\overset{\displaystyle O}{\|}}{}} \quad \xrightarrow{Et_2NOH} \quad \underset{\text{Ph-C-CHO}}{\overset{\overset{\displaystyle O}{\|}}{}} \qquad 78\%$$

JOC 42, 754 (1977)

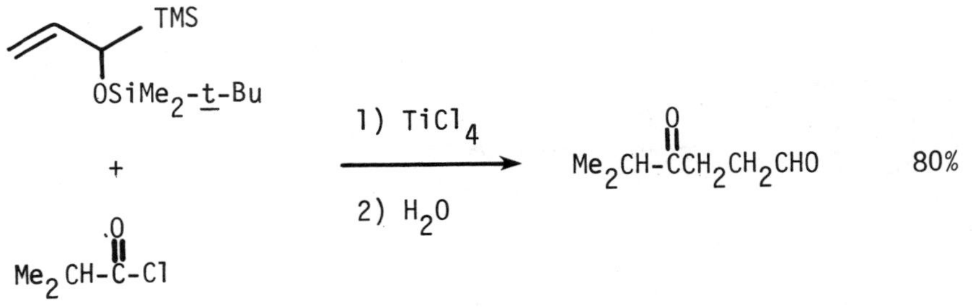

1) TiCl$_4$

2) H$_2$O

Me$_2$CH-C(=O)CH$_2$CH$_2$CHO   80%

JOC **43**, 2551 (1978)

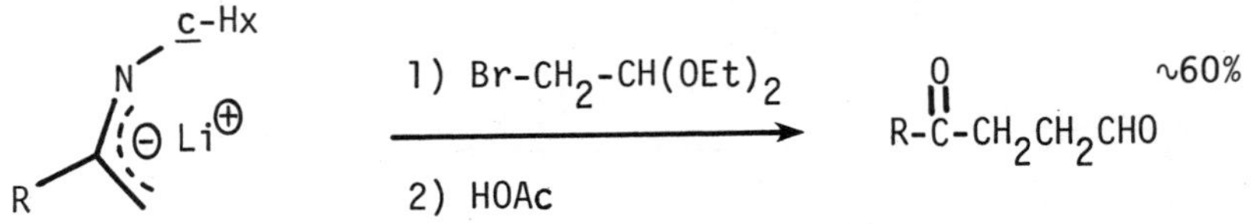

R = alkyl, Ph

1) Br-CH$_2$-CH(OEt)$_2$

2) HOAc

~60%

R-C(=O)-CH$_2$CH$_2$CHO

Tetrahedron **35**, 1745 (1979)

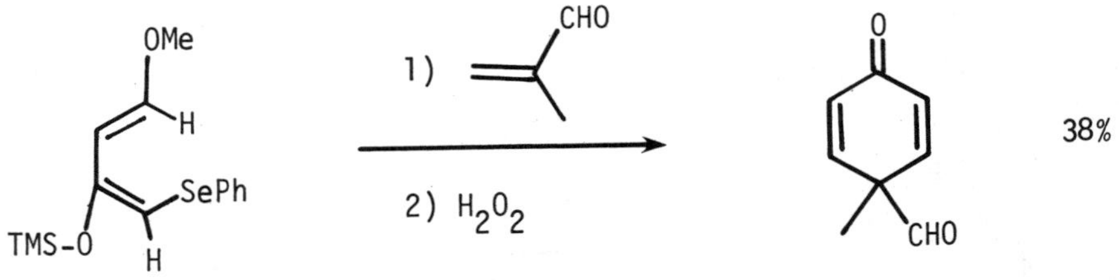

1) [CH$_2$=C(CH$_3$)CHO]

2) H$_2$O$_2$

38%

JOC **42**, 1819 (1977)

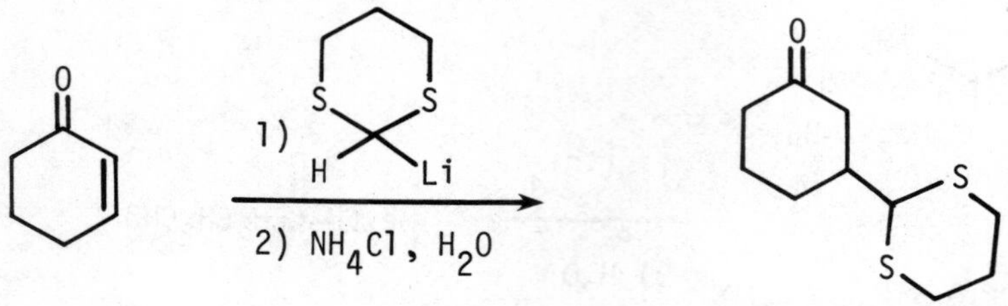

O
1)
S
S
H
Li
2) NH_4Cl, H_2O
O
S
S

95%

JCS Chem Comm, 100 (1979)

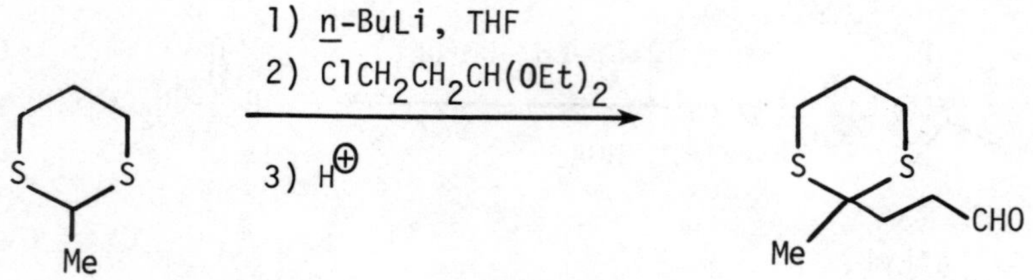

1) n-BuLi, THF
2) $ClCH_2CH_2CH(OEt)_2$
3) $H^{\oplus}$
S
S
Me
S
S
CHO
Me

57%

Synth Comm 9, 147 (1979)

Section 340 Aldehyde - Nitrile

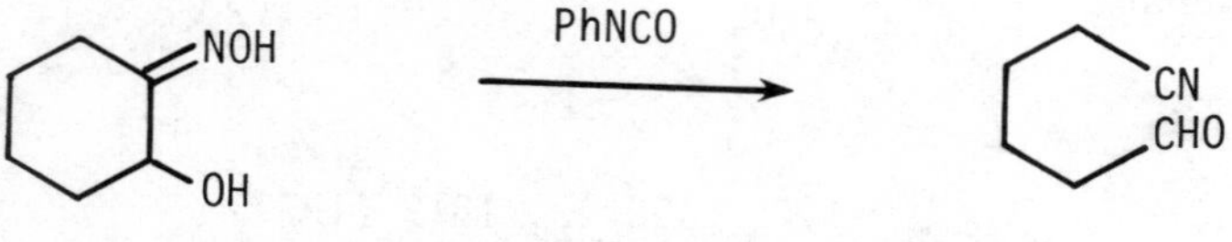

NOH
OH
PhNCO
CN
CHO

72%

Indian J Chem 18B, 175 (1979)

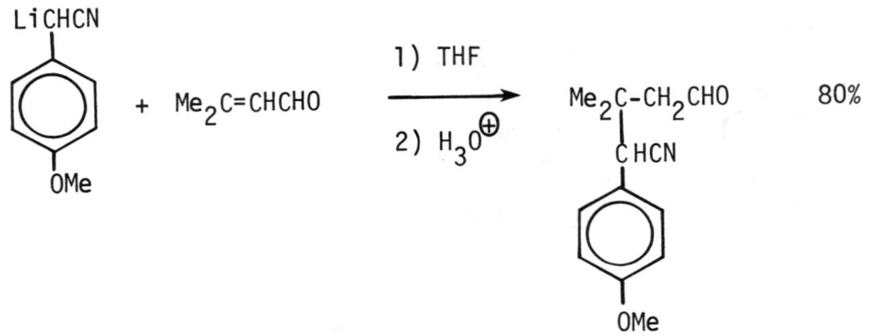
LiCHCN
OMe
+ Me_2C=CHCHO
1) THF
2) $H_3O^{\oplus}$
Me_2C-CH_2CHO
CHCN
OMe
80%

JCS Chem Comm, 779 (1979)

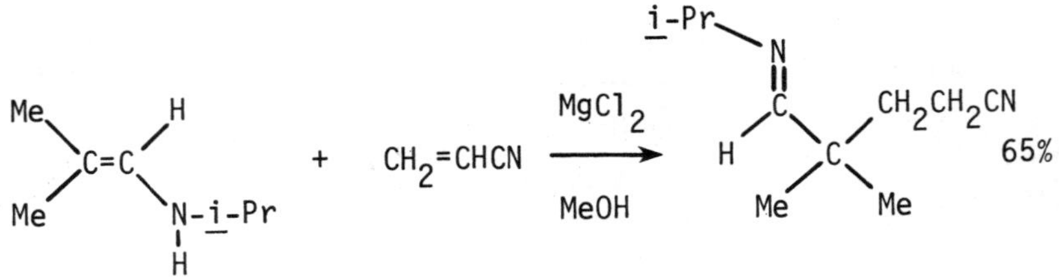
Me
Me
H
C=C
N-i-Pr
H
+ CH_2=CHCN
$MgCl_2$
MeOH
i-Pr
N
C
H
C
CH_2CH_2CN
Me
Me
65%

JCS Chem Comm, 565 (1977)

Section 341 Aldehyde - Olefin

For the oxidation of allylic alcohols to olefinic aldehydes see also Section 48 (Aldehydes from Alcohols).

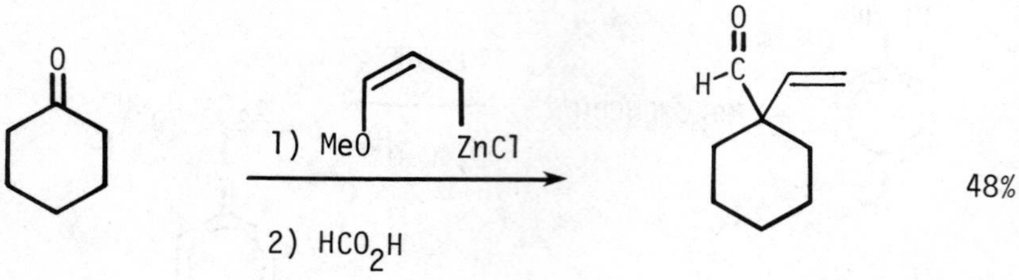

JACS <u>99</u>, 5453 (1977)

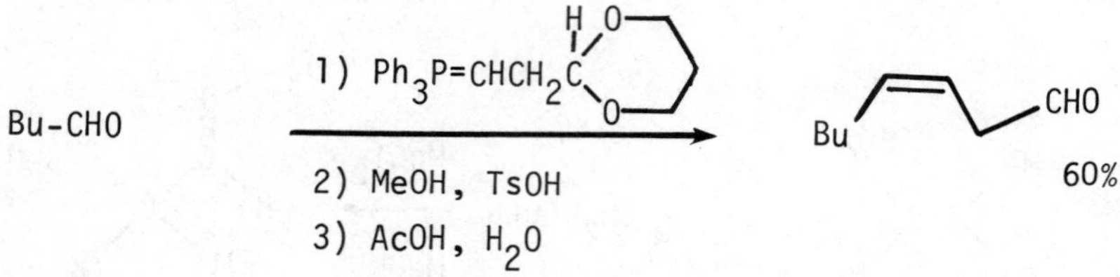

Synthesis, 132 (1979)

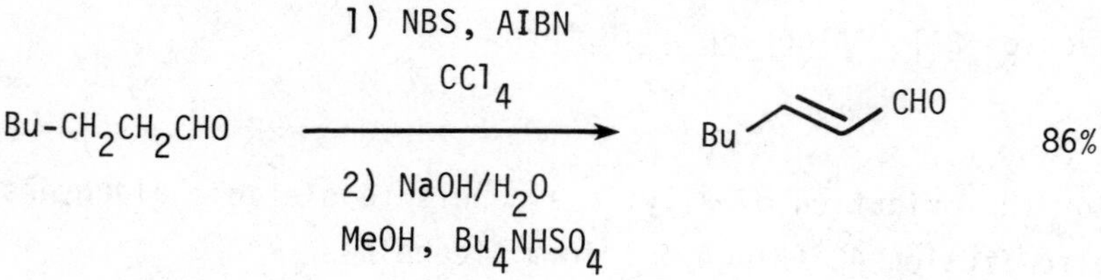

Synthesis, 507 (1979)

Ph $\diagup\!\!\!\diagdown$ CH$_2$OH $\quad$ [pyridinium reagent] $\quad$ $\overset{\ominus}{O}$Ts $\quad \xrightarrow{\text{DMSO}} \quad$ Ph $\diagup\!\!\!\diagdown$ CHO $\qquad$ 85%

Chem Lett, 369 (1978)

$\xrightarrow[\text{DMF}]{\text{POCl}_3}$ [cyclohexene-CHO/Cl] $\xrightarrow[\substack{\text{Pd/C} \\ \text{Et}_3\text{N}}]{\text{H}_2}$ [cyclohexene-CHO] $\qquad$ 85%

Tetr Lett, 2027 (1977)

1) Ph-$\overset{\overset{O}{\|}}{S}$-CH$_2$Cl, $\underline{t}$-BuOK

2) Δ

$\qquad$ 85%

Tetr Lett, 1377 (1977)

CH$_3$-(CH$_2$)$_7$-CHO $\quad \xrightarrow[\substack{\\ 2)\ H_3O^{\oplus}}]{1)\ Ph_3\overset{\oplus}{P}-\overset{\ominus}{CH}-CH(OEt)_2}$ $\quad$ CH$_3$(CH$_2$)$_7$ $\diagup$ C=C $\diagdown$ CHO

$\sim$40%

Angew Int Ed <u>18</u>, 687 (1979)

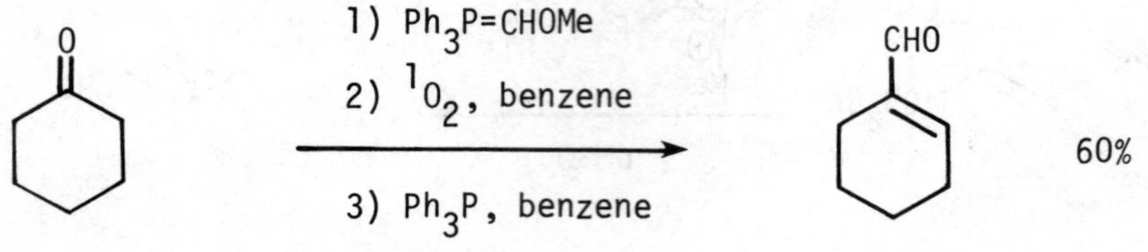

1) Ph₃P=CHOMe

2) ¹O₂, benzene

3) Ph₃P, benzene

60%

Synthesis, 67 (1978)

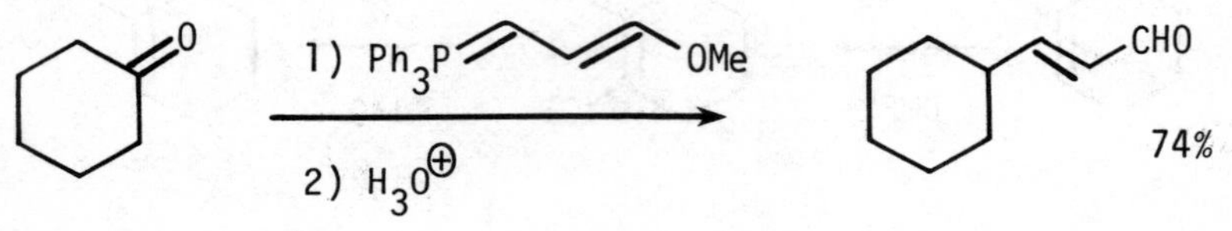

1) Ph₃P⌒⌒OMe

2) H₃O⊕

74%

Tetr Lett, 3875 (1977)

1) (EtO)₂P(=O)⌒⌒N-cyclohexyl, NaH

2) H₃O⊕

94%

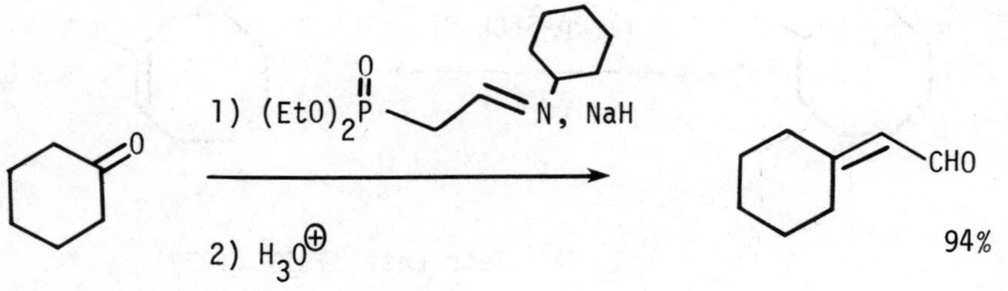

JOC 43, 3788 (1978)

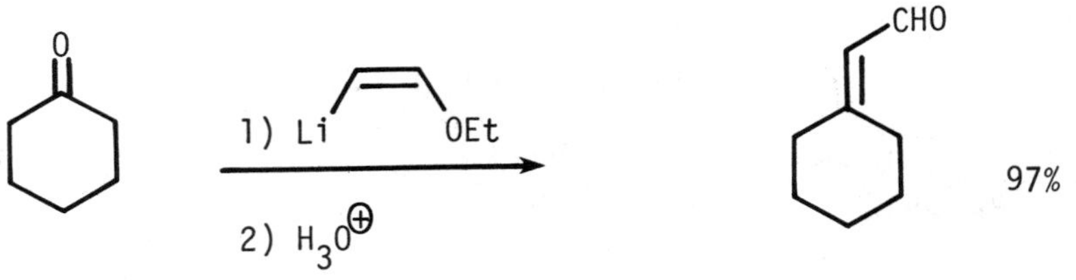

JACS $\underline{99}$, 7365 (1977)

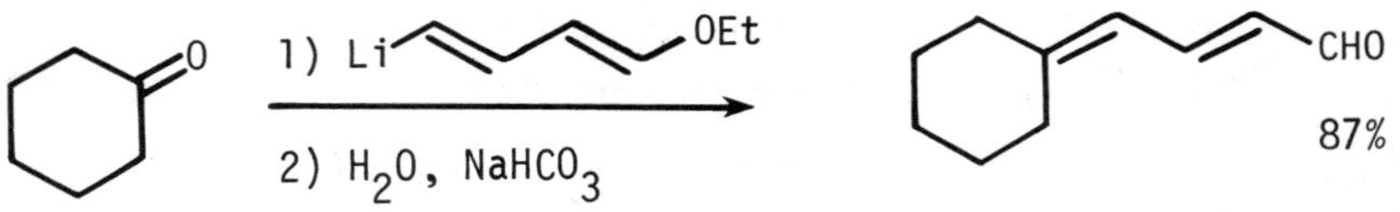

Tetr Lett, 717 (1978)

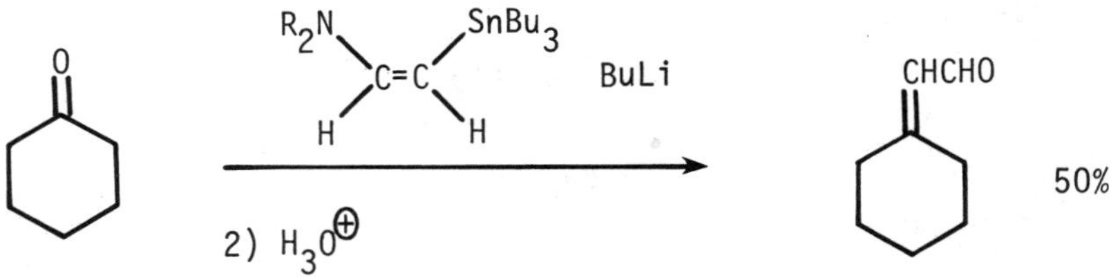

Tetr Lett, 3589 (1977)

1) TiCl$_4$, CH$_2$Cl$_2$

Pr-CH=C(TMS)(Pr)   CH$_3$OCHCl$_2$ →   Pr-CH=C(CHO)(Pr)   79%

2) H$_2$O

Synthesis, 721 (1977)

1. Cl$_2$CHOMe, TiCl$_4$ 79%

PrCH=C(Pr)TMS   →   PrCH=C(Pr)CHO

2. H$_2$O

Chem Lett, 859 (1978)

Me-CH=C-OTMS
        |
       TMS

        +   1) BF$_3$·OEt$_2$ Ph   Me
                                                    C=C   85%
                       2) Bu$_4$NOH   H   CHO

PhCH(OEt)$_2$

JACS **99**, 5827 (1977)

PhS—\\=/—CH$_2$OCH$_3$ $\xrightarrow[\text{2) } \underline{n}\text{-C}_6\text{H}_{13}\text{Br}]{\text{1) LDA, -78°}}$ PhS—\\(C$_6$H$_{13}$)—\\=/—OMe   73%

$\xrightarrow[\text{HCl-CH}_3\text{CN}]{\text{2 HgCl}_2}$

C$_6$H$_{13}$—\\=/—CHO (H)   76%

Chem Lett, 345 (1977)

$\underset{\parallel}{\overset{O}{Me-C}}$-CH=CHS-$\underline{t}$-Bu $\xrightarrow{\underline{n}\text{-BuLi}}$ $\underline{n}$-Bu$\underset{\overset{|}{CH_3}}{C}$=CHCHO

Tetr Lett, 2809 (1979)

1) LDA, allyl Br
2) MeOSO$_2$F

3) NaBH$_4$
4) AgNO$_3$

—CHO (with allyl and cyclohexene ring)   89%

Tetr Lett, 5 and 9 (1978)

Also via: β-Hydroxyaldehydes (Section 324)

Section 342   Amide - Amide

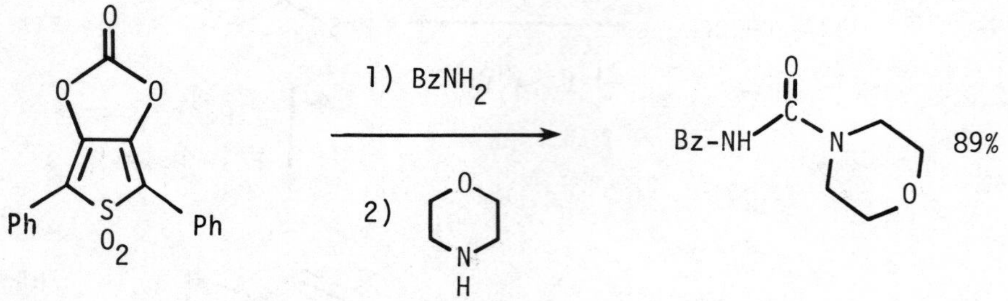

1) BzNH$_2$

2) [morpholine structure]

Bz-NH-C(=O)-N(morpholine)   89%

Several additional examples.

Chem Ber 112, 727 (1979)

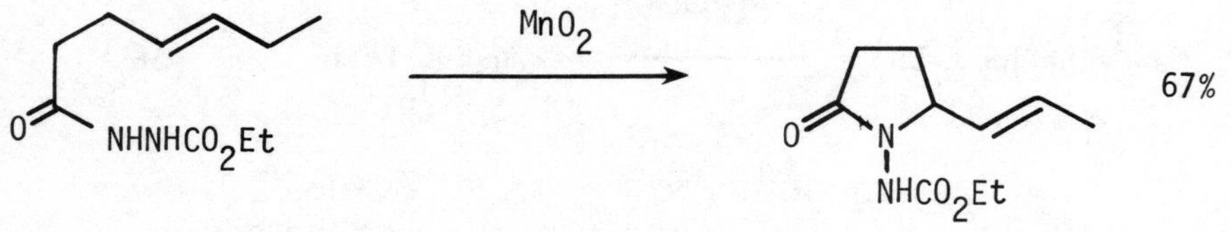

MnO$_2$

67%

Tetr Lett, 4185 (1979)

Also via: Dicarboxylic acids (Section 312)
            Diamines (Section 350)

Section 343   <u>Amide - Amine</u>

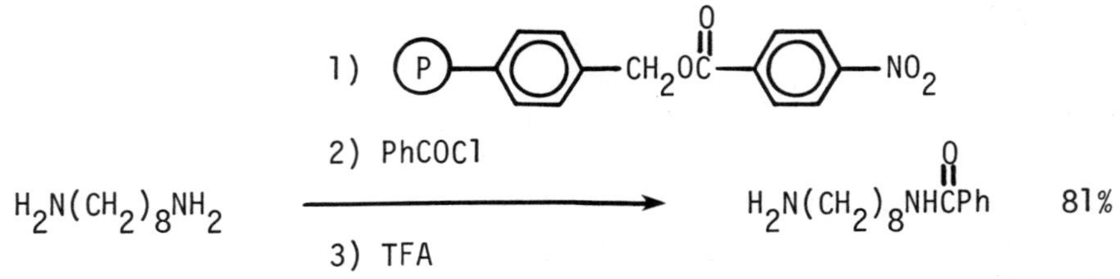

H$_2$N(CH$_2$)$_8$NH$_2$

1) (P)—⬡—CH$_2$OC(=O)—⬡—NO$_2$

2) PhCOCl

3) TFA

H$_2$N(CH$_2$)$_8$NHCPh   81%

Israel J Chem <u>17</u>, 248 (1979)

Section 344   <u>Amide - Ester</u>

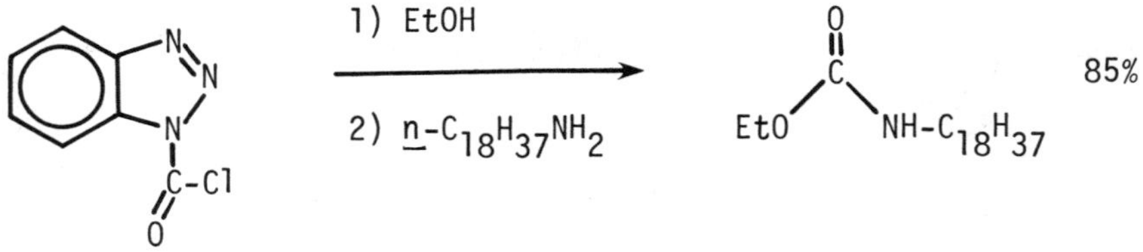

1) EtOH

2) <u>n</u>-C$_{18}$H$_{37}$NH$_2$

EtO—C(=O)—NH-C$_{18}$H$_{37}$ 85%

Synthesis, 704 (1977)

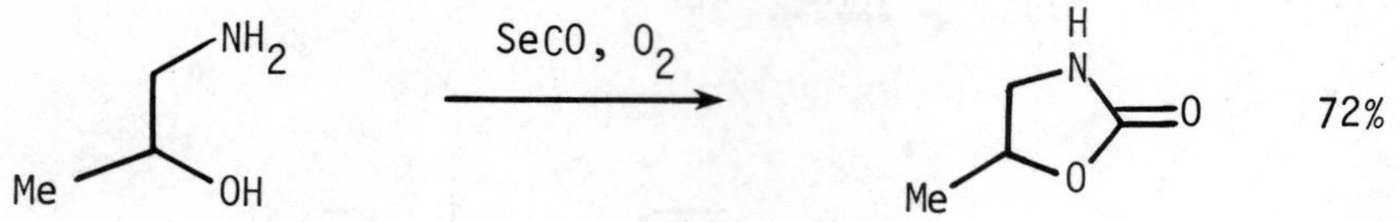

72%

Angew Int Ed 18, 692 (1979)

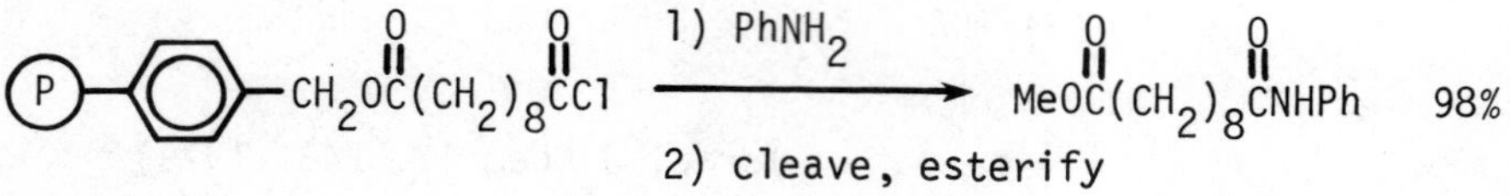

1) PhNH$_2$

2) cleave, esterify

$MeOC(CH_2)_8CNHPh$ 98%

Can J Chem 56, 1562 (1978)

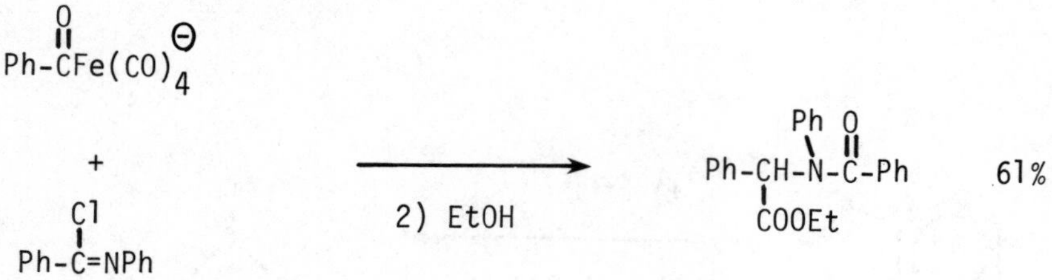

JACS 101, 4245 (1979)

$$Me_2C=CH_2 \quad + \quad \underset{NHCOPh}{\overset{MeOCHCO_2Me}{|}} \quad \xrightarrow[\text{benzene}]{\text{NSA}} \quad \underset{Me_2C=CH}{\overset{\overset{NHCOPh}{|}\;\overset{CH-CO_2Me}{|}}{}} \quad 52\%$$

(NSA = naphthalenesulfonic acid)

Tetrahedron <u>33</u>, 1533 (1977)

$$\underset{COOMe}{\overset{NHAc}{PhCH=C}} \quad \xrightarrow[\text{chiral Rh catalysts}]{H_2} \quad \underset{COOMe}{\overset{*\;NHAc}{PhCH_2-CH}} \quad \sim 100\%$$

55-96%ee

JACS <u>99</u>, 5946 (1977)
Tetr Lett, 1119 (1978)
Tetr Lett, 1635 (1978)
Chem Lett, 39 (1979)

Related methods: Section 315 (Acid-Amide)
                    Section 316 (Acid-Amine)
                    Section 351 (Amine-Ester)

Section 345   <u>Amide - Epoxide</u>

No additional examples

Section 346   <u>Amide - Halide</u>

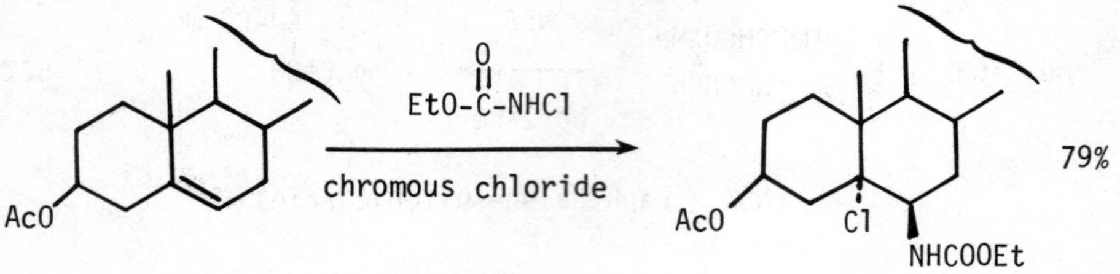

$$\overset{\overset{\textstyle O}{\|}}{EtO-C-NHCl}$$

chromous chloride

79%

Can J Chem <u>55</u>, 700 (1977)

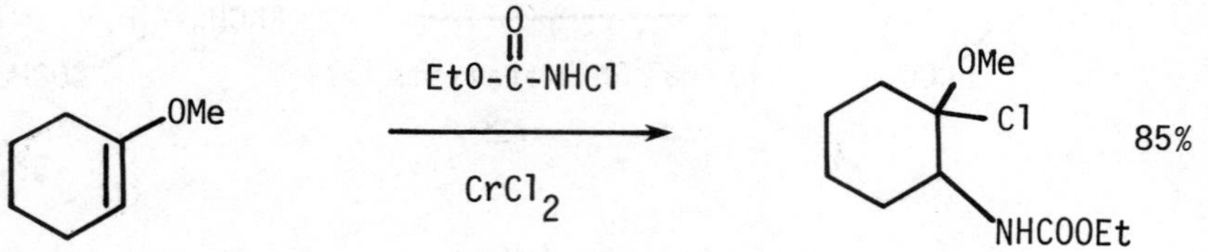

$$\overset{\overset{\textstyle O}{\|}}{EtO-C-NHCl}$$

$CrCl_2$

85%

Can J Chem <u>56</u>, 119 (1978)

Chromous chloride

MeOH

92%

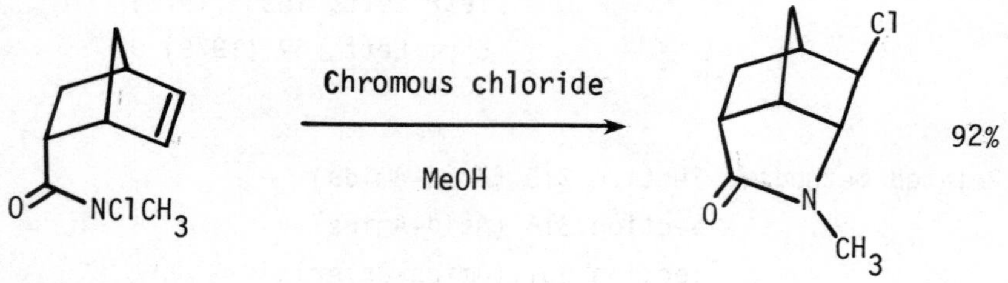

JOC <u>43</u>, 3750 (1978)

Section 347   Amide - Ketone

No additional examples

Section 348   Amide - Nitrile

No additional examples

Section 349   Amide - Olefin

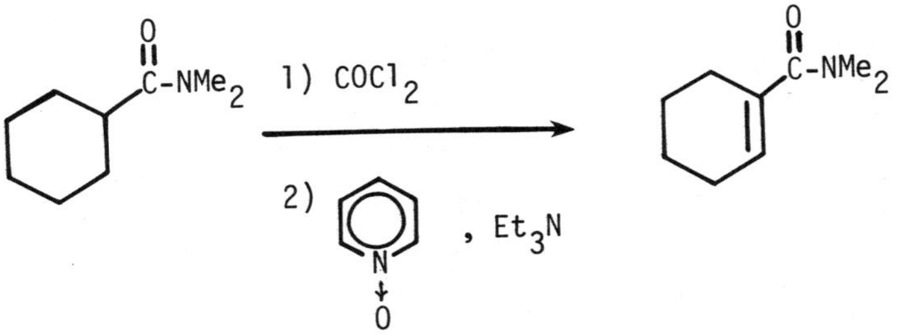

JACS 101, 4381 (1979)

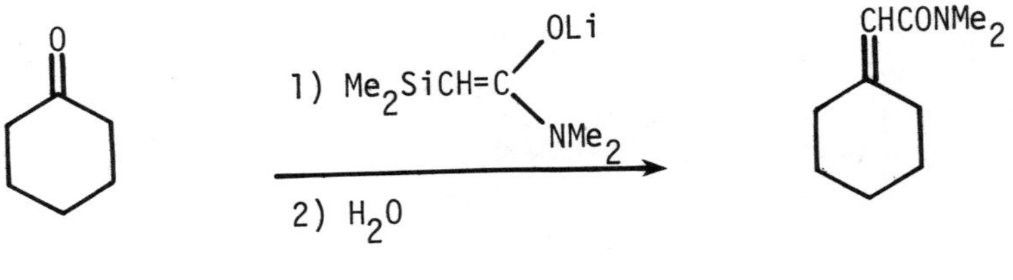

JOC 43, 1947 (1978)

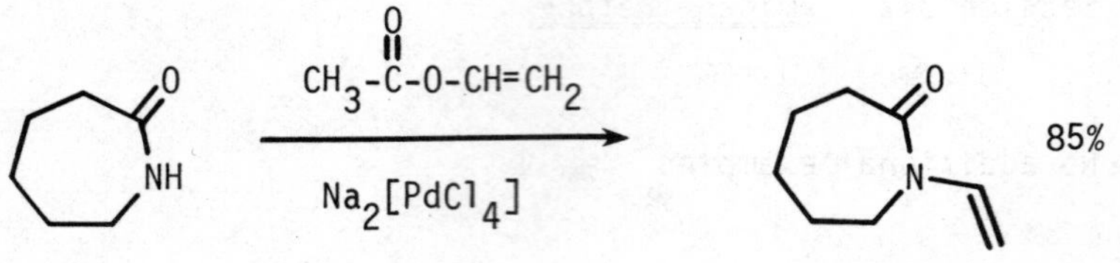

Angew Int Ed <u>18</u>, 533 (1979)

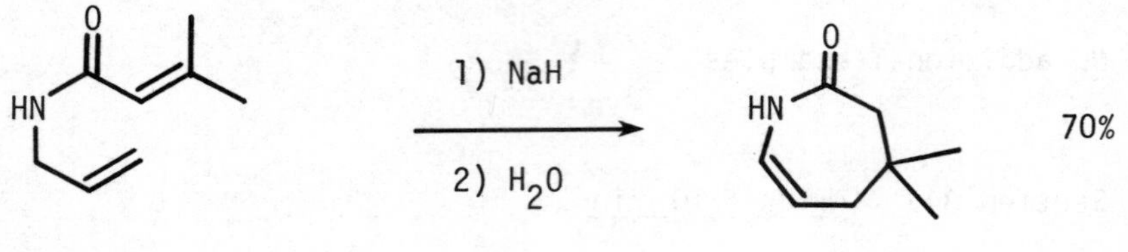

Tetr Lett, 2289 (1977)

Also via: Olefinic acids (Section 322)

Section 350   <u>Amine - Amine</u>

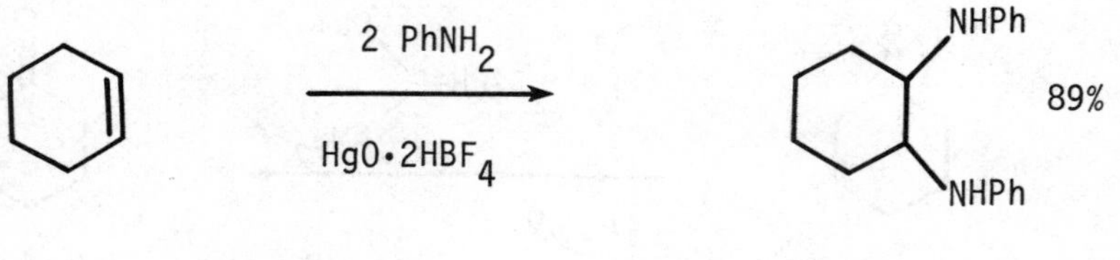

Synthesis, 962 (1979)

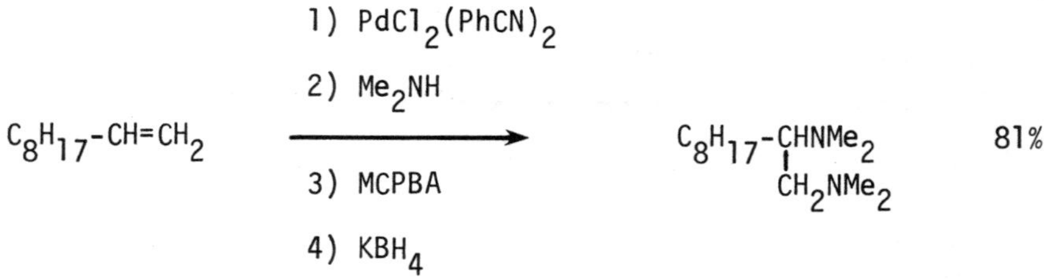

1) $PdCl_2(PhCN)_2$
2) Me_2NH
3) MCPBA
4) KBH_4
C_8H_{17}-CH=CH_2
C_8H_{17}-CHNMe$_2$
CH_2NMe_2
81%

Tetr Lett, 163 (1978)

CH_2NH_2
H_2N
Me_2CHBr
EtOH
Me_2CHNH
$CH_2NHCHMe_2$
80%

J Prakt Chem 321, 680 (1979)

N
N
$NaBH_4$
N
N
85%

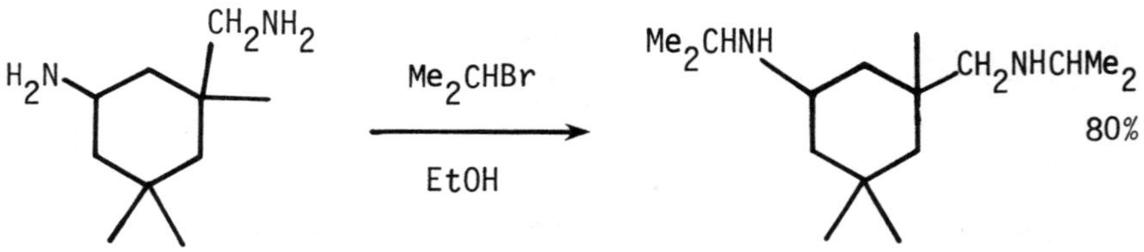

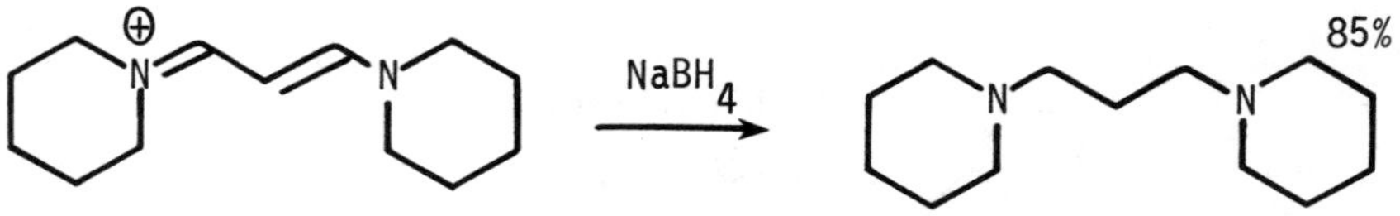

Chem Ber 110, 1259 (1977)

Section 351   <u>Amine - Ester</u>

$$\underset{\text{Z-NH-CH-COOH}}{\overset{R}{|}} \xrightarrow[\text{R}_4\text{NCl, NaHCO}_3]{\text{BzBr, CH}_2\text{Cl}_2} \underset{\text{Z-NH-CH-C-OBz}}{\overset{R \quad O}{| \quad ||}}$$

$$\underset{\text{Z-NH-CH-COOH}}{\overset{R}{|}} \xrightarrow[\text{NaHCO}_3\text{, DMF}]{\text{EtBr}} \underset{\text{Z-NH-CH-C-OEt}}{\overset{R \quad O}{| \quad ||}}$$

Synthesis, 957 and 961 (1979)

$$\underset{\text{H}_3\text{N-CH-COO}^\ominus}{\overset{\oplus \quad R}{ |}} \xrightarrow[\text{reflux}]{\text{TsOH, EtOH}} \text{TsO}^\ominus \; \underset{\text{H}_3\text{N-CH-COOEt}}{\overset{\oplus \quad R}{ |}}$$

Bull Chem Soc Japan <u>52</u>, 1879 (1979)

1) LiNH(<u>t</u>-Bu)

2) H$^\oplus$, MeOH

47%

JOC <u>42</u>, 2653 (1977)

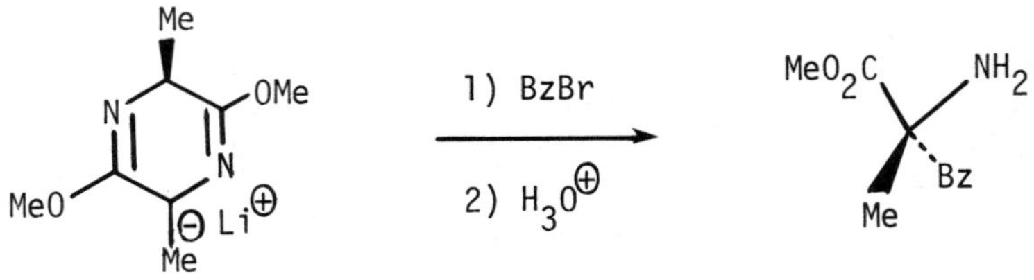

Angew Int Ed <u>18</u>, 863 (1979)

oxalic acid

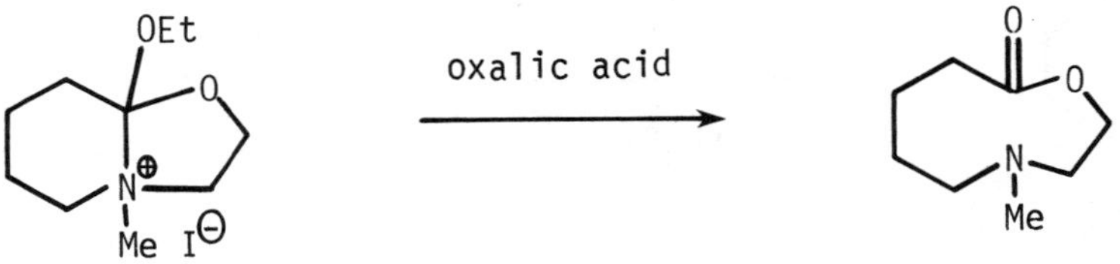

68%

Tetr Lett, 809 (1979)

Related methods: Section 315 (Acid-Amide)
                   Section 316 (Acid-Amine)
                   Section 344 (Amide-Ester)

Section 352   <u>Amine - Ether</u>

No additional examples

Section 353   <u>Amine - Halide</u>

Ph-CHF-CMe$_2$ ∼80%
    |
    NH$_2$

Tetr Lett, 3247 (1978)

80%

Synth Comm <u>8</u>, 549 (1978)

Section 354   <u>Amine - Ketone</u>

$Me_2N=CH_2$ $CF_3CO_2^{\ominus}$

90%

Tetrahedron <u>35</u>, 613 (1979)

$(H_2C=CH-CH_2\rightarrow)_2NH$

+

$HC\equiv C-CH_2OH$

$\xrightarrow[Cd(OAc)_2]{Zn(OAc)_2}$

$(H_2C=CH-CH_2-)_2N-CH_2-\overset{\overset{O}{\|}}{C}-CH_3$ 45%

Tetr Lett, 3523 (1979)

Section 355 Amine - Nitrile

O
(EtO)2PCN,
N
H
THF
N
CN
88%

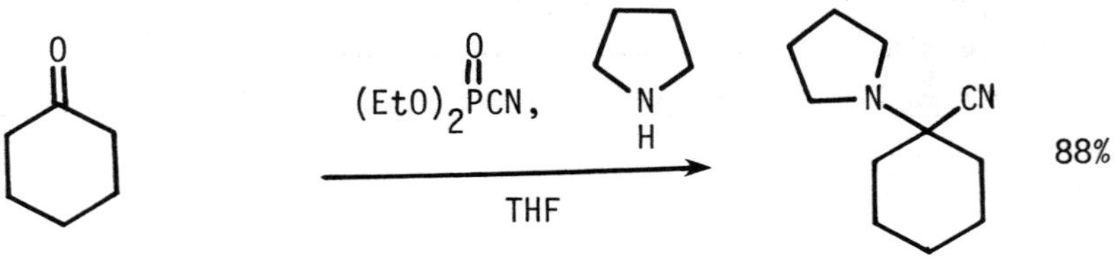

Tetr Lett, 4663 (1979)

Section 356 Amine - Olefin

H
N
N-NO2
N
+
N
81%

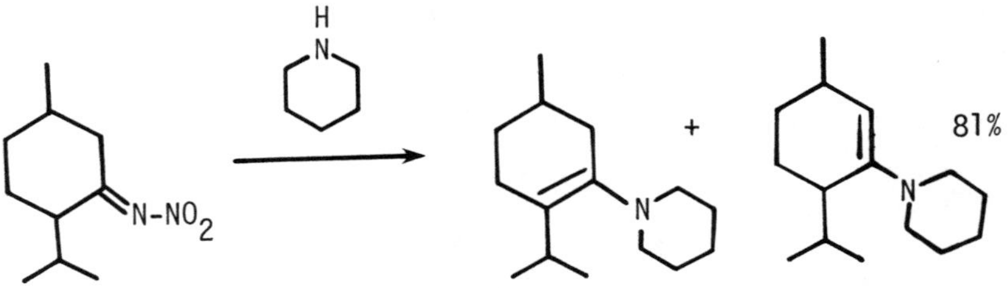

Synthesis, 830 (1979)

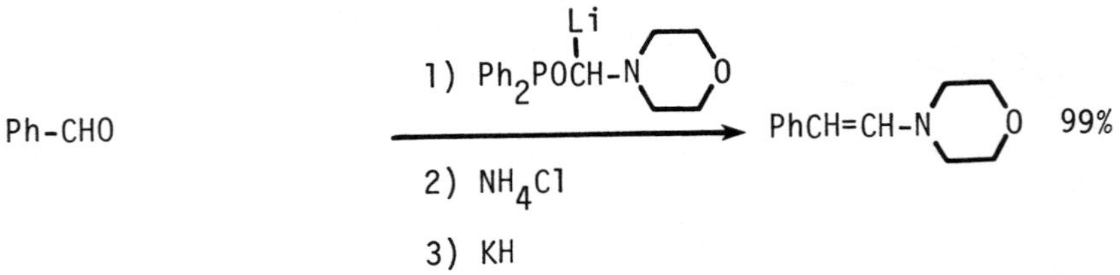

Ph-CHO
Li
1) Ph2POCH-N O
2) NH4Cl
3) KH
PhCH=CH-N O
99%

Tetr Lett, 2433 (1979)

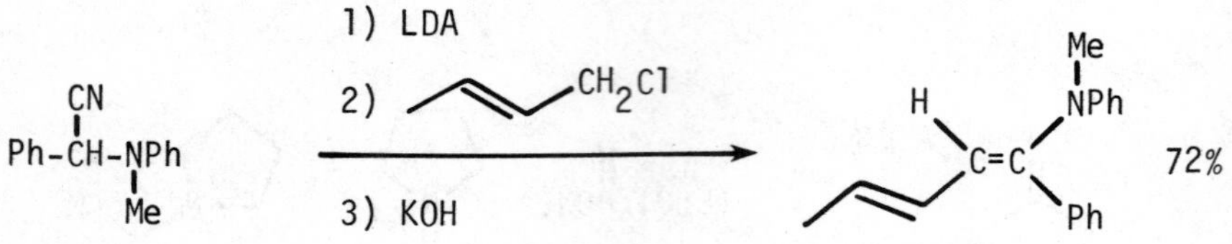

1) LDA

2) ⌇⌇CH₂Cl

3) KOH

Ph-CH-NPh with CN and Me →

structure with Me, NPh, H, C=C, Ph   72%

Synthesis, 127 (1979)

2-aminopyridine + H₂N-C(=CHCOOEt)(Ph) → pyridine-NH-C(Ph)=CHCOOEt

J Het Chem 14, 1419 (1977)

structure with Br + morpholine (H-N-O ring) + hexene →

Pd(OAc)₂

P(O-C₆H₄-CH₃)₃

product morpholine-N-CH₂-C(CH₃)=CH-CH₂CH₂CH₂CH₃   64%

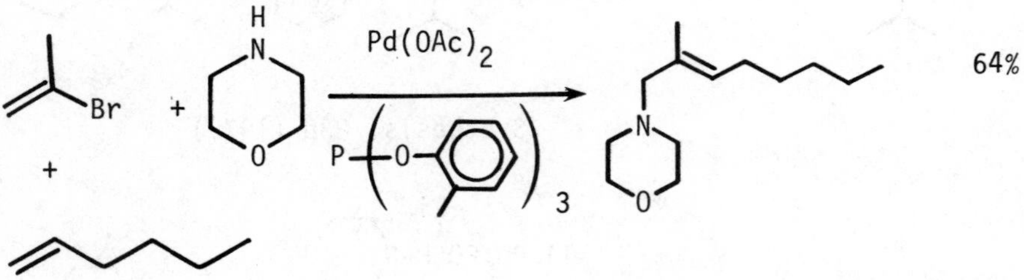

JOC 43, 3898 (1978)

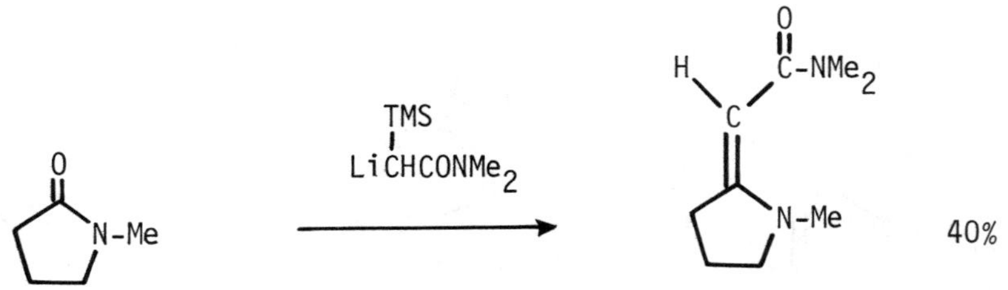
O
N-Me
TMS
LiCHCONMe2
H
O
C-NMe2
C
N-Me
40%

Tetr Lett, 709 (1978)

Section 357 Ester - Ester

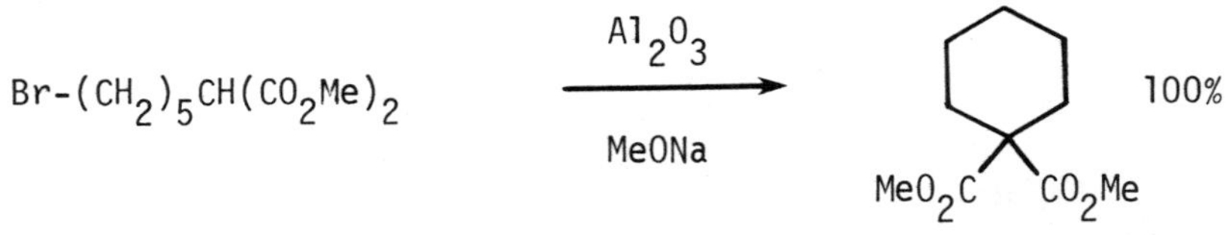
Br-(CH2)5CH(CO2Me)2
Al2O3
MeONa
MeO2C
CO2Me
100%

JCS Chem Comm, 522 (1979)

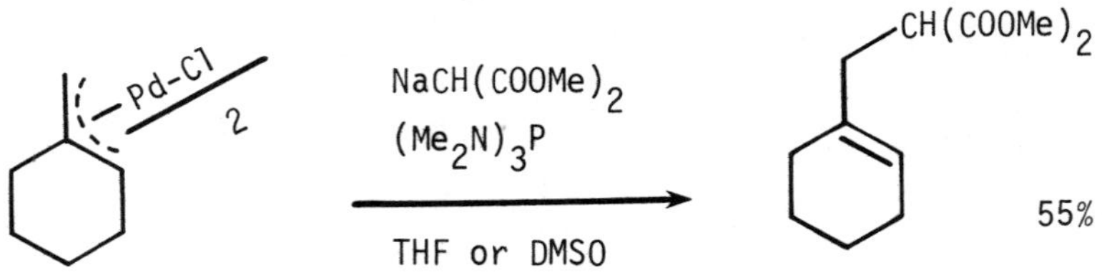
Pd-Cl
2
NaCH(COOMe)2
(Me2N)3P
THF or DMSO
CH(COOMe)2
55%

JACS 100, 3416 (1978)

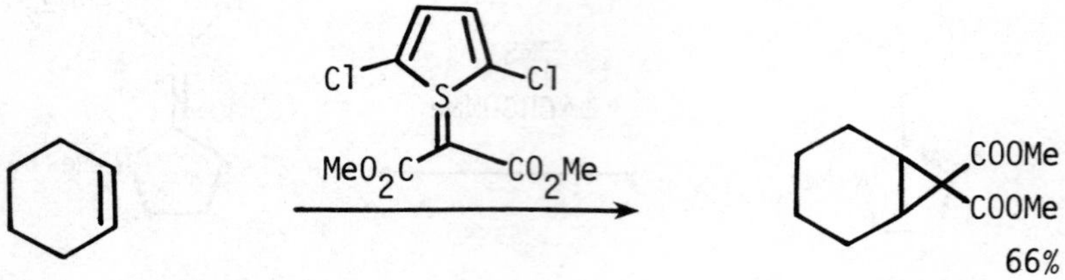

JCS Chem Comm, 641 (1978)

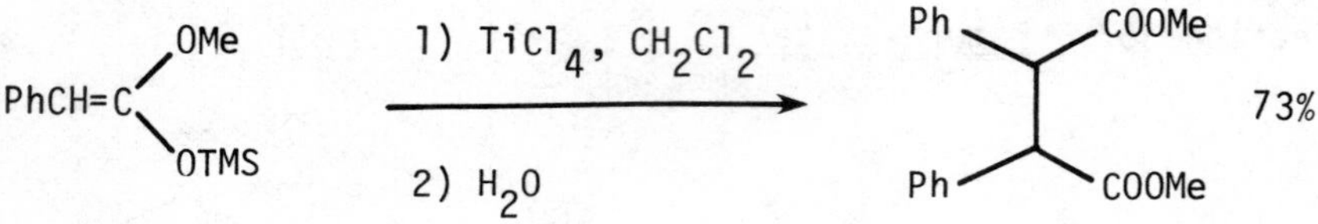

Tetr Lett, 2009 (1977)

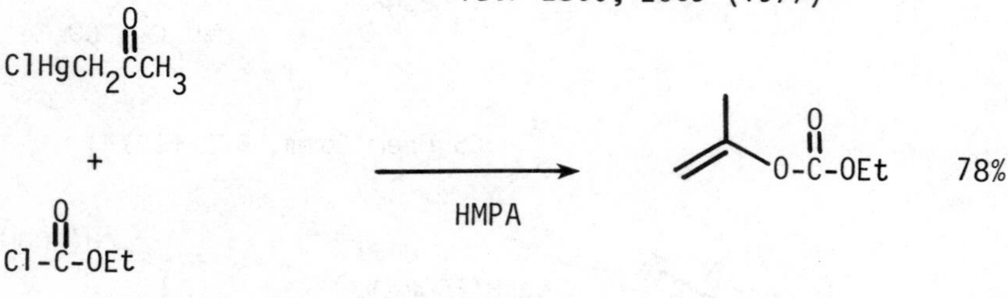

Tetr Lett, 3737 (1978)

Also via: Dicarboxylic acids (Section 312)
           Hydroxyesters (Section 327)
           Diols (Section 323)

Section 358   Ester - Ether, Epoxide

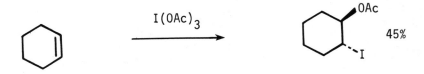

1) LiCH$_2$COOEt

2) 2 LDA

3) I$_2$

43%

Tetr Lett, 4575 (1977)

CH(COOEt)$_2$

CH$_2$OH

CF$_3$SO$_2$Cl

DBU

COOEt
COOEt

100%

Tetr Lett, 3645 (1979)

Section 359   Ester - Halide

I(OAc)$_3$

OAc
I

45%

JCS Perkin I, 2231 (1977)

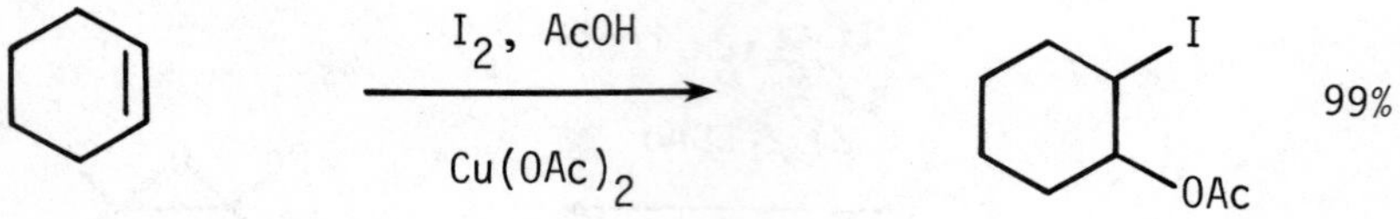

Synthesis, 402 (1978)

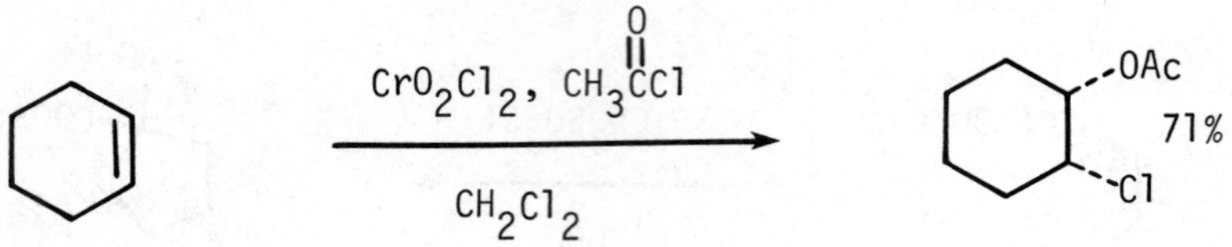

Tetr Lett, 3523 (1977)

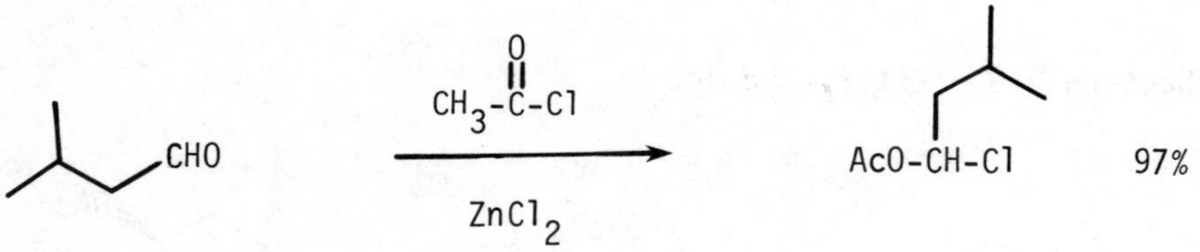

Helv Chim Acta 61, 2047, 2059,
                2165, 2381 (1978)

Synthesis, 593 (1978)

$$\underset{\substack{| \\ Me_2C-C-OMe}}{\overset{\substack{H \quad O \\ | \quad \|}}{}} \qquad \xrightarrow[\text{2) } CBr_4, \text{ THF}]{\text{1) LDA}} \qquad \underset{\substack{| \\ Me_2C-C-OMe}}{\overset{\substack{Br \quad O \\ \backslash \quad \|}}{}} \qquad 79\%$$

JOC <u>43</u>, 3687 (1978)

1-Decene + CCl_3COOH $\xrightarrow{\text{Ru(II) complex}}$ 85%

Chem Lett, 363 (1978)

$$\underset{\substack{| \\ OH}}{Ph-CH-CH_2COOEt} \qquad \xrightarrow[\text{2) } PhPF_4]{\substack{\text{1) } Me_3SiCl, \\ \text{pyridine}}} \qquad \underset{\substack{| \\ F}}{Ph-CH-CH_2COOEt} \qquad 92\%$$

Tetr Lett, 4507 (1978)

$\xrightarrow{\substack{TMS \\ | \\ Cl-C-CO_2\underline{t}\text{-Bu} \\ | \\ Li}}$ 44%

Tetr Lett, 515 (1978)

$$Cl_3C-\overset{\overset{O}{\|}}{P}(OEt)_2 \quad \xrightarrow{\begin{array}{l}1)\ BuLi\\ 2)\ ClCOOEt\\ \\ 3)\ \end{array}}$$

1) BuLi
2) ClCOOEt
3) cyclohexanone

cyclohexylidene C(Cl)-COOEt   71%

Synthesis, 31 (1978)

Review: "Synthesis and Synthetic Utility of Halolactones"

Chem Soc Rev 8, 171 (1979)

Also via: Haloacids (Section 319)
             Halohydrins (Section 329)

Section 360   Ester - Ketone

$$(CH_3)_2CH-C\equiv C-OMe \quad \xrightarrow[OsO_4,\ H_2O]{KClO_3} \quad$$

isopropyl -CO-CO-OMe   80%

JOC 43, 4245 (1978)

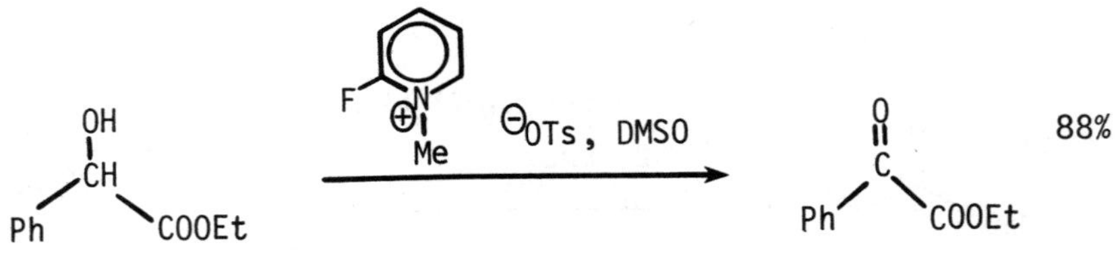

Chem Lett, 369 (1978)

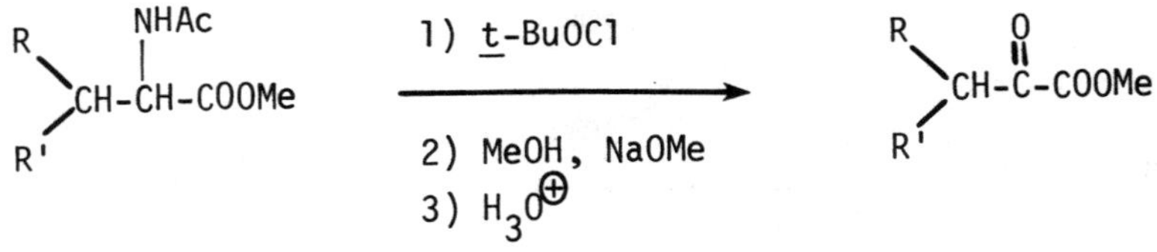

R,R' = H, alkyl

Chem Ber 111, 3136 (1978)

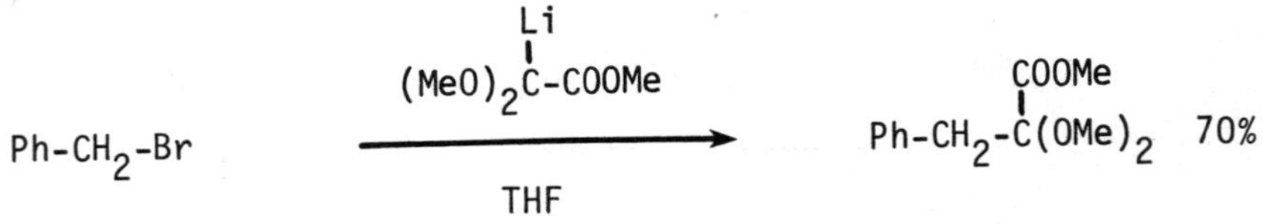

Synthesis, 33 (1979)

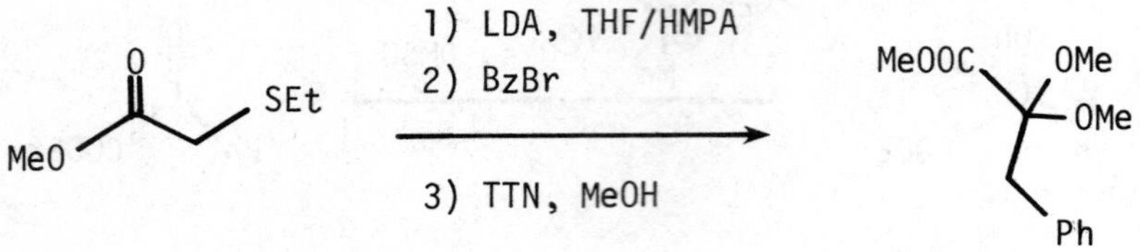

Tetr Lett, 4115 (1978)

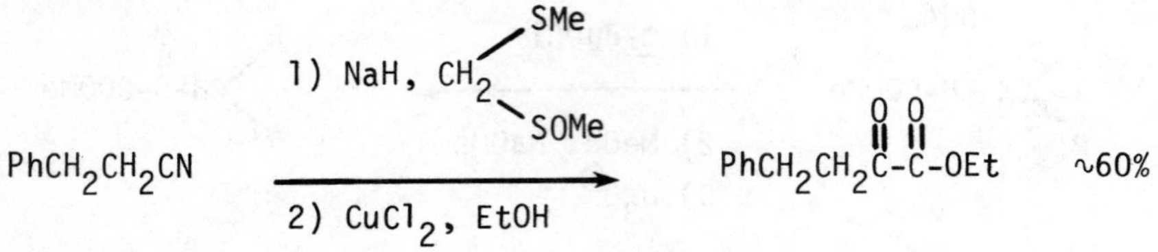

Tetr Lett, 375 (1978)

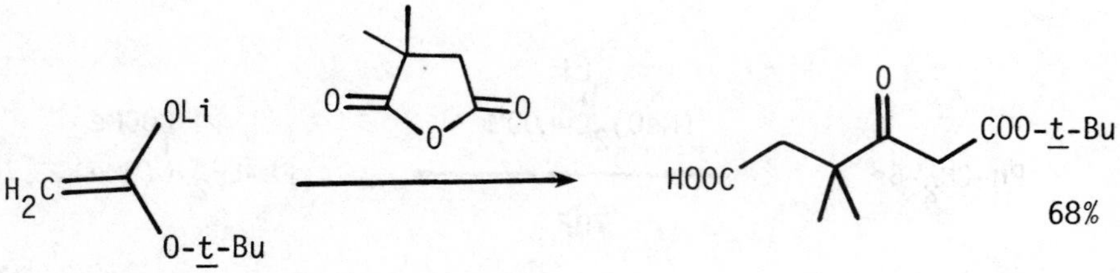

Angew Int Ed 18, 632 (1979)

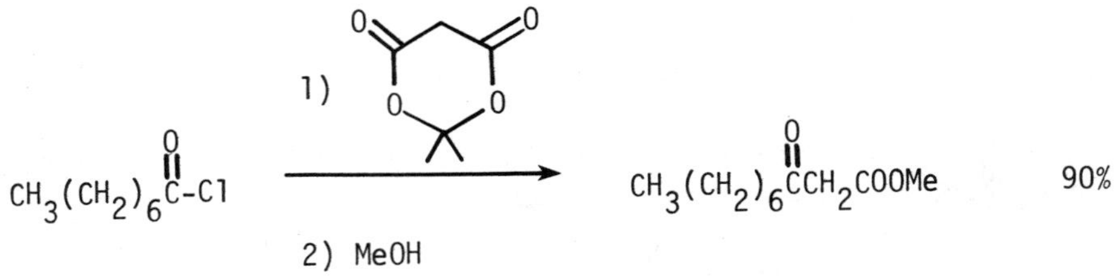

$CH_3(CH_2)_6\overset{O}{\overset{\|}{C}}-Cl$

1)

2) MeOH

$CH_3(CH_2)_6\overset{O}{\overset{\|}{C}}CH_2COOMe$ 90%

JOC **43**, 2087 (1978)

$HOOC-CH_2COOEt$

1) BuLi

2) Me_2CHCH_2COCl

3) $H_3O^{\oplus}$

$Me_2CHCH_2\overset{O}{\overset{\|}{C}}CH_2COOEt$ 98%

JOC **44**, 310 (1979)

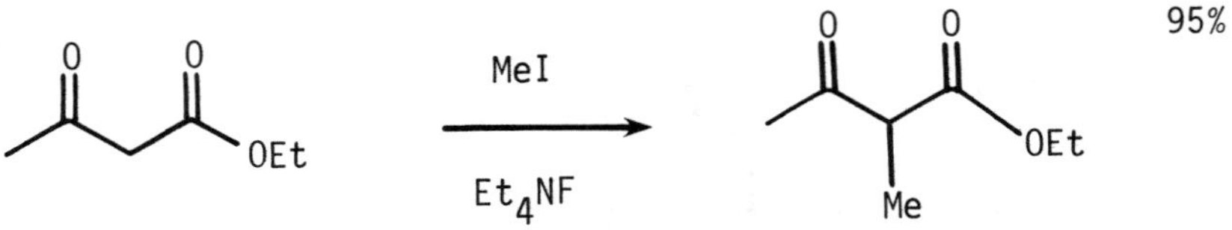

95%

MeI

Et_4NF

JCS Chem Comm, 64 (1977)

1) Ph$_3$P=CHCOOMe

2) Δ

$C_5H_{11}\overset{\overset{O}{\|}}{C}-Cl$ $\xrightarrow{\hspace{3cm}}$ $C_5H_{11}\overset{\overset{O}{\|}}{C}CH_2CO_2Me$ ~60%

3) piperidine ⟨NH⟩

4) H$_3$O$^{\oplus}$

Liebigs Ann, 282 (1977)

$\xrightarrow{(EtO\overset{\overset{O}{\|}}{C})_2\overset{\ominus}{C}H\overset{\oplus}{N}a}$

70%

Synth Comm 8, 59 (1978)

2) H$_3$O$^{\oplus}$

74%

Synthesis, 142 (1978)

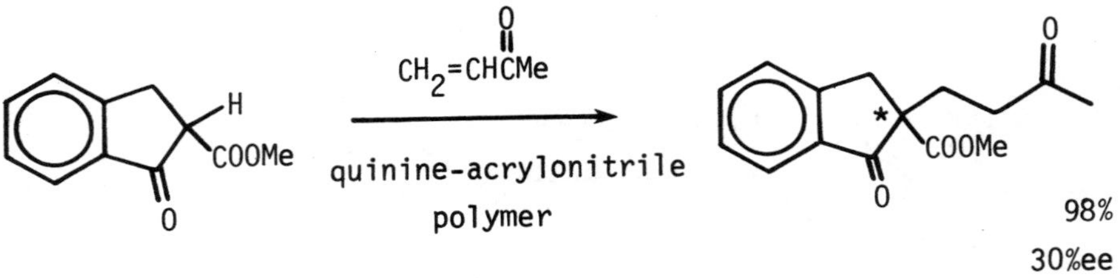
H
COOMe
O
CH_2=CHCMe
quinine-acrylonitrile
polymer
COOMe
*
98%
30%ee

JACS 100, 7071 (1978)

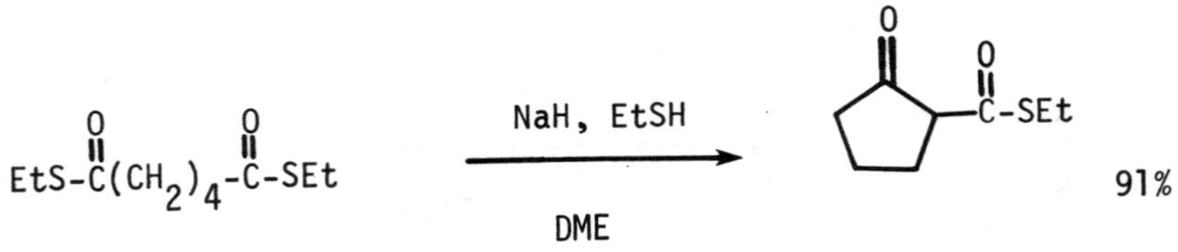
EtS-C(CH_2)$_4$-C-SEt
NaH, EtSH
DME
C-SEt
91%

Tetr Lett, 1193 (1979)

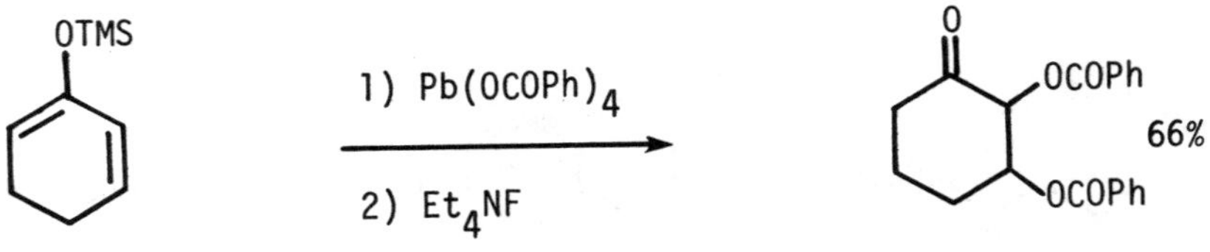
OTMS
1) Pb(OCOPh)$_4$
2) Et_4NF
OCOPh
OCOPh
66%

JOC 42, 1051 (1977)

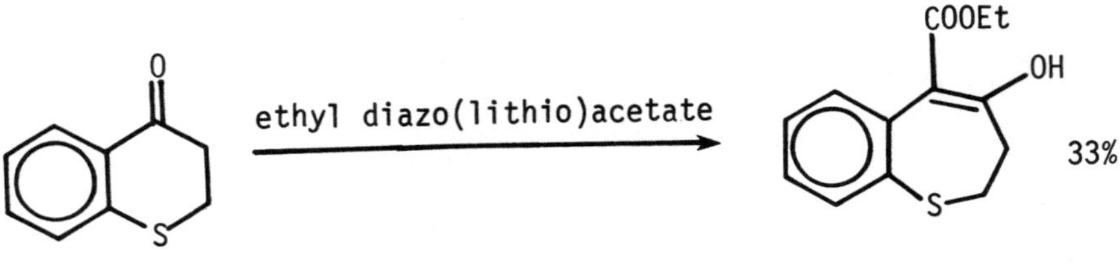
O
S
ethyl diazo(lithio)acetate
COOEt
OH
33%

JCS Perkin I, 1822 (1977)

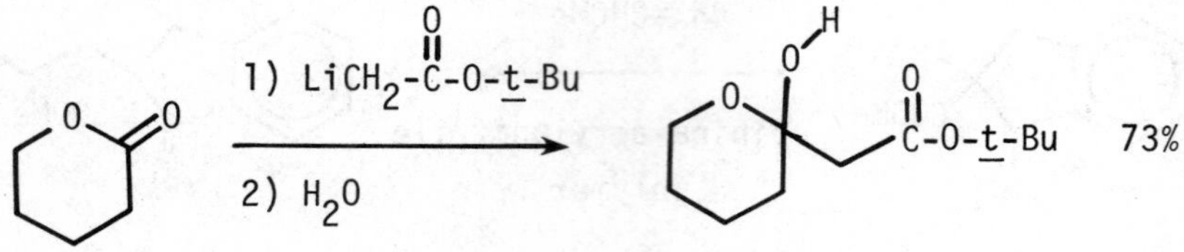

Tetr Lett, 4323 (1978)

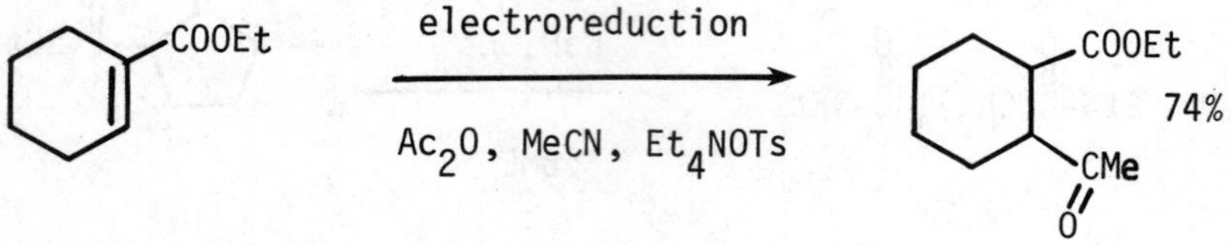

JACS 99, 7396 (1977)

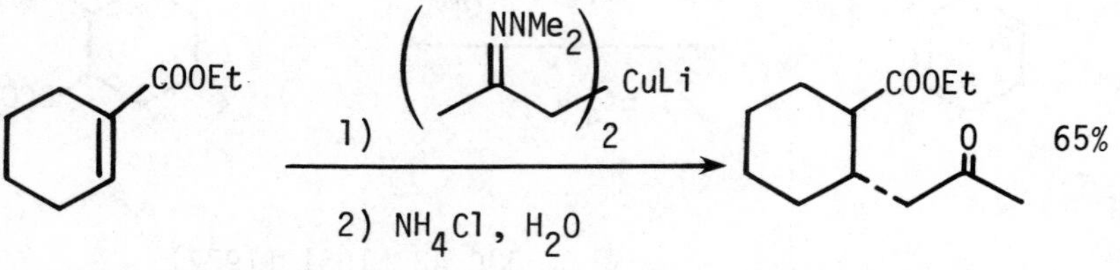

Tetr Lett, 4597 (1978)

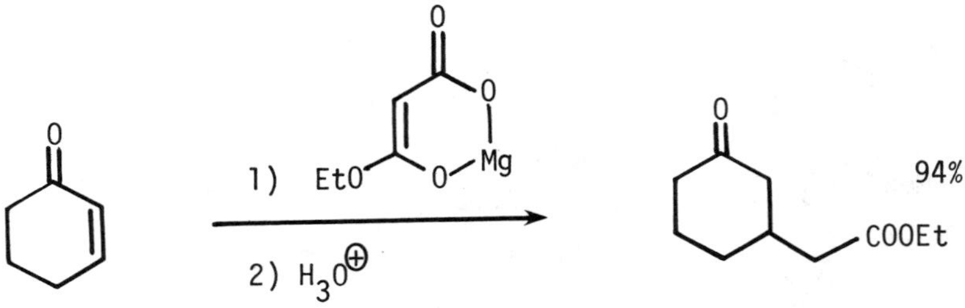

Synth Comm <u>8</u>, 53 (1978)

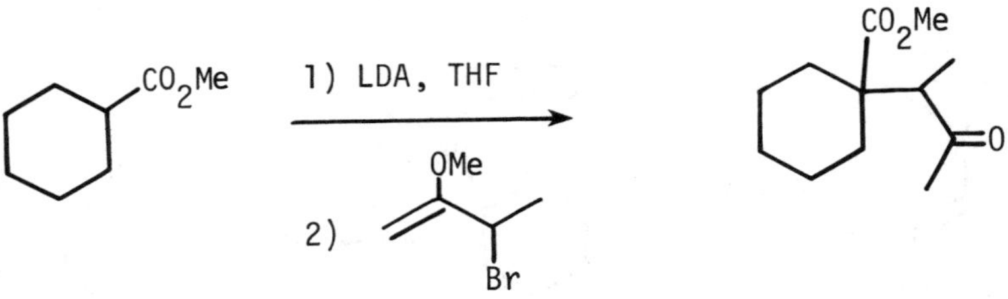

JOC <u>43</u>, 4650 (1978)

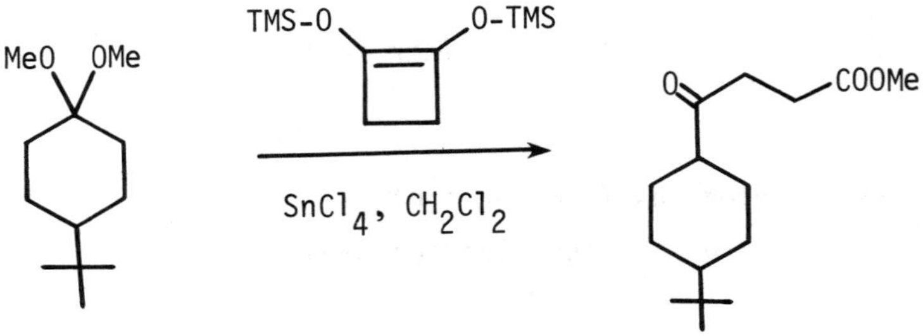

JOC <u>42</u>, 4166 (1977)

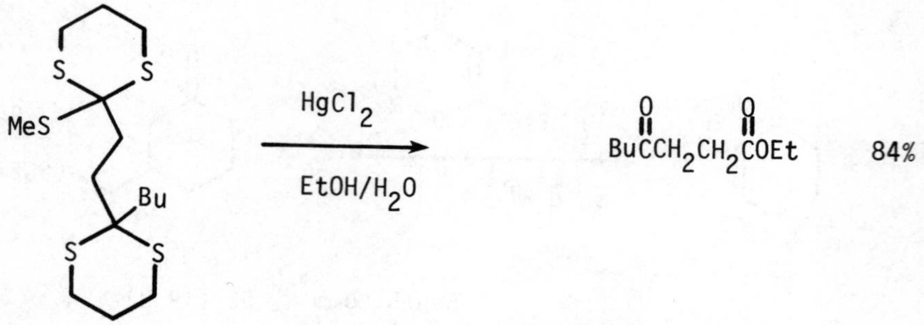

$$\text{BuCCH}_2\text{CH}_2\text{COEt} \quad 84\%$$

Synth Comm $\underline{8}$, 279 (1978)

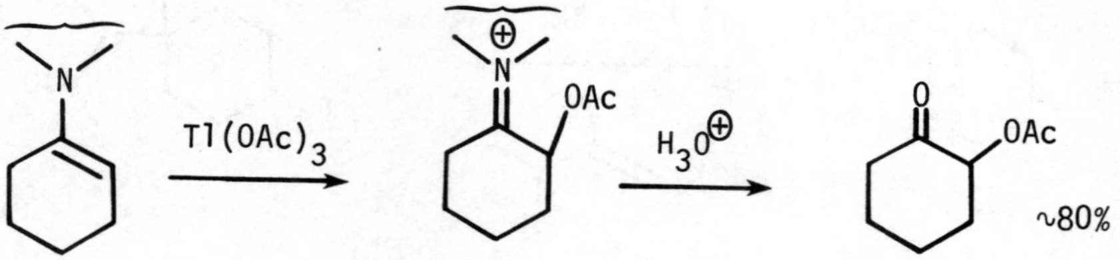

~80%

Tetr Lett, 23 (1977)

Also via: Ketoacids (Section 320)

Hydroxyketones (Section 330)

Section 361   Ester - Nitrile

CH$_2$CN (on cyclohexane)

1) LDA

2) ClCOOEt

COOEt / CHCN (on cyclohexane)   86%

JOC $\underline{42}$, 2009 (1977)

EtOOC-CH$_2$-CN

+

CH$_3$-(CH$_2$)$_7$-CH=CH$_2$

CuO, 85°

CH$_3$-(CH$_2$)$_9$-CH with CN and COOEt   75%

Synthesis, 454 (1977)

OMe (benzene ring)

+

$\overset{H}{\underset{Cl}{}} C=C \overset{COOEt}{\underset{CN}{}}$

AlCl$_3$

HC=C with COOEt and CN (on benzene ring with OMe)   62%

Chem Ber $\underline{110}$, 86 (1977)

Ph$_3$P=CHCN

+ $\longrightarrow$ NC-CH=C(COOEt)$_2$ 88%

O=C(COOEt)$_2$

Synthesis, 626 (1977)

Section 362   Ester - Olefin

This section contains syntheses of enol esters and esters of
unsaturated acids.

1) LDA

2) C$_6$H$_{13}$CHO

3) POCl$_3$, Et$_3$N

~48%

Tetr Lett, 2693 (1978)

$(EtO)_2\overset{O}{\overset{\|}{P}}-S-\overset{Me}{\underset{\ominus}{C}}-COOEt$

90%

Chem Lett, 1039 (1979)

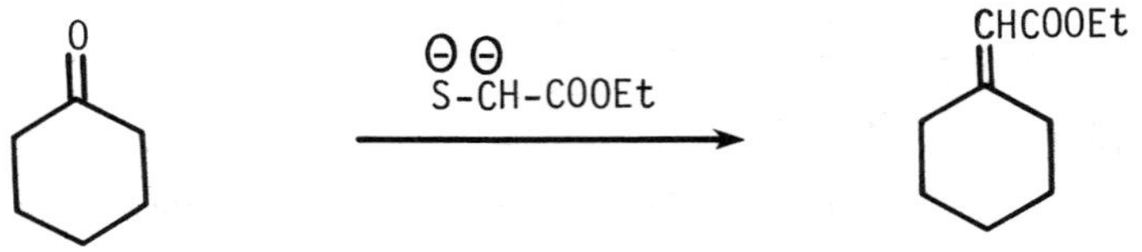

Chem Lett, 471 (1977)

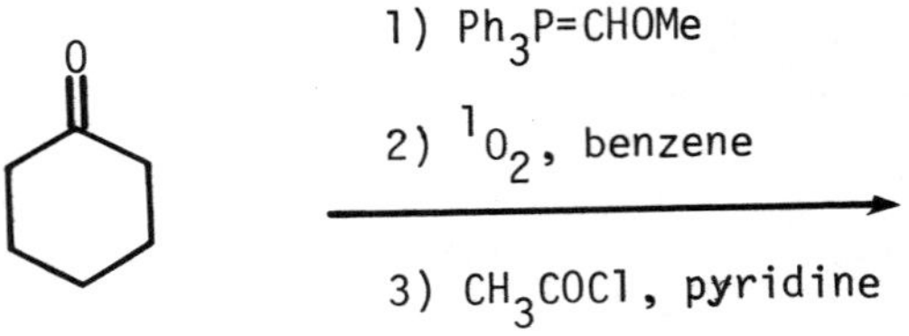

Synthesis, 67 (1978)

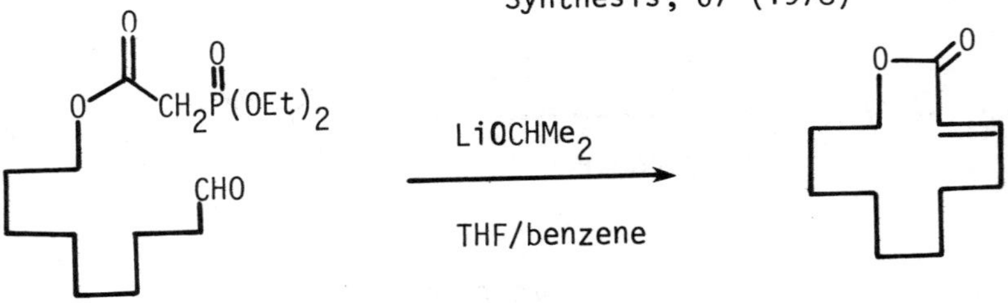

JOC 44, 4010 (1979)
JOC 44, 4011 (1979)

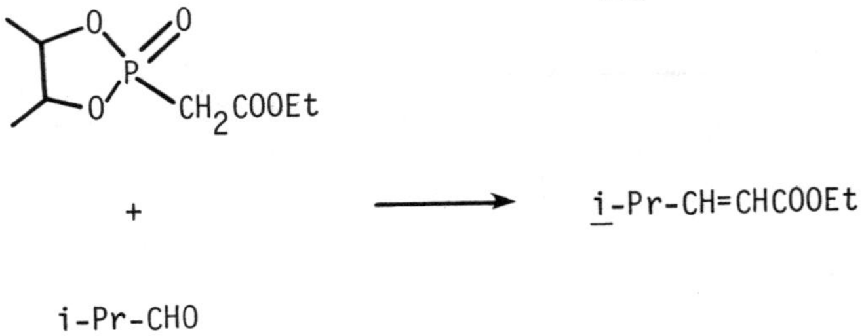

i-Pr-CH=CHCOOEt   ∿70%

i-Pr-CHO

Tetrahedron 34, 997 (1978)

$$[Pr_3BMe]Cu \; + \; \underset{H}{\overset{Br}{>}}C=C\underset{COOEt}{\overset{H}{<}} \quad \xrightarrow[\text{H}_2\text{O}_2]{\overset{2)\;\ominus OH,}{}} \quad \underset{H}{\overset{Pr}{>}}C=C\underset{COOEt}{\overset{H}{<}} \qquad 80\%$$

Tetr Lett, 3369 (1977)

$$\xrightarrow[\text{AlCl}_3]{\text{HC}\equiv\text{CCO}_2\text{CH}_3}$$

62%

JACS <u>101</u>, 5283 (1979)

$$C_5H_{11}-C\equiv CH \quad \xrightarrow[\text{2) ClCOOEt}]{\text{1) Me}_3\text{Al-Cl}_2\text{ZrCp}_2} \quad \underset{Me}{\overset{C_5H_{11}}{>}}C=C\underset{COOEt}{\overset{H}{<}} \qquad 86\%$$

Tetr Lett, 2357 (1978)

$(CH_2=CHCH_2)_2CuLi$

+

$Bu-C\equiv C-COOEt$

$$\xrightarrow{}$$

90%

Tetr Lett, 1811 (1979)

$Ph-C\equiv C-CO_2Et$ $\xrightarrow{BuCu\cdot BEt_3}$

Ph, COOEt
C=C
Bu, H

96%

JOC **44**, 1744 (1979)

$[Bu_3BMe]Cu$

+

$HC\equiv C-COOEt$

$\xrightarrow{DME}$

Bu, H
C=C
H, COOEt

61%

BCS Japan **50**, 3431 (1977)

CHSCONMe$_2$ $\xrightarrow[\substack{2)\ MeSSMe \\ 3)\ H_2O,\ HgCl_2}]{1)\ LDA}$ CHCO$_2$Et

57%

Tetr Lett, 2895 (1978)

$\left[\begin{array}{c} CH_3(CH_2)_5 \quad\quad H \\ C=C \\ H \quad\quad SiF_5 \end{array} \right] K_2$ $\xrightarrow[\substack{PdCl_2,\ AcONa}]{CO,\ MeOH}$

91%

$CH_3(CH_2)_5$, H
C=C
H, COOMe

Tetr Lett, 619 (1979)

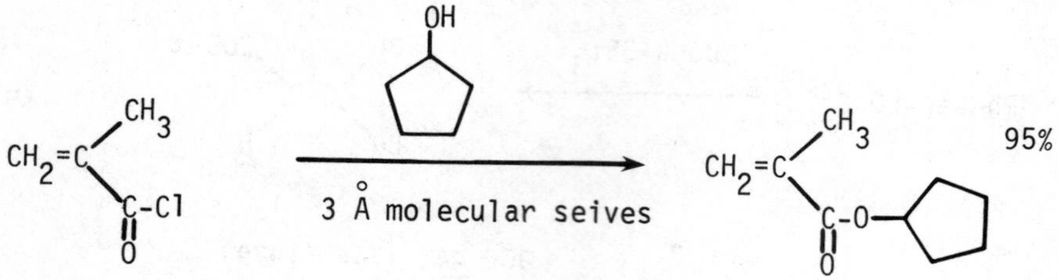

JOC 42, 3965 (1977)

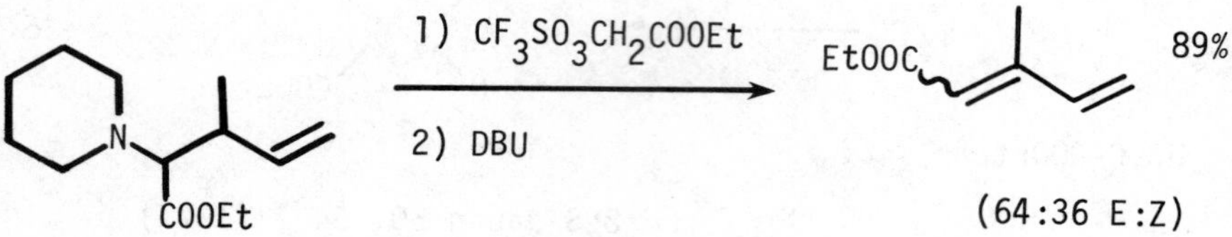

Tetr Lett, 1241 (1977)

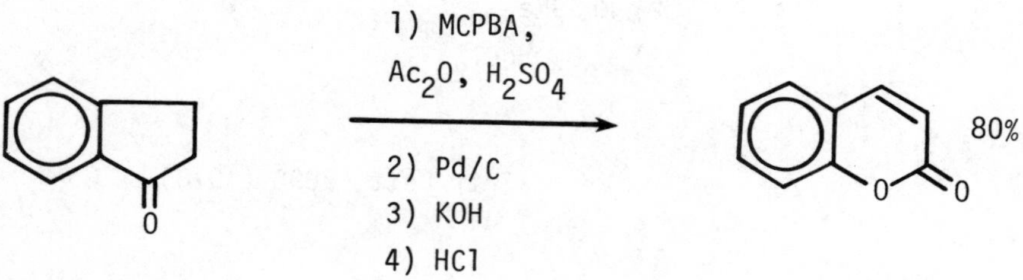

Indian J CHem 15B, 214 (1977)

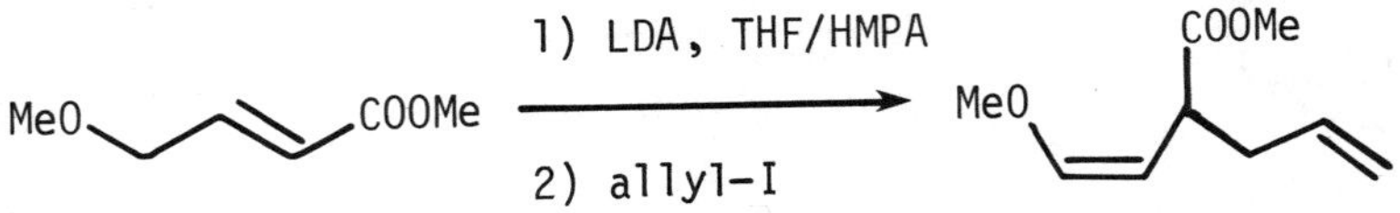

Synth Comm <u>7</u>, 189 (1977)

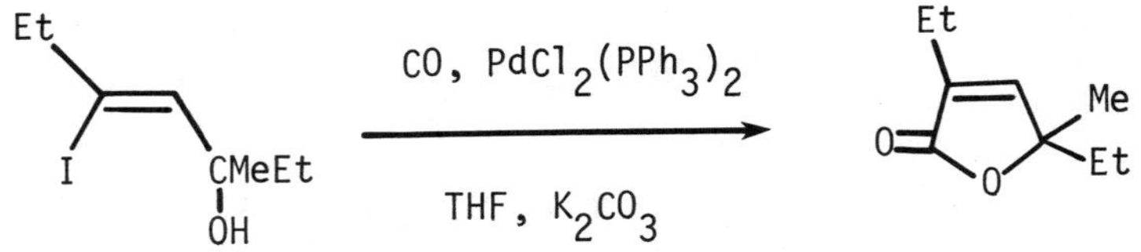

Tetr Lett, 133 (1979)

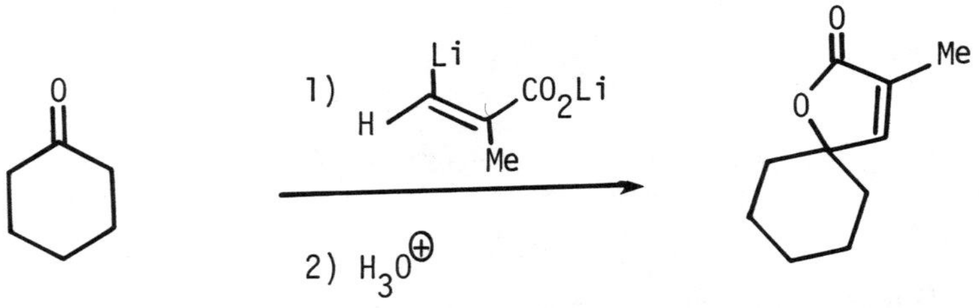

Tetr Lett, 5167 (1978)

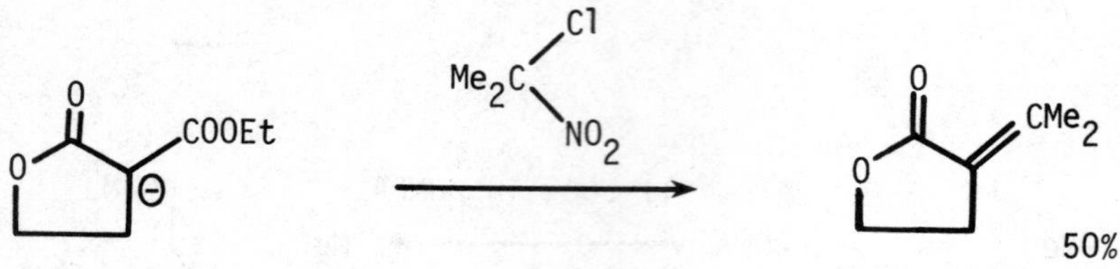

Cl
Me_2C
NO_2
COOEt
O
CMe_2
50%

Chem Lett, 189 (1977)

S
O
+
2) ClCOOEt
43%

Chem Lett, 653 (1978)

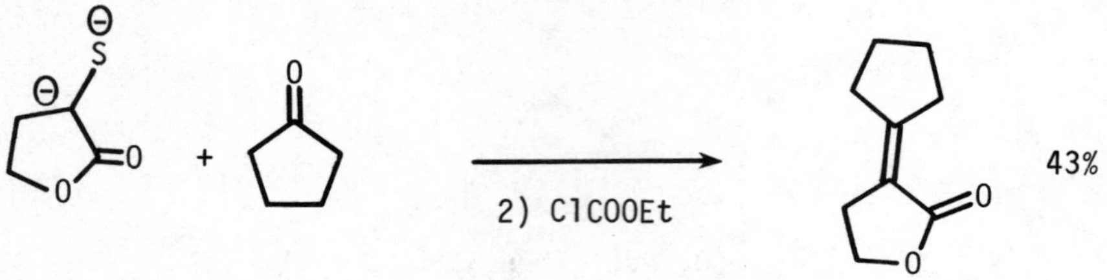

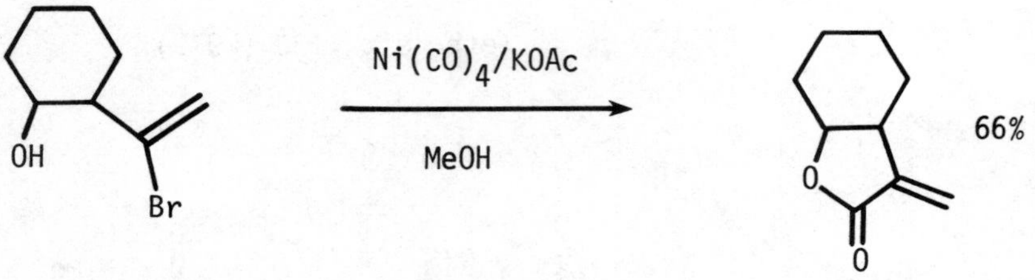

OH
Br
$Ni(CO)_4$/KOAc
MeOH
O
66%

Chem Lett, 773 (1978)

Review: "Recent Methods for the Synthesis of Conjugated Lactones"

Aldrichimica Acta 10, 64 (1977)

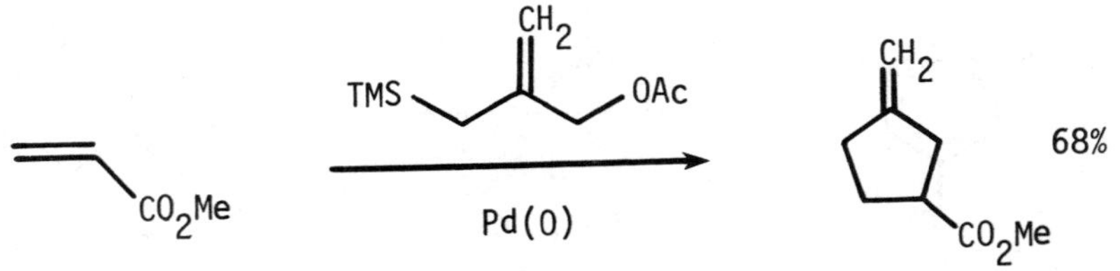

JACS <u>101</u>, 6429 (1979)

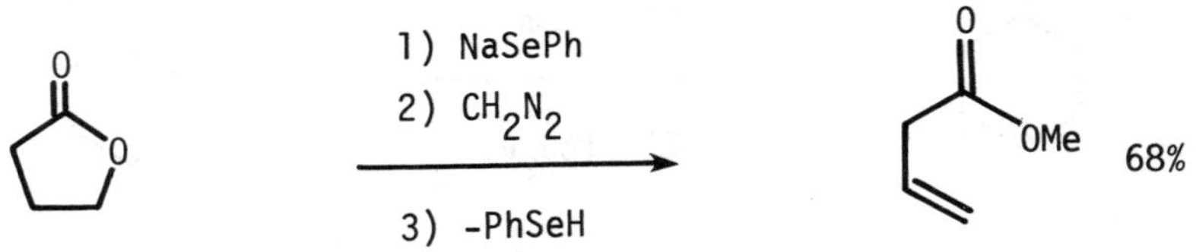

1) NaSePh

2) CH_2N_2

3) -PhSeH

68%

Tetr Lett, 4361 and 4369 (1977)

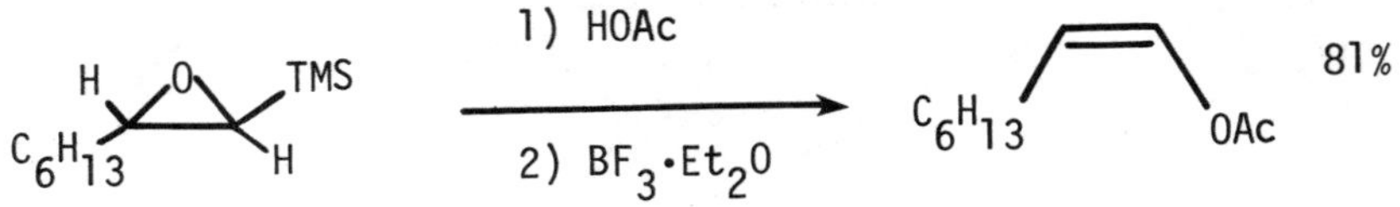

1) HOAc

2) $BF_3 \cdot Et_2O$

81%

JACS <u>99</u>, 1993 (1977)

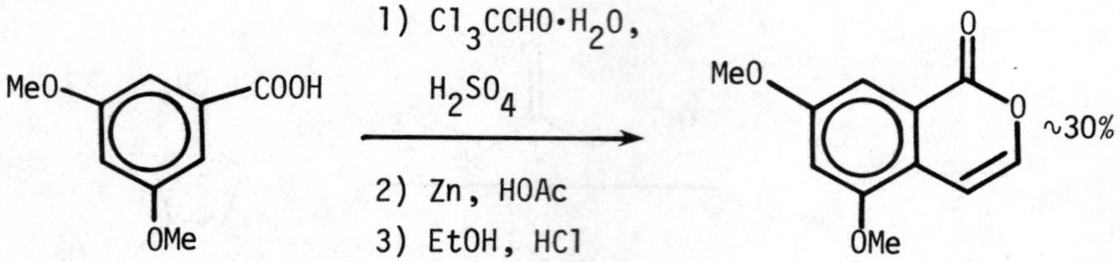

Indian J Chem 15B, 103 (1977)

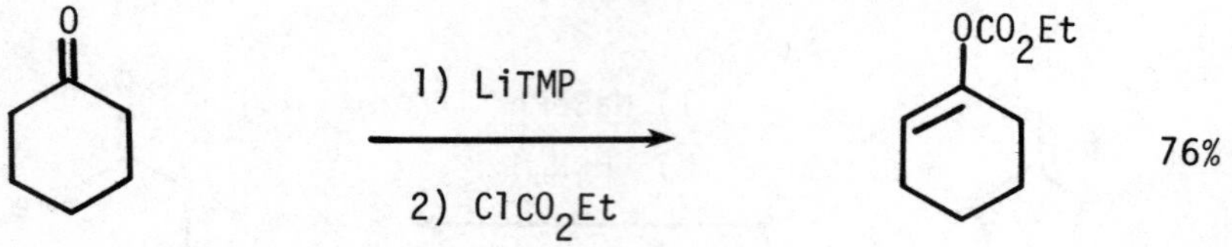

JOC 43, 2073 (1978)

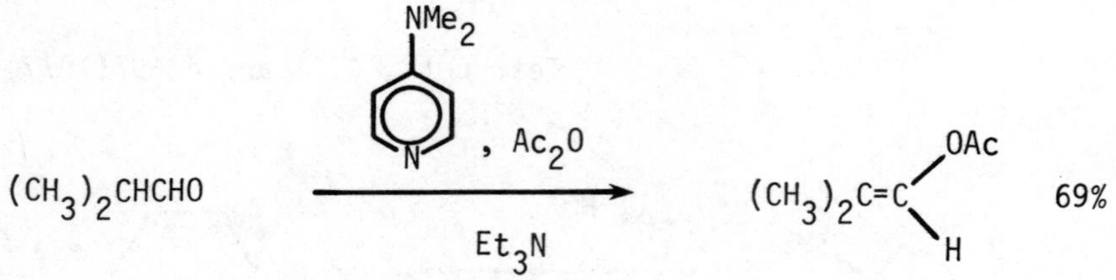

Synth Comm 9, 157 (1979)

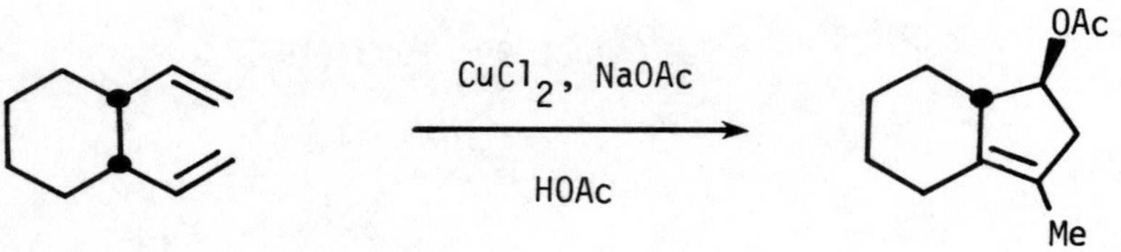

Angew Int Ed 18, 866 (1979)

Also via: Acetylenic esters (Section 306)
             Olefinic acids (Section 322)
             β-Hydroxyesters (Section 327)

Section 363   Ether - Ether

See Section 60A (Protection of Aldehydes) and Section 180A
(Protection of Ketones) for reactions involving the formation of
acetals and ketals.

$$CH(OEt)_2 \quad \xrightarrow[\text{THF}]{LiAlH_4-TiCl_4} \quad 98\%$$

BCS Japan 51, 2059 (1978)

Section 364   Ether - Halide

$$\xrightarrow[Cu(OAc)_2, \; AcOH]{I_2, \; BzOH} \quad 90\%$$

Synthesis, 402 (1978)

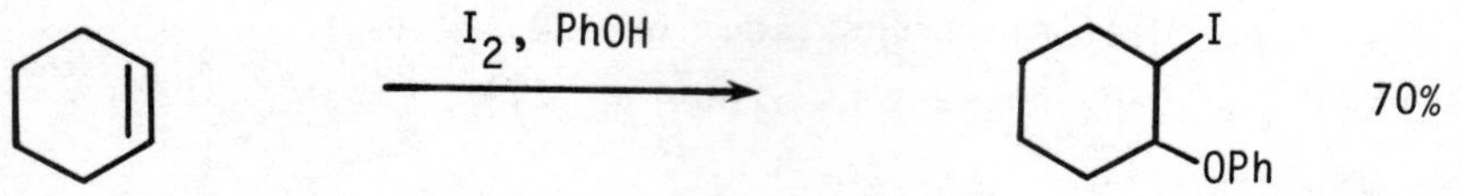

I_2, PhOH

70%

Synthesis, 67 (1979)

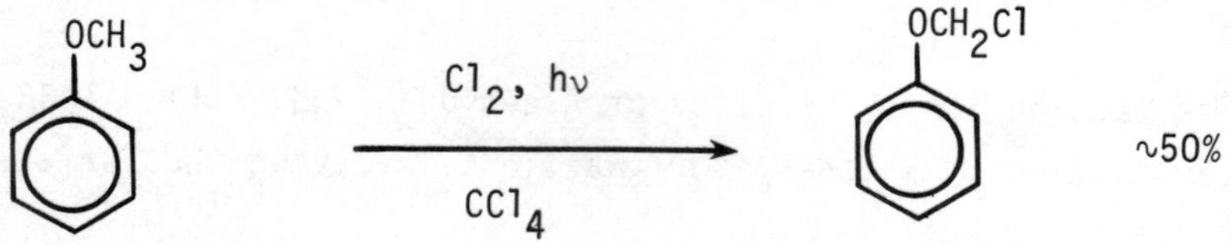

Cl_2, hν

CCl_4

~50%

Chem and Ind, 127 (1977)

Section 365   Ether, Epoxide - Ketone

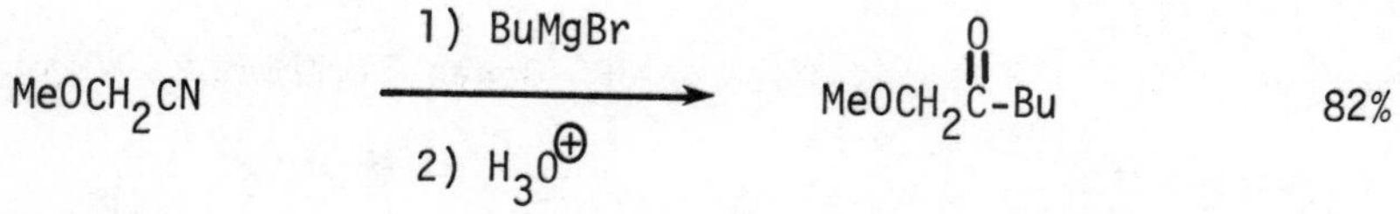

$MeOCH_2CN$

1) BuMgBr

2) $H_3O^{\oplus}$

$MeOCH_2\overset{O}{\overset{\|}{C}}-Bu$

82%

Tetr Lett, 23 (1977)

$AlCl_3 \cdot NaCl$

99%

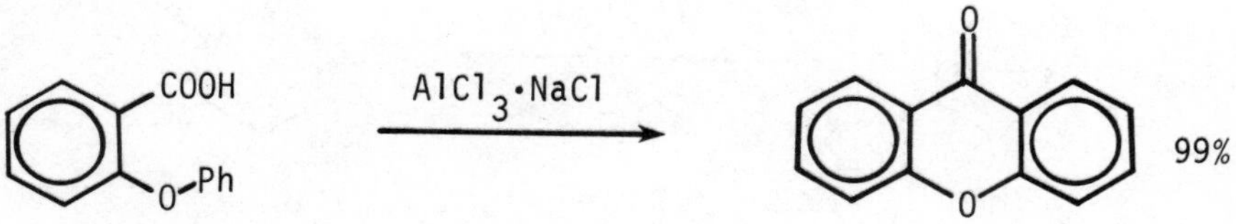

JOC 44, 3724 (1979)

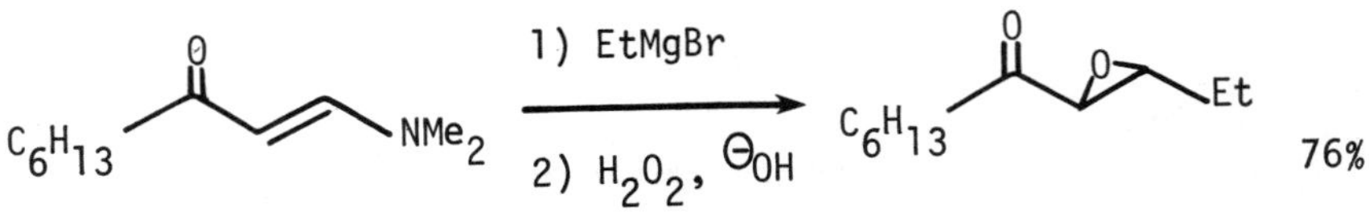

JOC (USSR) 13, 1062 (1977)

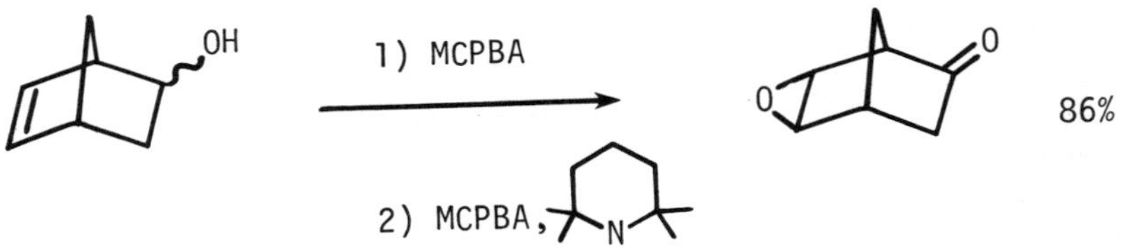

JOC 42, 2077 (1977)

Section 366   Ether, Epoxide - Nitrile

No additional examples

Section 367   Ether - Olefin

Related methods: Protection of Ketones (Section 180A).

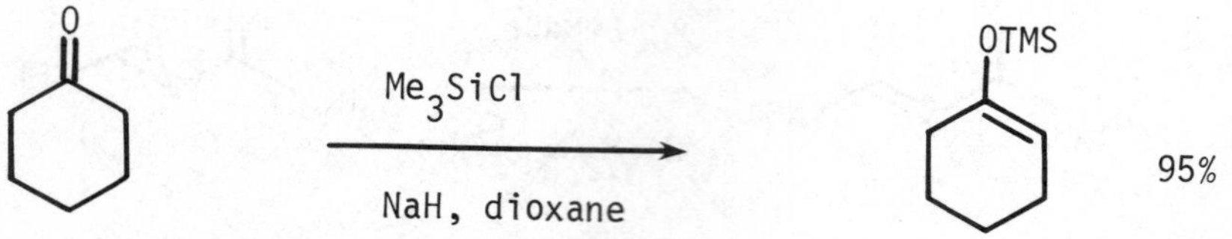

JOC 43, 3861 (1978)

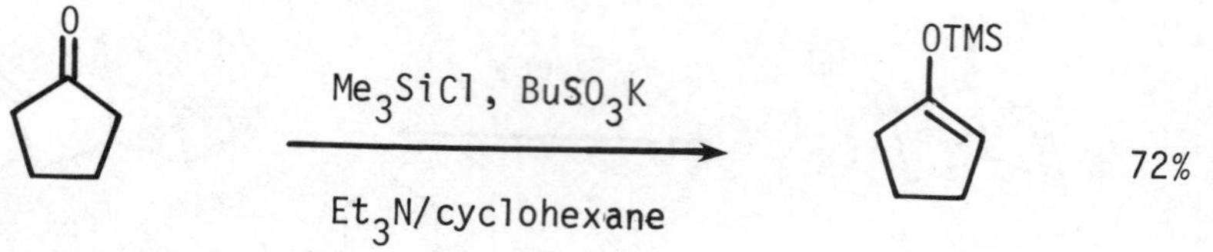

Synthesis, 34 (1979)

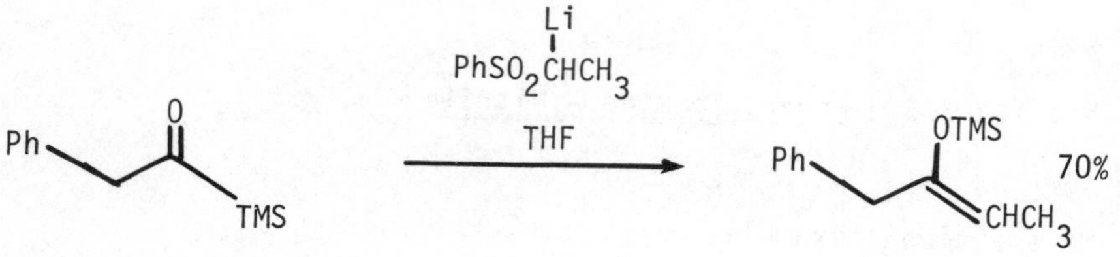

JACS 101, 2225 (1979)

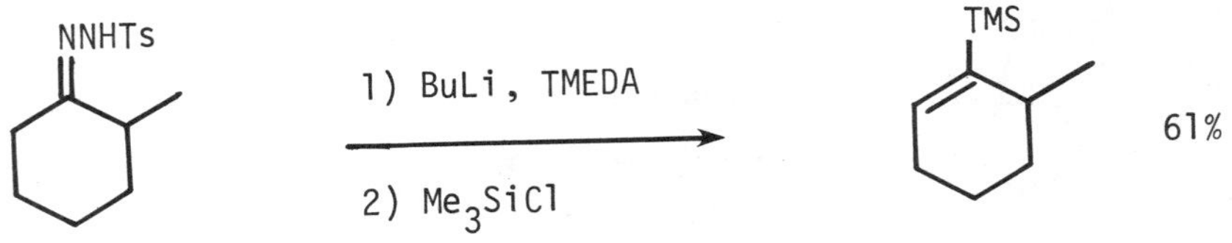
NNHTs
1) BuLi, TMEDA
2) Me3SiCl
TMS
61%

Tetr Lett, 159 (1977)

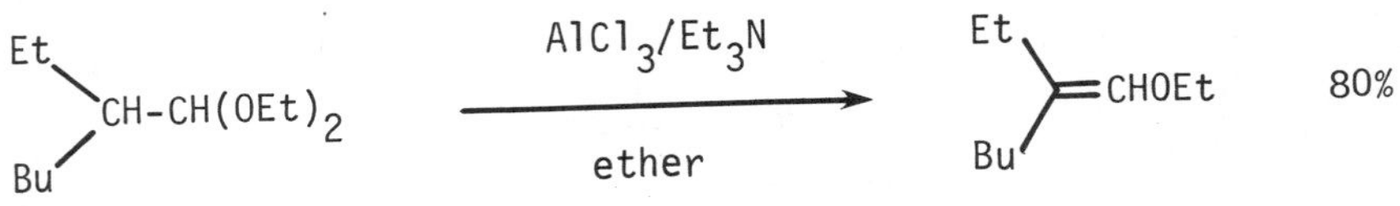
Et
CH-CH(OEt)2
Bu
AlCl3/Et3N
ether
Et
CHOEt
Bu
80%

Helv Chim Acta 62, 1451 (1979)

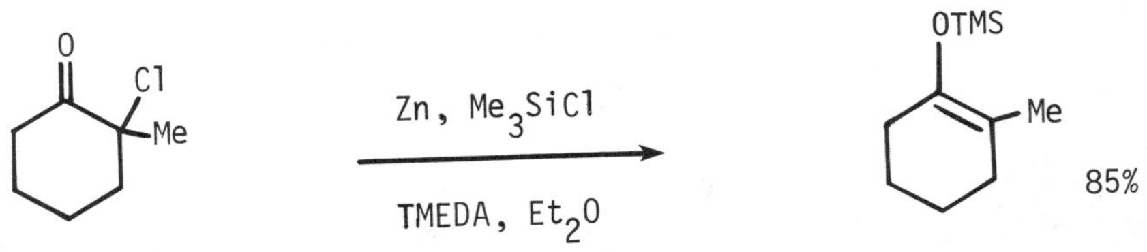
O
Cl
Me
Zn, Me3SiCl
TMEDA, Et2O
OTMS
Me
85%

Synth Comm 7, 327 (1977)

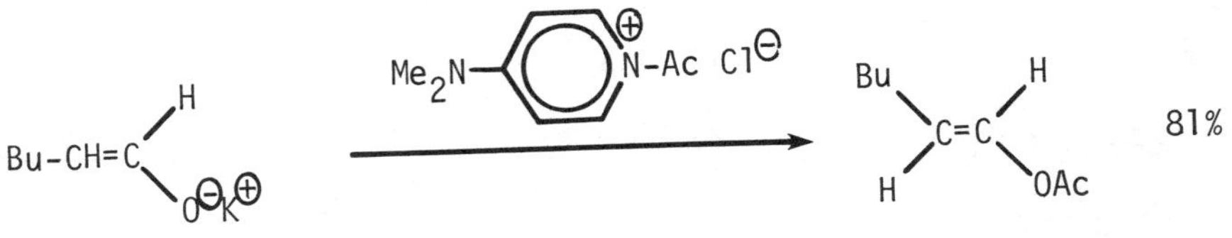
Bu-CH=C
H
O K
Me2N
N-Ac Cl
Bu
H
C=C
H
OAc
81%

Synthesis, 504 (1979)

Bz-O-CH-COOH
|
—OH

$\xrightarrow[\text{pyridine}]{\text{PhSO}_2\text{Cl}_2}$

H OBz
\ /
C

93%

Synthesis, 388 (1979)

$\xrightarrow[\text{2) NaH, THF}]{\text{1) Ph}_2\overset{\text{O}}{\overset{\|}{\text{P}}}-\text{CH}_2\text{OMe}}$

HCOMe

57%

JCS Chem Comm, 314 (1977)

$\text{Ph}_2\overset{\text{O}}{\overset{\|}{\text{P}}}-\underset{\text{OMe}}{\text{CH}}-\text{Li}$ + $\underset{\text{Me}}{\overset{\text{O}}{\overset{\|}{\text{C}}}}\text{Ph}$ $\xrightarrow{\text{2) NaH}}$

MeO   H
\   /
C
‖
C
/   \
Me   Ph

89%

JCS Perkin I, 3099 (1979)

$\xrightarrow[\substack{\text{2) PhCOONa} \\ \text{3) }\underline{t}\text{-AmONO}}]{\text{1) Ph}_3\text{P}}$

—CH₂Br

—CH₂Br

Ph

75%

Tetr Lett, 2149 (1979)

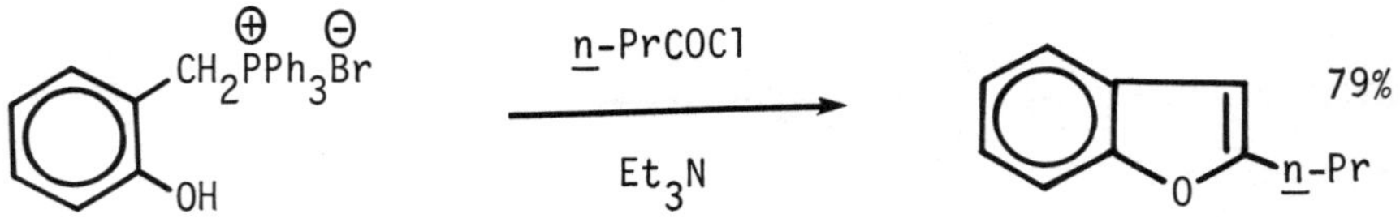

Tetr Lett, 2145 (1979)

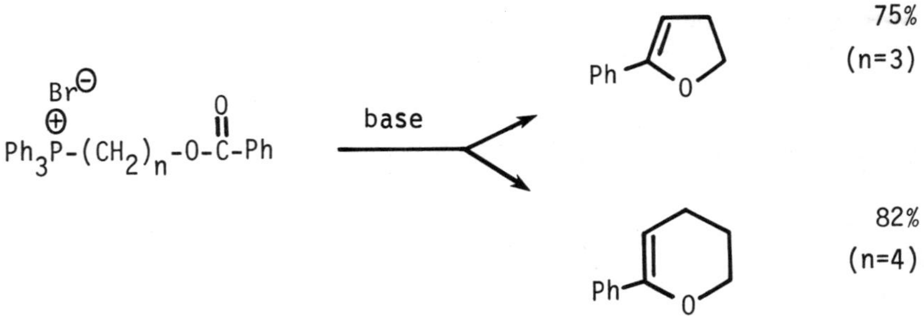

Tetr Lett, 5 (1979)

Tetr Lett, 2995 (1979)

$$ClCH_2CH_2OH \xrightarrow[\text{Hg(OAc)}_2]{\text{CH}_2\text{=CHOAc, H}_2\text{SO}_4} ClCH_2CH_2OCH=CH_2 \quad \sim45\%$$

JOC (USSR) <u>13</u>, 606 (1977)

Section 368   <u>Halide - Halide</u>

Halocyclopropanations are found in Section 74 (Alkyls from Olefins).

$$Ph-CHO \xrightarrow[\text{DMF}]{\text{SOCl}_2} Ph-\overset{\text{Cl}}{\underset{\text{H}}{C}}-Cl \quad 90\%$$

JOC <u>43</u>, 4367 (1978)

$$\xrightarrow{\text{Et}_2\text{NSF}_3}$$

up to ~90%

JCS Perkin I, 1354 (1979)

Et-C≡C-Et $\dfrac{\text{pyridine} \cdot (HF)_n}{THF}$ (difluorohexane, F F)   75%

JOC <u>44</u>, 3872 (1979)

(cyclohexene) $\dfrac{\text{HCl, } H_2O_2}{\text{phase-transfer cat.}}$ (1,2-dichlorocyclohexane, Cl Cl)   76%

Synthesis, 676 (1977)

Ph-CH=CHCH$_3$ $\xrightarrow{XeF_2}$ Ph-CH-CH-CH$_3$ (F F)   60%

JOC <u>42</u>, 1559 (1977)

(cyclohexene oxide) $\dfrac{\text{2-chloro-3-ethyl-benzoxazolium } BF_4^{\ominus}}{Et_4N^{\oplus}Cl^{\ominus}, \ Et_3N}$ (trans-1,2-dichlorocyclohexane, Cl Cl)   58%

Chem Lett, 1013 (1977)

Section 369   Halide - Ketone

JCS Perkin I, 501 (1977)

$$Me-\overset{O}{\overset{\|}{C}}-CH_2Me \xrightarrow{NBS} Me-\overset{O}{\overset{\|}{C}}-CBr_2Me \qquad 95\%$$

$$Me-\overset{O}{\overset{\|}{C}}-\underset{\underset{Cl}{|}}{C}H-Me \xrightarrow{NBS} Me-\overset{O}{\overset{\|}{C}}-\underset{\underset{Cl}{|}}{C}Br-Me \qquad 95\%$$

JOC 42, 3527 (1977)

$$Ph-\overset{O}{\overset{\|}{C}}-CH_2Ph \xrightarrow[\text{hv, benzene}]{PhICl_2} Ph-\overset{O}{\overset{\|}{C}}-\underset{\underset{Cl}{|}}{C}HPh \qquad 80\%$$

JOC (USSR) 14, 1414 (1978)

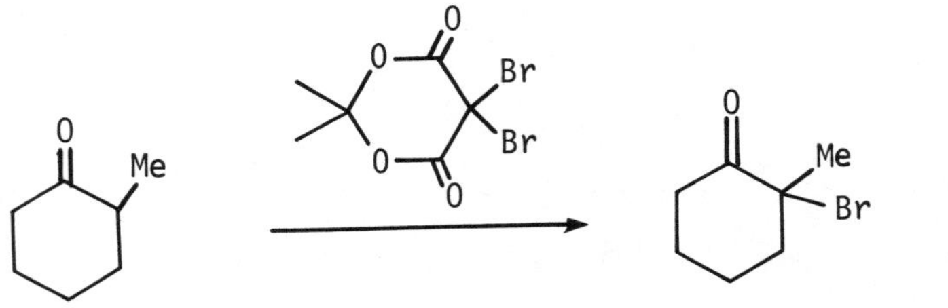

Synthesis, 140 (1978)

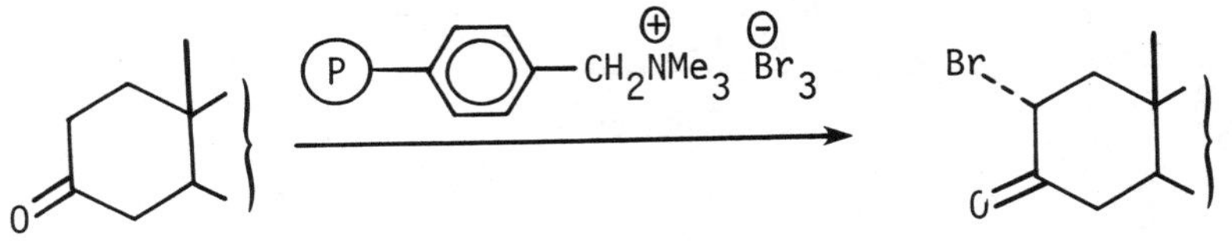

Synthesis, 64 (1979)

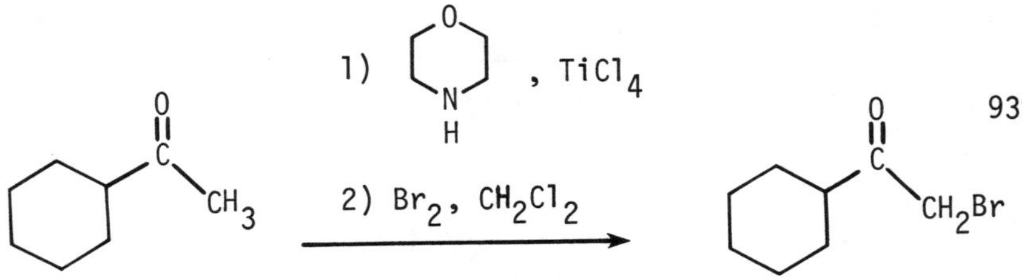

Acta Chem Scand B32, 646 (1979)

$$\underset{\substack{\| \quad \quad \| \\ \text{Me-C-CH}_2\text{-C-Me}}}{\overset{\text{O} \quad \quad \text{O}}{}} \quad \xrightarrow[\text{DBU}]{\text{CF}_3\text{SO}_2\text{Cl}} \quad \underset{\substack{\| \quad \quad \| \\ \text{Me-C-CCl}_2\text{-C-Me}}}{\overset{\text{O} \quad \quad \text{O}}{}} \quad 100\%$$

Tetr Lett, 3643 (1979)

CF$_3$COONa

F$_2$, Freon

85%

JCS Chem Comm, 479 (1979)

Tetr Lett, 725 (1979)

1) CF$_3$CF$_2$OF

2) Zn, HOAc

85%

JACS **101**, 2782 (1979)

I$_2$, TlOAc

HOAc

46%

JCS Perkin I, 126 (1978)

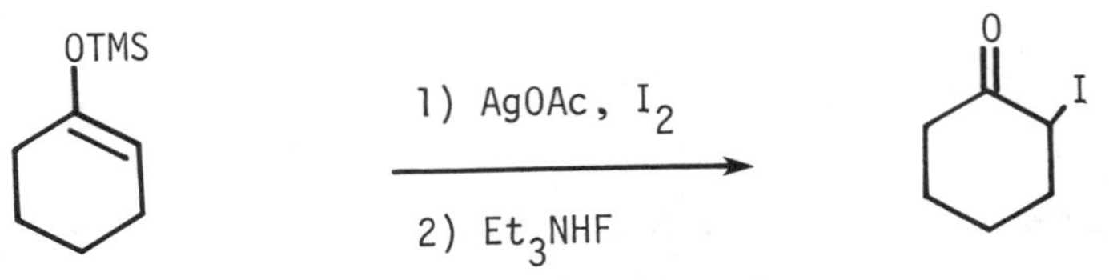

1) AgOAc, I$_2$

2) Et$_3$NHF

84%

JOC $\underline{44}$, 1731 (1979)

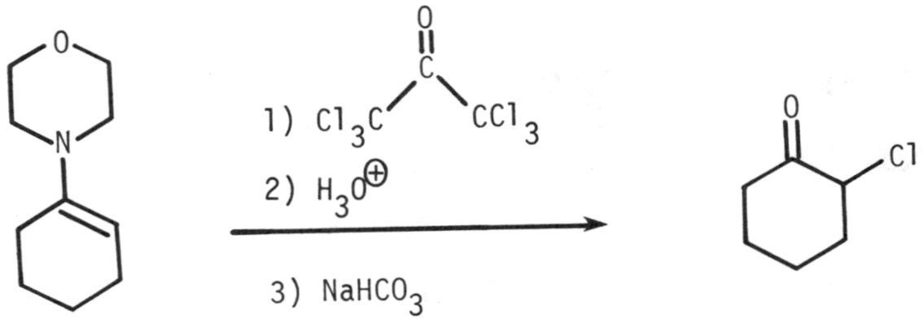

1) Cl$_3$C—C(=O)—CCl$_3$

2) H$_3$O$^{\oplus}$

3) NaHCO$_3$

65-90%

JACS $\underline{99}$, 6672 (1977)
Tetr Lett, 759 (1977)

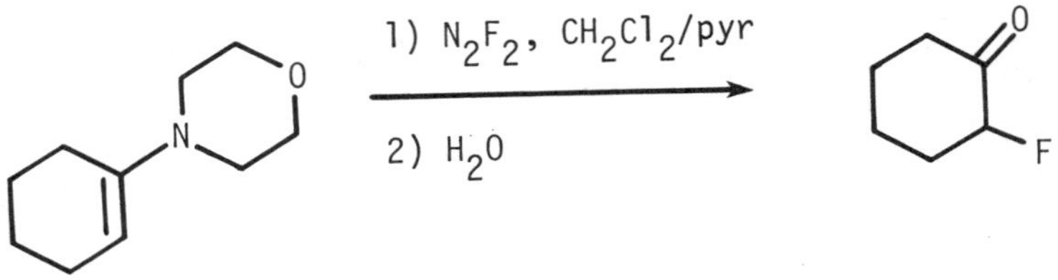

1) N$_2$F$_2$, CH$_2$Cl$_2$/pyr

2) H$_2$O

53%

Tetr Lett, 2797 (1977)

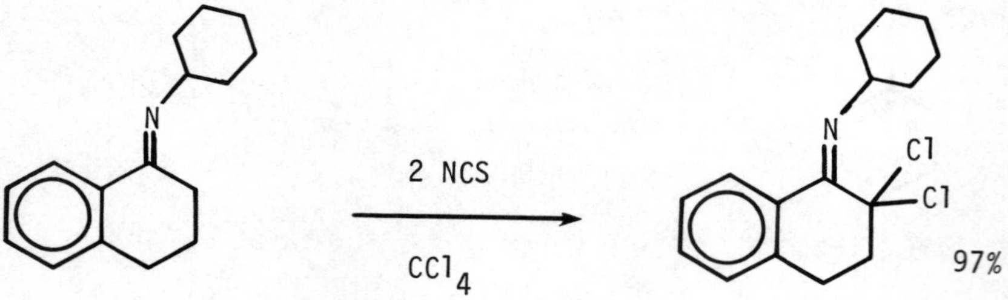

Synth Comm <u>8</u>, 75 (1978)

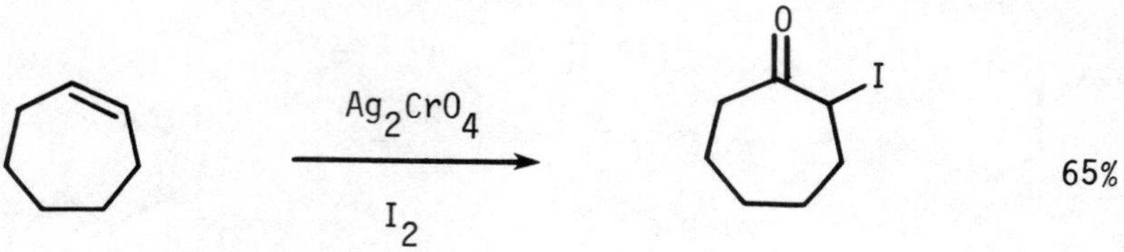

JOC <u>42</u>, 4268 (1977)

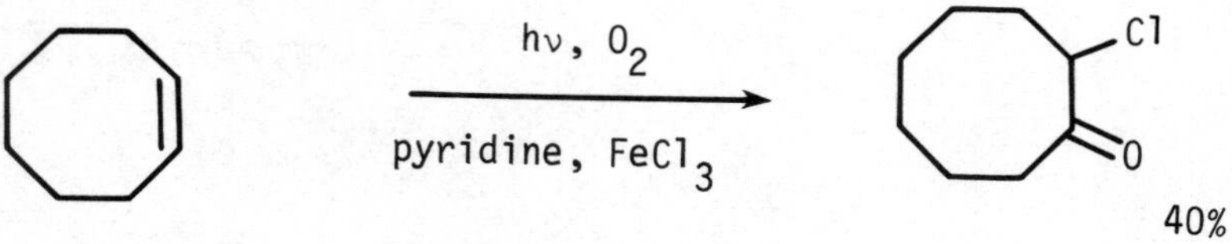

Chem Lett, 161 (1978)

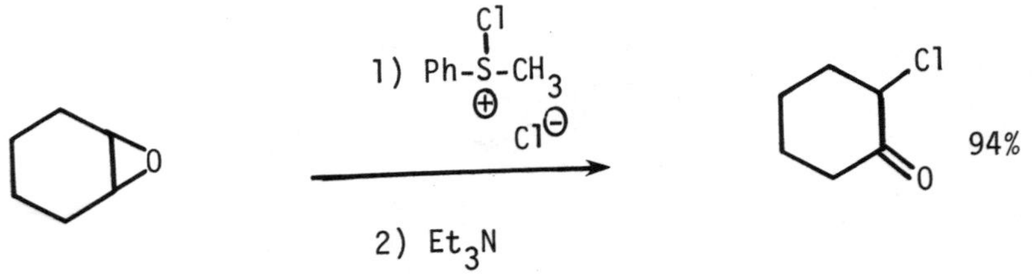

Chem Lett, 995 (1977)

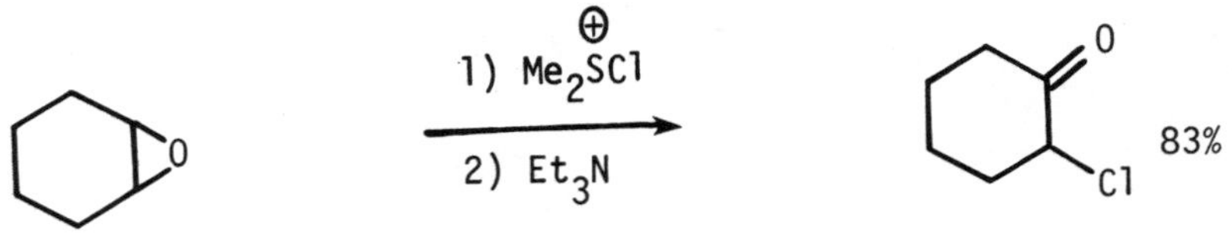

Tetr Lett, 3653 (1979)

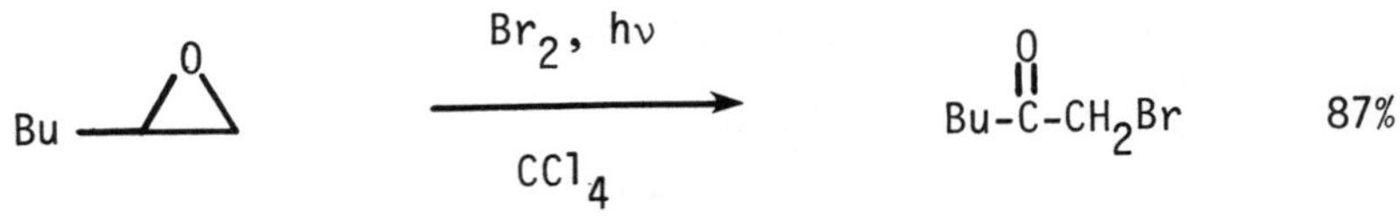

Synthesis, 139 (1978)

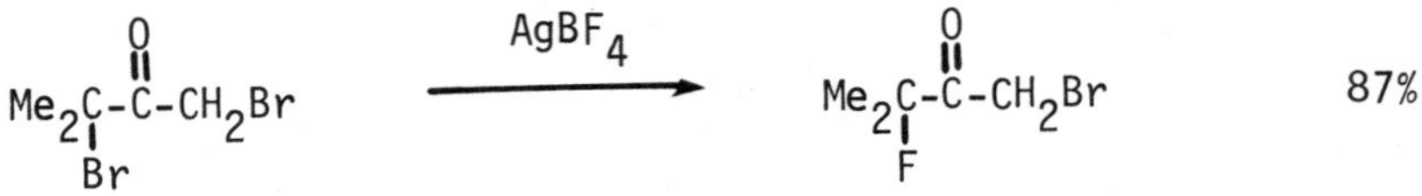

Tetr Lett, 3357 (1979)

70%

JOC <u>43</u>, 2879 (1978)

85%

JOC <u>42</u>, 459 (1977)

$$Ph-\overset{O}{\overset{||}{C}}-Cl \quad \xrightarrow[\;(CF_3)_2CFZnI\;]{\text{pyridine/THF, } ZnF_2} \quad Ph-\overset{O}{\overset{||}{C}}-CF(CF_3)_2 \qquad 89\%$$

Also works with anhydrides.

Chem Lett, 81 (1977)

$$Bu-CHO \quad \xrightarrow[\;2)\;\Delta\;]{1)\; Ph-\overset{O}{\overset{||}{S}}-\overset{Li}{\underset{|}{C}}HX} \quad Bu-\overset{O}{\overset{||}{C}}-CH_2Cl \qquad 81\%$$

(X=Cl, Br)

Tetr Lett, 1225 (1977)
Chem Lett, 209 (1979)

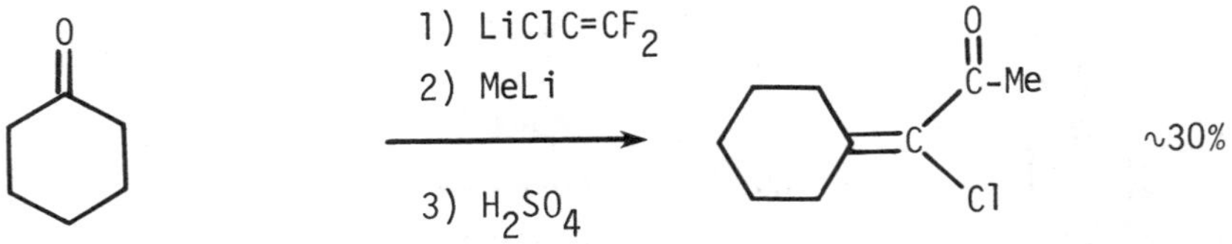

Synthesis, 458 (1978)

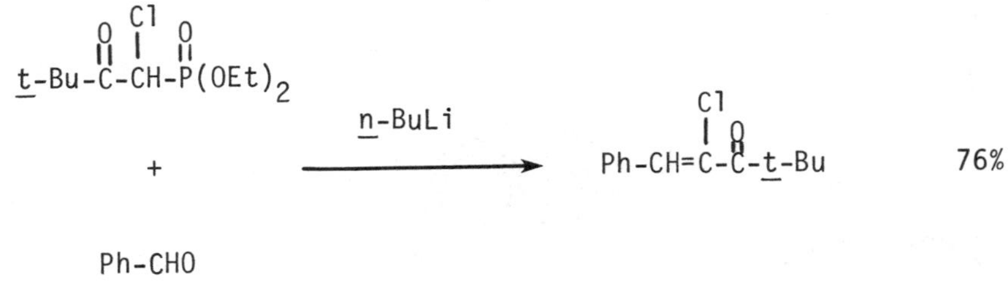

$$t\text{-Bu-C-CH-P(OEt)}_2$$

with Cl and O substituents shown, + Ph-CHO, n-BuLi →

$$\underset{\text{Cl}}{\text{Ph-CH=C}}\text{-C-}\underline{t}\text{-Bu}$$ 76%

Synthesis, 29 (1978)

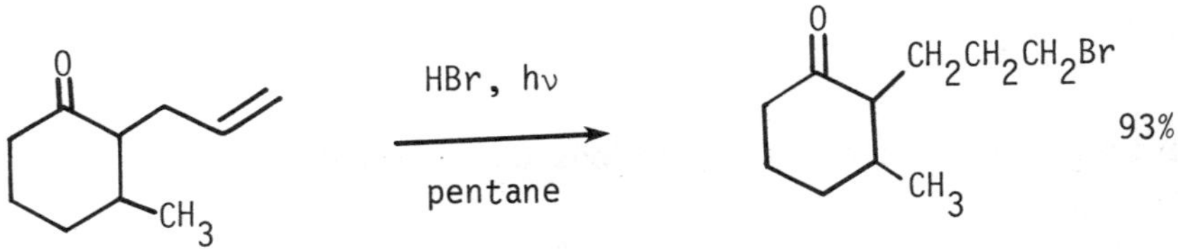

HBr, hν

pentane

→   93%

Synthesis of several additional vinyl ketones and their conversion to ω-bromoketones are also presented.

JOC **42**, 1709 (1977)

Section 370   Halide - Nitrile

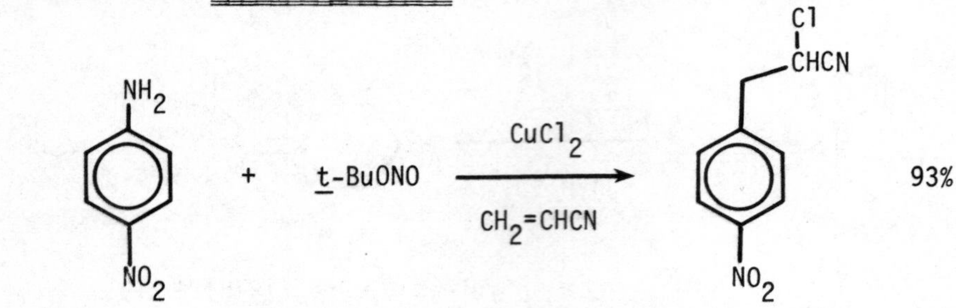

JOC 42, 2431 (1977)

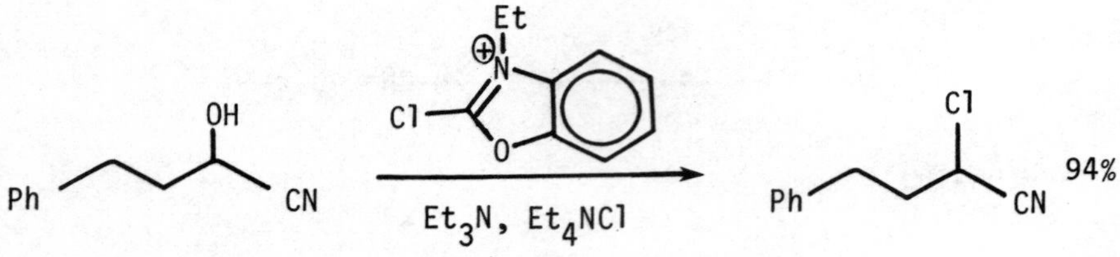

Chem Lett, 1117 (1979)

Review: "Preparation and Reactions of Chloro-derivatives of Nitriles"

Russ Chem Rev 48, 282 (1979)

Section 371   Halide - Olefin

Ph-C≡C-Et $\xrightarrow[\text{MeCN}]{\text{CuCl}_2\text{-LiCl}}$ Ph, Cl / Cl, Et C=C   94%

$\xrightarrow[\text{MeCN}]{\text{CuCl}_2\text{-I}_2}$ Ph, I / Cl, Et C=C   100%

JCS Perkin I, 676 (1977)

Ph-C≡CH $\xrightarrow[\text{2) NBS}]{\text{1) [EtCuBr]MgBr}}$ Ph, H / Et, Br C=C   >90%

Rec Trav Chim 96, 168 (1977)

Bu-C≡CH $\xrightarrow[\text{2) I}_2\text{, THF}]{\text{1) Cp}_2\text{ZrCl}_2\text{, Me}_3\text{Al}}$ Bu, H / Me, I C=C   85%

Synthesis, 501 (1979)

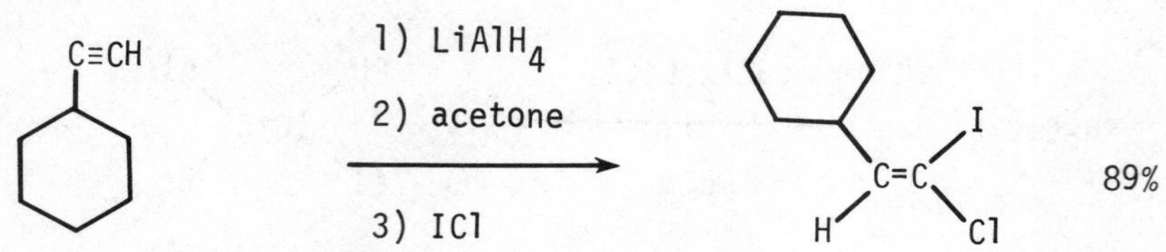

JACS 101, 5101 (1979)

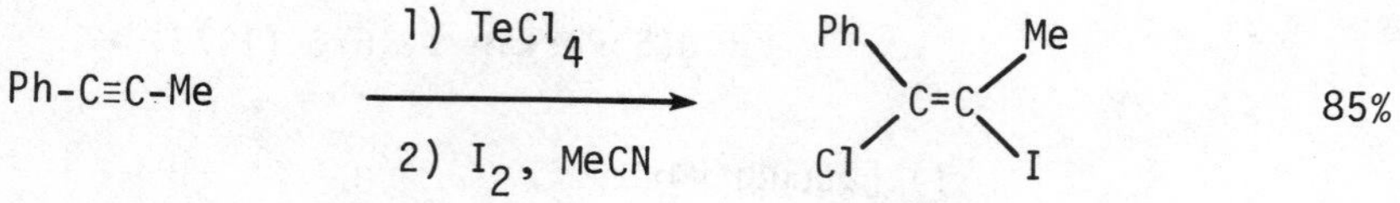

Chem Lett, 1357 (1979)

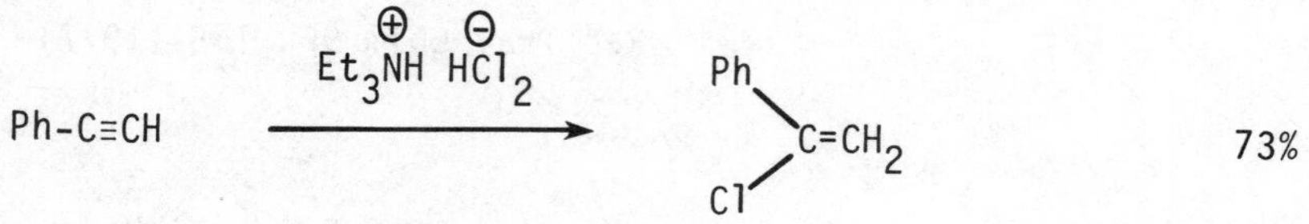

JCS Perkin I, 1797 (1977)

$$\underset{\substack{\text{Et} \\ \cdot \text{MgBr}}}{\overset{\text{Bu}}{\diagdown}} C=C \overset{\text{H}}{\underset{\text{CuEt}}{\diagup}} \quad \xrightarrow[\text{THF}]{I_2} \quad \underset{\text{Et}}{\overset{\text{Bu}}{\diagdown}} C=C \overset{\text{H}}{\underset{\text{I}}{\diagup}} \qquad >90\%$$

Rec Trav 96, 168 (1977)

$$\underset{\text{Et}}{\overset{\text{Pr}}{\diagdown}} C=C \overset{\text{Cu} \cdot \text{MgBr}}{\underset{\text{H}}{\diagup}} \quad \xrightarrow[\text{or NCS (x=Cl)}]{\text{NBS (X=Br)}} \quad \underset{\text{Et}}{\overset{\text{Pr}}{\diagdown}} C=C \overset{\text{X}}{\underset{\text{H}}{\diagup}} \qquad \sim 90\%$$

Tetr Lett, 3545 (1977)

$$\text{Bu-CH=CHSiMe}_3 \quad \xrightarrow[\text{2) alumina, pentane}]{\text{1) Br}_2} \quad \text{BuCH=CHBr} \qquad 74\%$$

Synth Comm 7, 475 (1977)

$$\underset{\text{H}}{\overset{\underline{t}\text{-Bu}}{\diagdown}} C=C \overset{\text{TMS}}{\underset{\text{H}}{\diagup}} \quad \xrightarrow[\substack{\text{2) KF, Me}_2\text{SO} \\ \text{or NaOMe, MeOH} \\ \text{or Al}_2\text{O}_3\text{, Pentane}}]{\text{1) Br}_2} \quad \underset{\text{H}}{\overset{\underline{t}\text{-Bu}}{\diagdown}} C=C \overset{\text{Br}}{\underset{\text{H}}{\diagup}} \qquad 70\%$$

JOC 43, 4424 (1978)

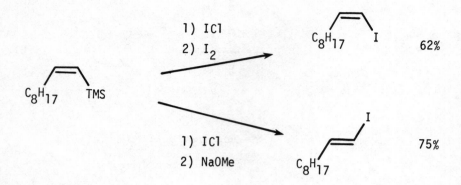

1) ICl
2) I$_2$
C$_8$H$_{17}$ I   62%

1) ICl
2) NaOMe
C$_8$H$_{17}$ I   75%

Tetr Lett, 1073 (1979)

C$_6$H$_{13}$-CH=CH$_2$

1) PhSeBr
2) O$_3$
3) i-Pr$_2$NH

C$_6$H$_{13}$-C=CH$_2$ 85%
Br

Tetr Lett, 3909 (1977)

CH=CH-CMe$_3$

1) CH$_3$CNHBr, HF
2) KOH

C=CH-CMe$_3$ 74%
F

OMe

OMe

Synthesis, 217 (1978)

90%

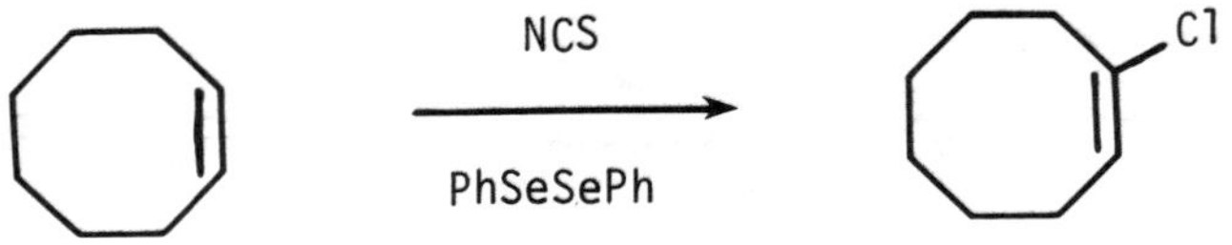

Several other cases give allylic chlorides.

JOC 44, 4204 (1979)

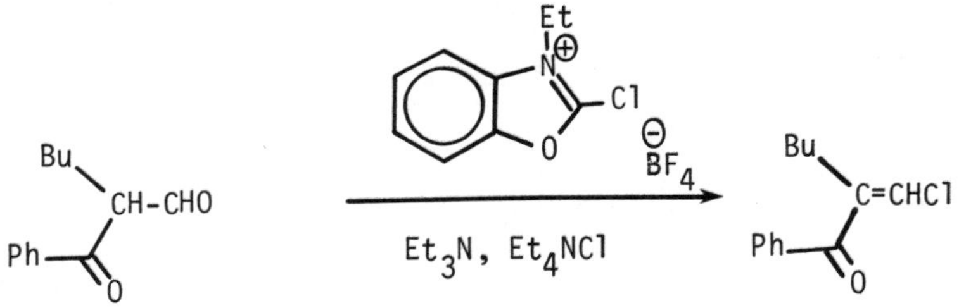

82%

Chem Lett, 465 (1978)

$$C_5H_{11}CHO \xrightarrow{(Me_2N)_3P=CCl_2} C_5H_{11}CH=CCl_2$$

94%

Tetr Lett, 1239 (1977)

94%

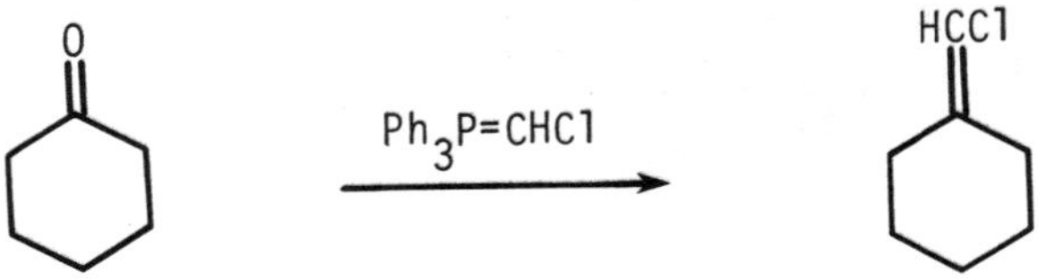

JCS Chem Comm, 446 (1978)

$$n\text{-}C_6H_{13}CHO \xrightarrow[\text{triglyme}]{CBr_2F_2-(Me_2N)_3P} n\text{-}C_6H_{13}CH=CF_2 \qquad 80\%$$

Chem Lett, 983 (1979)

$$Ph\text{-}CHO \xrightarrow{[Ph_3\overset{\oplus}{P}-\overset{\ominus}{C}(CF_3)_2] \text{ in situ}} PhCH=C(CF_3)_2 \quad 50\%$$

Tetr Lett, 3397 (1979)

$$CH_3\text{-}(CH_2)_{14}\text{-}CH_2Br \xrightarrow[\text{2) } \Delta,\text{ xylene}]{\text{1) } Ph\overset{O}{\overset{\|}{S}}-\overset{Li}{\overset{|}{C}}H\text{-}Cl} CH_3(CH_2)_{14}\text{-}CH=CHCl$$

80%

Tetr Lett, 617 (1979)

85%

JACS 99, 1993 (1977)

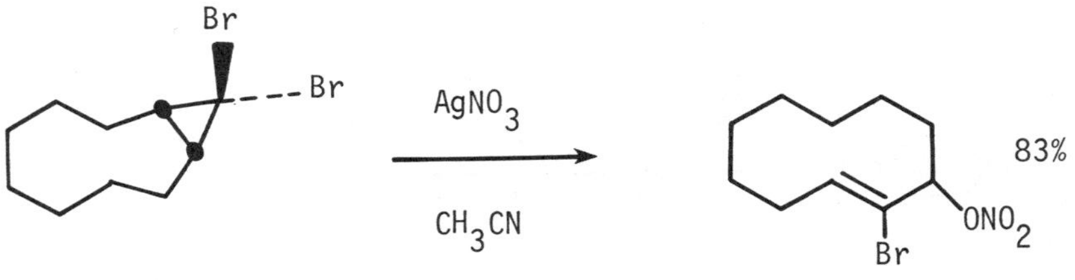

Br
Br
AgNO3
CH3CN
ONO2
Br
83%

JOC 42, 418 (1977)

O
Cl3C-C-CCl3
Ph3P
CH2OH
CH2Cl
100%

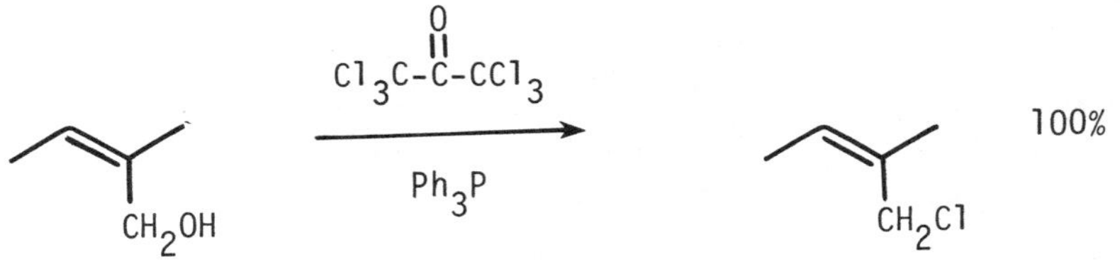

JOC 44, 359 (1979)

CH2=CH-CH-CH3
OH
PPh3
Cl3C-C-CCl3
O
CH2=CH-CH-CH3
Cl
94%

Tetr Lett, 2999 (1977)

CCl4
Ru(II)complex
Cl
Cl
83%

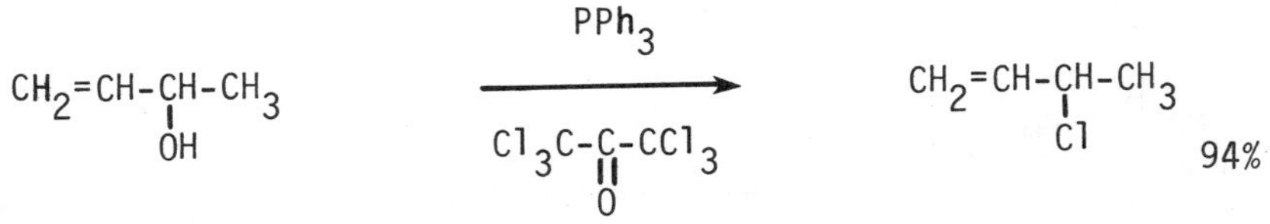

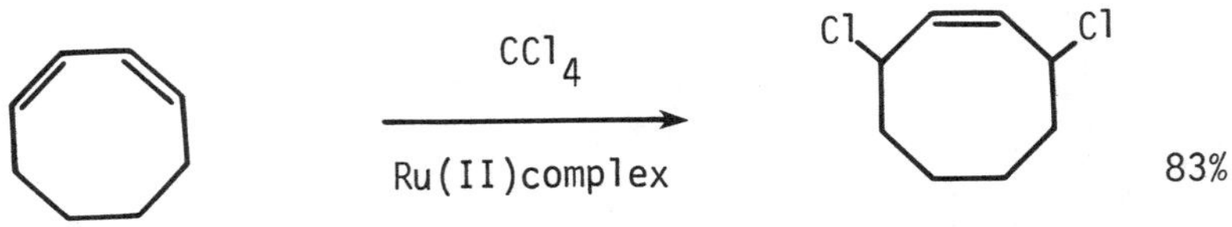

Chem Lett, 115 (1978)

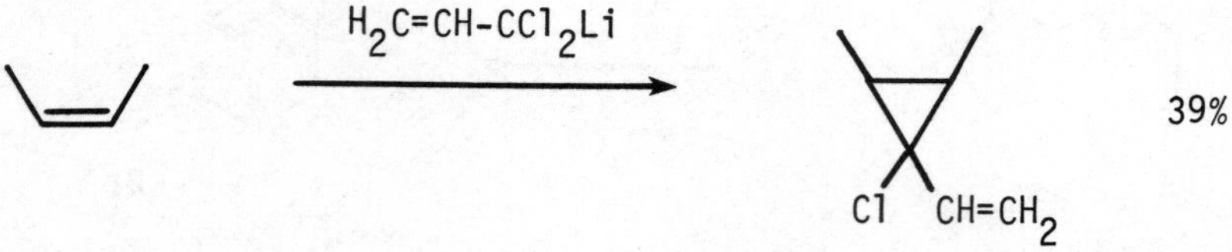

$$H_2C=CH-CCl_2Li$$

39%

Synthesis, 425 (1979)

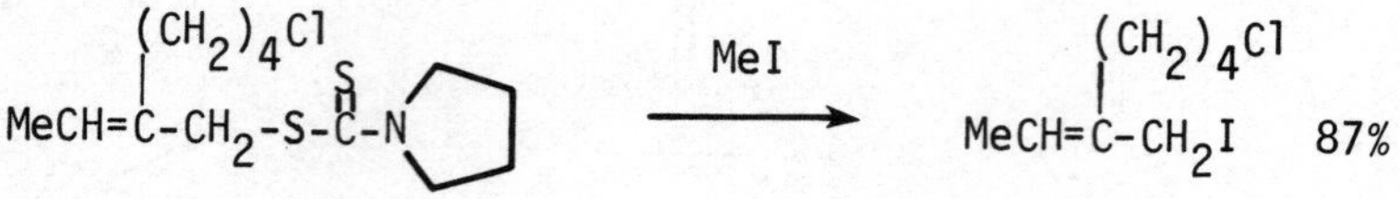

MeI

87%

Synthesis, 370 (1978)

ClCOOEt

91%

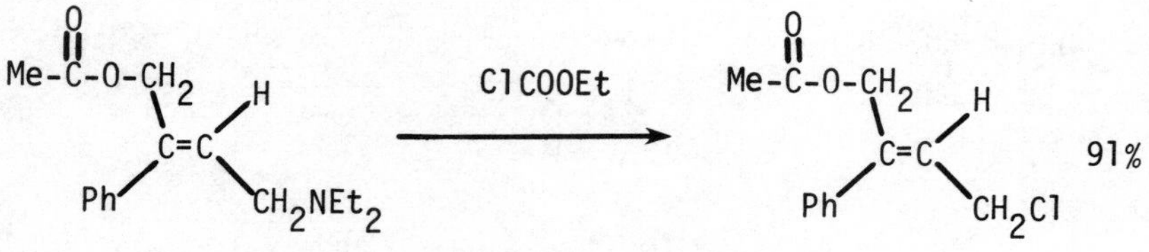

Synthesis, 786 (1977)

Section 372   Ketone - Ketone

$$\underset{\text{Ph-C-CH-Ph}}{\overset{\text{O}\quad\text{OH}}{}} \quad \xrightarrow[\text{NaOH, MeOH}]{\text{K}_3\text{Fe(CN)}_6} \quad \underset{\text{Ph-C-C-Ph}}{\overset{\text{O}\quad\text{O}}{}} \qquad 95\%$$

Chem and Ind, 741 (1977)

$$\underset{\text{CH}_3\text{CHBrC-C}_8\text{H}_{17}}{\overset{\text{O}}{}} \quad \xrightarrow[\text{2) HCl}]{\text{1) PhNHSO}_2\text{CF}_3, \text{ K}_2\text{CO}_3} \quad \underset{\text{CH}_3\text{C-C-C}_8\text{H}_{17}}{\overset{\text{O}\quad\text{O}}{}} \qquad 92\%$$

JOC 44, 1835 (1979)

$$\underset{\text{Ph-C-CH}_2\text{Ph}}{\overset{\text{O}}{}} \quad \xrightarrow[\text{3) HCl/H}_2\text{O}]{\begin{array}{l}\text{1) SOCl}_2, \text{ pyr} \\ \text{2) Morpholine}\end{array}} \quad \underset{\text{Ph-C-C-Ph}}{\overset{\text{O}\quad\text{O}}{}} \qquad 80\%$$

Tetr Lett, 695 (1977)

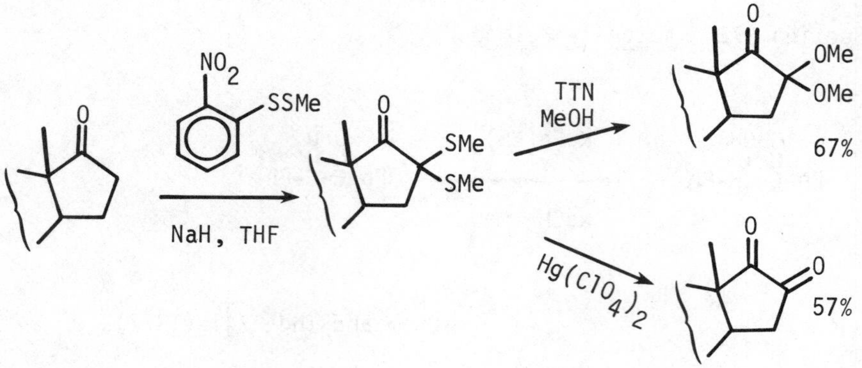

Tetr Lett, 5021 (1978)

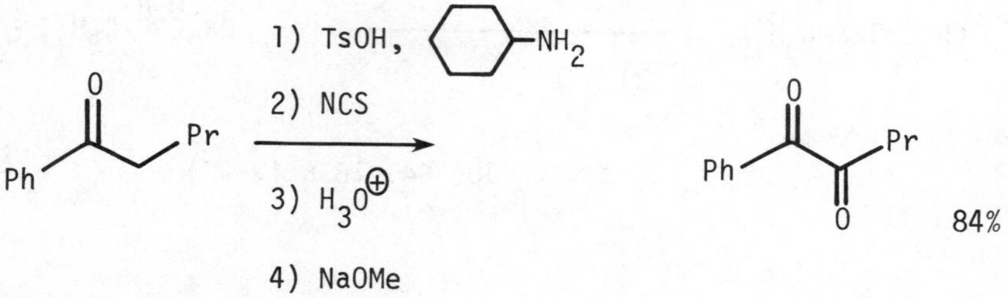

JOC 43, 2933 (1978)

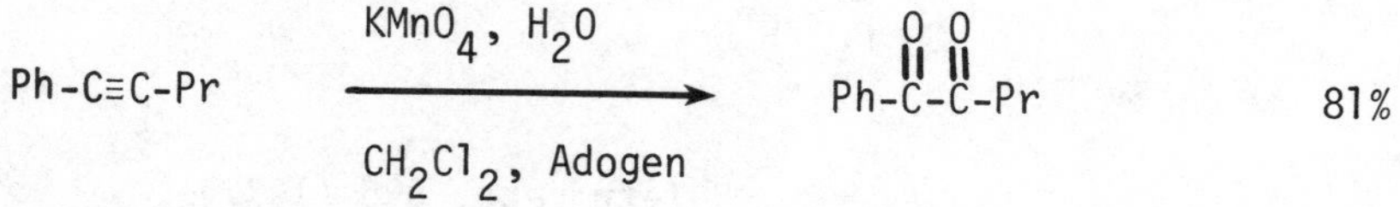

Synthesis, 462 (1978)

81%

$$CH_3(CH_2)_6-C\equiv C-(CH_2)_6CH_3 \xrightarrow[\text{acetone}]{KMnO_4} CH_3(CH_2)_6-\overset{O}{\overset{\|}{C}}-\overset{O}{\overset{\|}{C}}-(CH_2)_6CH_3$$

JOC **44**, 1574 (1979)

$$\begin{array}{c} Et \\ C \\ H \end{array} \begin{array}{c} S \\ S \end{array} \xrightarrow[\substack{3) \ EtI \\ 4) \ H_2O}]{\substack{1) \ BuLi \\ 2) \ Fe(CO)_5}} Et-\overset{O}{\overset{\|}{C}}-\overset{O}{\overset{\|}{C}}-Et \qquad 70\%$$

JCS Chem Comm, 691 (1977)

$$Ts-\underset{Ph}{CH}-N=C \xrightarrow[\substack{3) \ H_3O^{\oplus}}]{\substack{1) \ BuLi \\ 2) \ MeCOCl}} Ph-\overset{O}{\overset{\|}{C}}-\overset{O}{\overset{\|}{C}}-Me \qquad 73\%$$

Tetr Lett, 4233 (1977)

1) NaH, THF/HMPT
2) MeI

3) TTN, MeOH

53%

Tetr Lett, 4115 (1978)

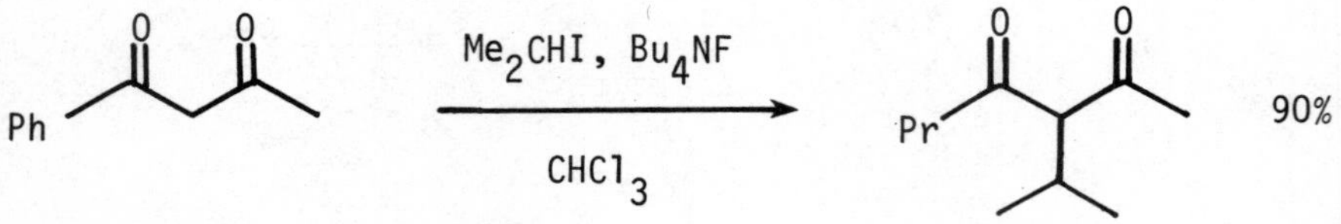

JCS Perkin I, 1743 (1977)

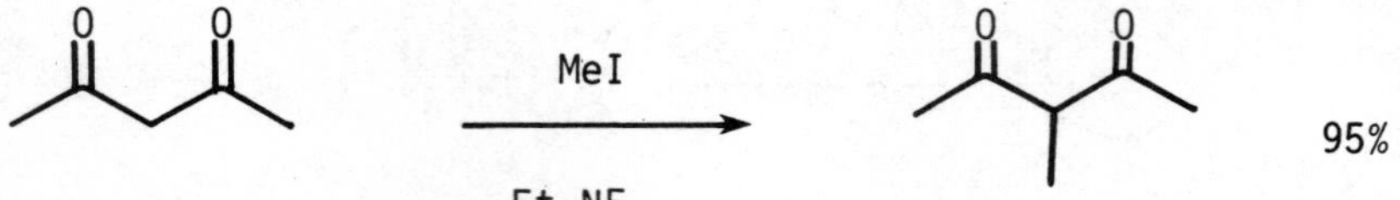

JCS Chem Comm, 64 (1977)

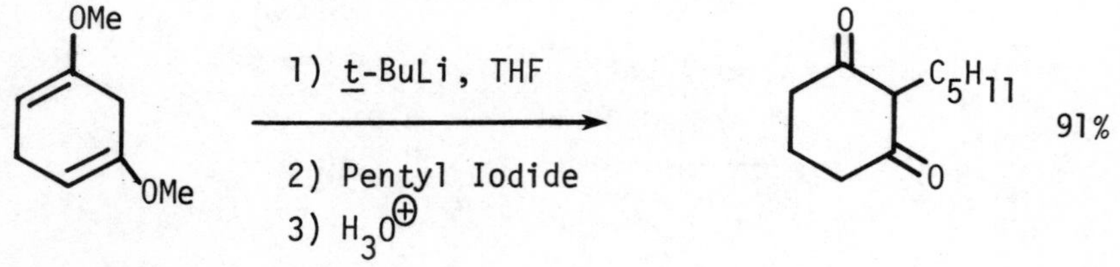

JOC 42, 3755 (1977)

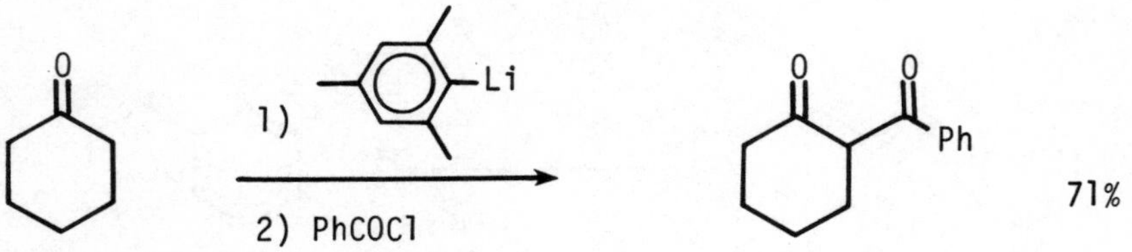

Tetr Lett, 1187 (1977)

81%

$$CH_3(CH_2)_6-C{\equiv}C-(CH_2)_6CH_3 \xrightarrow[\text{acetone}]{\text{KMnO}_4} CH_3(CH_2)_6-\overset{O}{\overset{\|}{C}}-\overset{O}{\overset{\|}{C}}-(CH_2)_6CH_3$$

JOC 44, 1574 (1979)

1) BuLi
2) Fe(CO)$_5$

$$Et-\overset{O}{\overset{\|}{C}}-\overset{O}{\overset{\|}{C}}-Et$$ 70%

3) EtI
4) H$_2$O

JCS Chem Comm, 691 (1977)

1) BuLi
2) MeCOCl

$$Ts-\underset{\underset{Ph}{|}}{C}H-N{=}C \xrightarrow{} Ph-\overset{O}{\overset{\|}{C}}-\overset{O}{\overset{\|}{C}}-Me$$ 73%

3) H$_3$O$^{\oplus}$

Tetr Lett, 4233 (1977)

1) NaH, THF/HMPT
2) MeI

53%

3) TTN, MeOH

Tetr Lett, 4115 (1978)

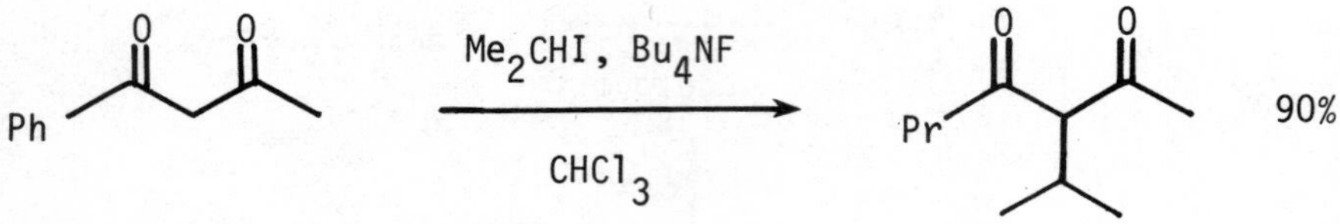

JCS Perkin I, 1743 (1977)

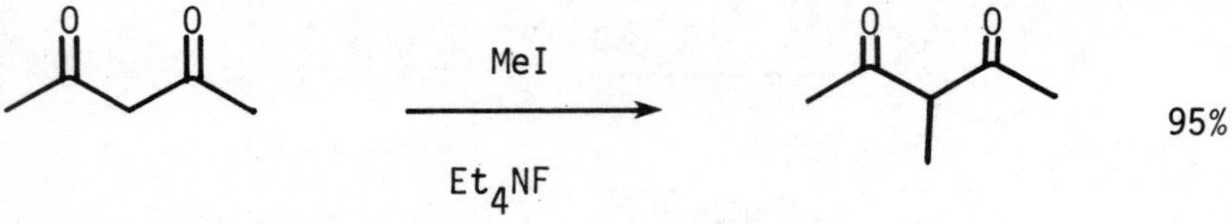

JCS Chem Comm, 64 (1977)

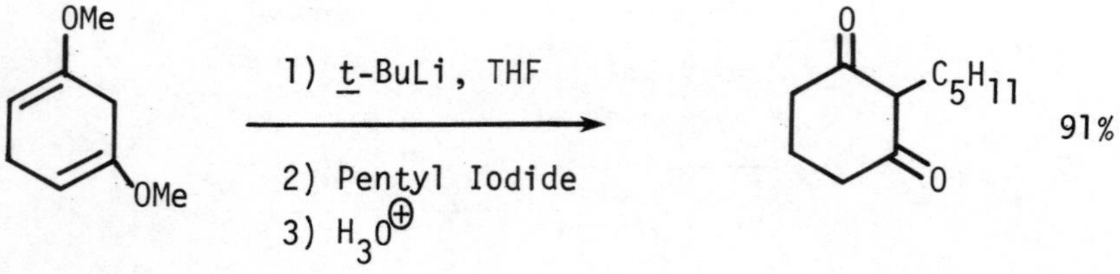

JOC 42, 3755 (1977)

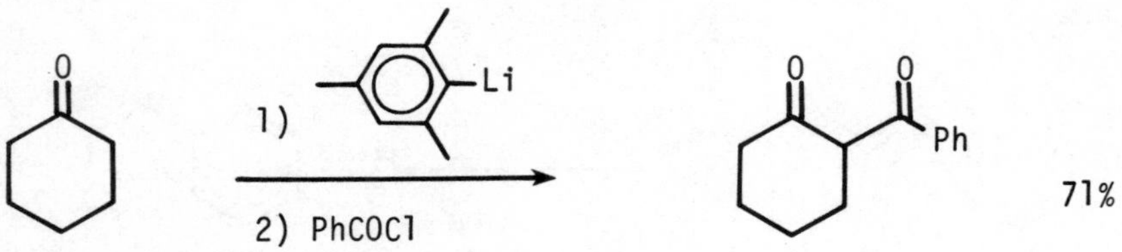

Tetr Lett, 1187 (1977)

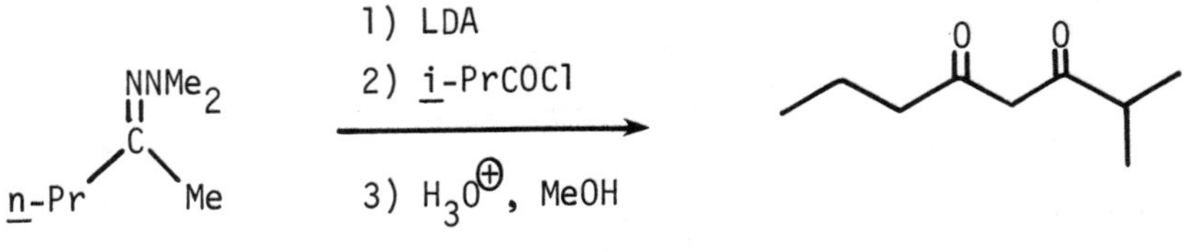
NNMe2
n-Pr
Me
1) LDA
2) i-PrCOCl
3) H3O⊕, MeOH

60%

Tetr Lett, 2853 (1978)

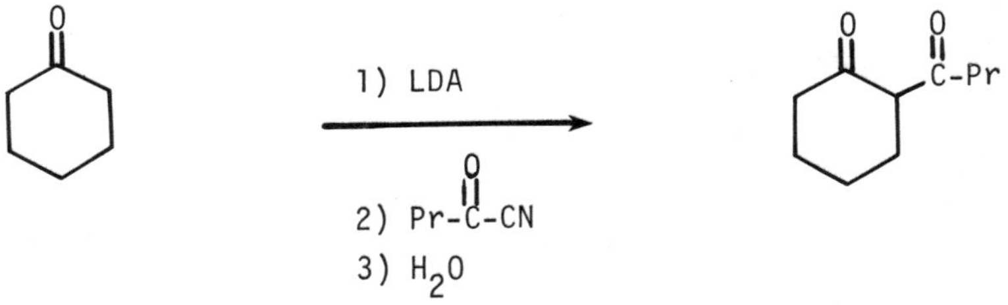
1) LDA
2) Pr-C-CN
3) H2O
C-Pr

94%

Tetr Lett, 1339 (1979)

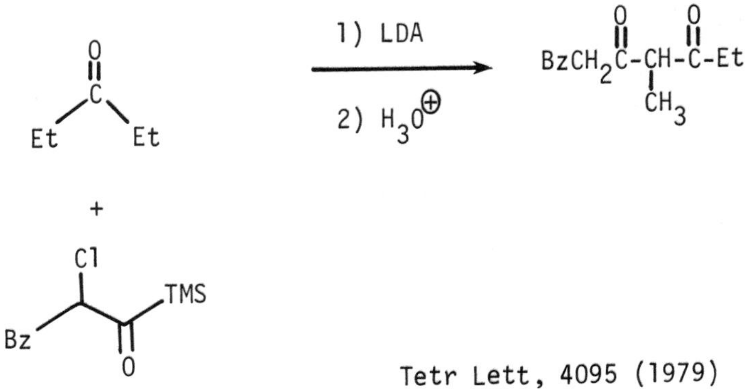
Et
Et
+
Cl
TMS
Bz
1) LDA
2) H3O⊕
BzCH2C-CH-C-Et
CH3

93%

Tetr Lett, 4095 (1979)

RCOSeK

\+

BrCH$_2$COR'

R, R' = subst. Ph, alkyl

$$\xrightarrow{\text{2) } \underline{t}\text{-C}_5\text{H}_{11}\text{OK}}$$

R-$\overset{\text{O}}{\overset{\|}{\text{C}}}$-CH$_2$-$\overset{\text{O}}{\overset{\|}{\text{C}}}$-R'   ~30-80%

Chem Lett, 1007 (1978)

\+

67%

Angew Int Ed <u>17</u>, 204 (1978)

$$\xrightarrow[\text{2) } \text{H}^{\oplus}]{\text{1) BuFe(CO)}_4^{\ominus}}$$

78%

JACS <u>99</u>, 5222 (1977)

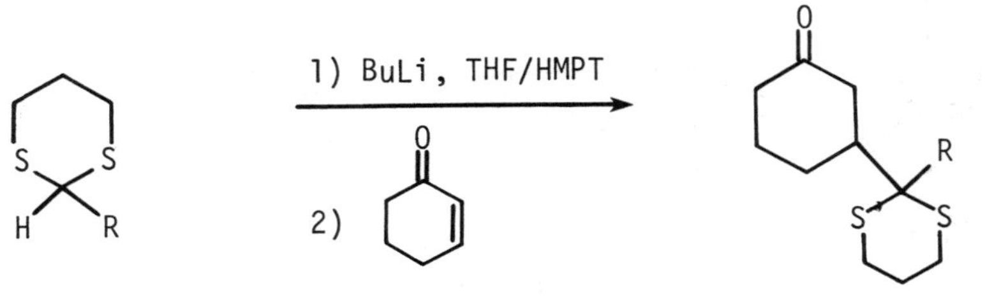

1) BuLi, THF/HMPT

2)

~70-95%

Tetr Lett, 3549 (1977)
Tetr Lett, 2695 (1979)

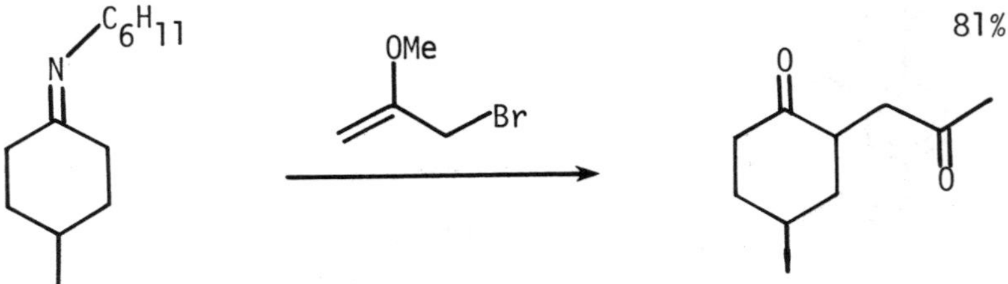

81%

Can be used to form cyclopentenones

JOC 42, 2545 (1977)

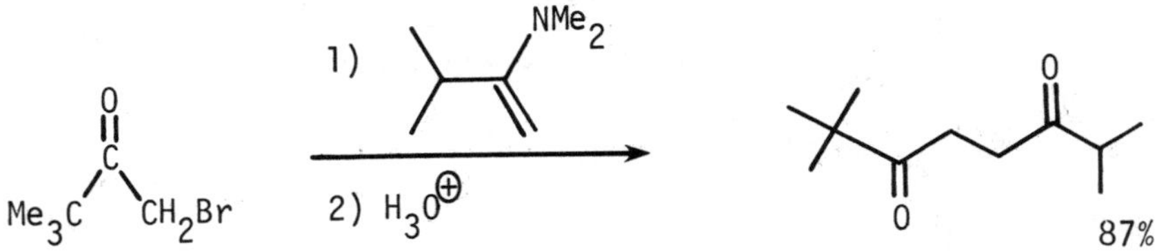

1)

2) $H_3O^{\oplus}$

87%

Acta Chem Scand (B) 30, 1000 (1976)

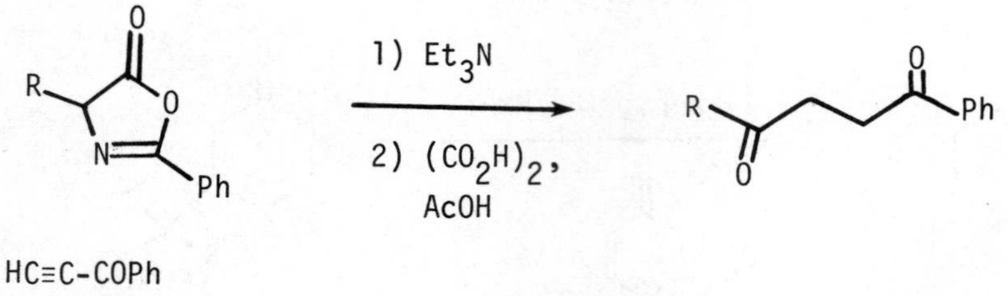

HC≡C-COPh

Chem Ber 112, 3221 (1979)

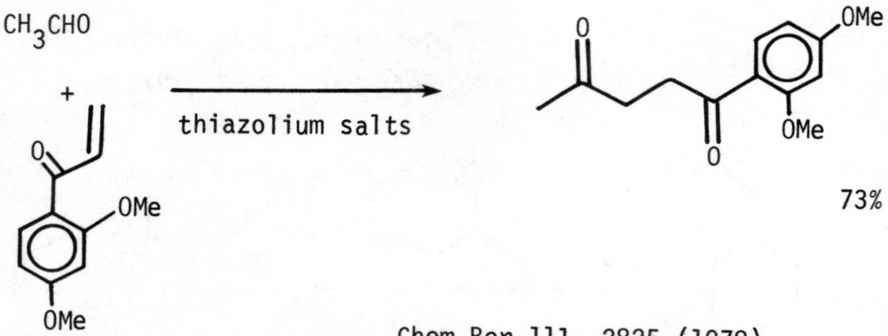

73%

Chem Ber 111, 2825 (1978)
Chem Ber 112, 84 (1979)

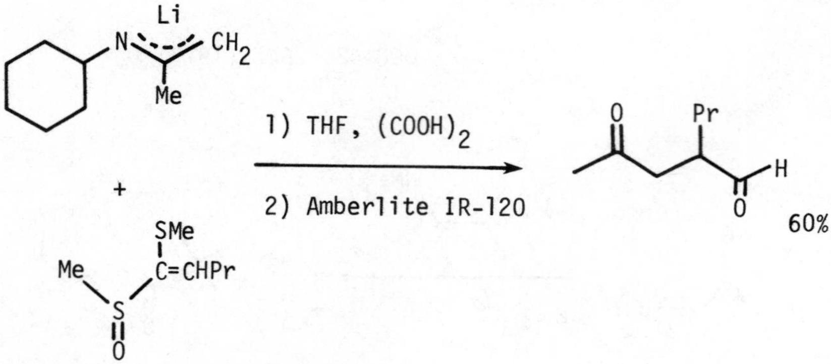

60%

J Chem Research (S), 68 (1978)

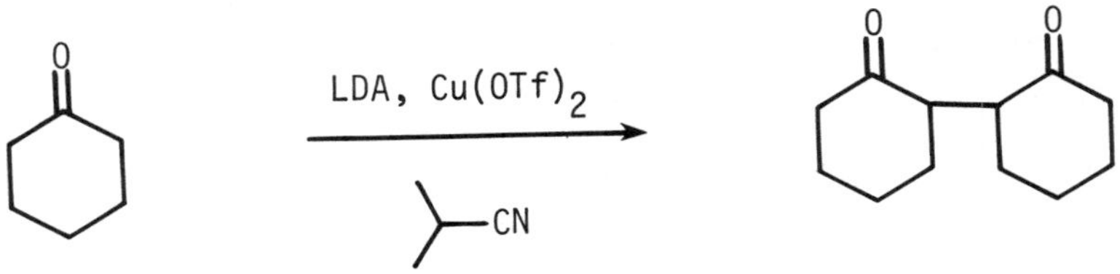

Tetr Lett, 3741 (1977)

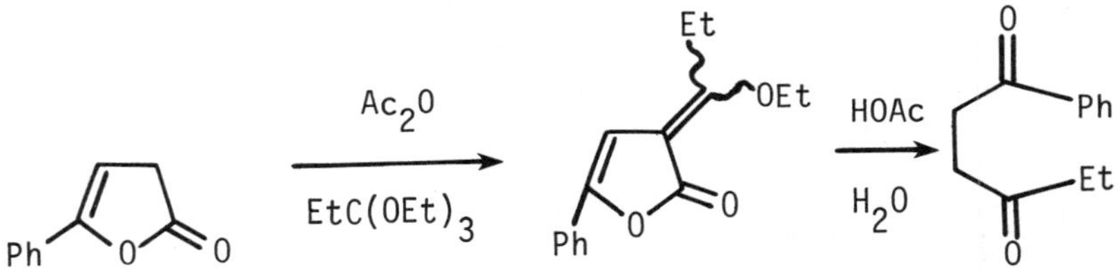

Chem Lett, 171 (1977)

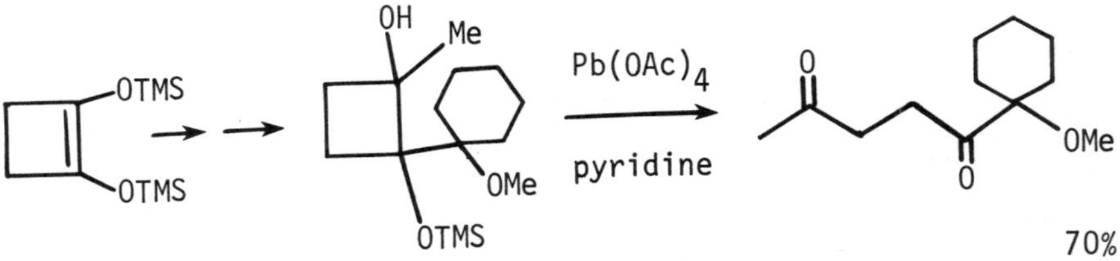

Chem Lett, 1433 (1978)

Review: "Synthesis of Polyketide-type Aromatic Natural Products by Biogenetically Modeled Routes"

Tetrahedron 33, 2159 (1977)

Section 373   Ketone - Nitrile

1) Et_2NLi, THF

$$\overset{\overset{\displaystyle O}{\|}}{2) \; Pr\text{-}C\text{-}O\text{-}CO_2Et}$$

$Et\text{-}CH_2CN$ $\xrightarrow{\hspace{3cm}}$ $Pr\text{-}\overset{\overset{\displaystyle O}{\|}}{C}\text{-}\underset{\underset{\displaystyle Et}{|}}{CH}\text{-}CN$ 74%

3) $H_3O^{\oplus}$

Tetr Lett, 1585 (1979)

$$\overset{\displaystyle OH}{Me_2\overset{|}{C}CN, \; KCN}$$

$\xrightarrow{\hspace{3cm}}$ 86%

18-crown-6

benzene

Tetr Lett, 1117 (1977)

2) CF_3COOH

+   NC Li $\xrightarrow{\hspace{3cm}}$ 70%

3) NH_4OH

$Cr(CO)_3$ 4) HCl

JOC **44**, 3275 (1979)

PhSCH$_2$CN

LDA/THF

90%

Tetr Lett, 1121 (1979)

Et-CHBr-C
         ‖
         O
          \
           Br

Me$_3$SiCN

Me$_3$SiCl

Et-CHBr-C-CN
         ‖
         O

64%

Synthesis, 204 (1979)

Section 374   Ketone - Olefin

For the oxidation of allylic alcohols to olefinic ketones, see
Section 168 (Ketones from Alcohols and Phenols).

For the oxidation of allylic methylene groups (C=C-CH$_2$ → C=C-CO),
see Section 170, Vol. 1 and 2 (Ketones from Alkyls and Methylenes).

For the alkylation of olefinic ketones, see also Section 177, Vol.
1 and 2 (Ketones from Ketones) and Section 74 (Alkyls from Olefins),
Vols. 3 and 4 for conjugate alkylations.

$$\text{cholestane} \xrightarrow[\text{HCl, 80°}]{\text{t-BuOH, PdCl}_2} \qquad 80\%$$

Synthesis, 773 (1977)

$$\xrightarrow{\text{PhSe-O-Se-Ph}} \qquad 39\text{-}92\%$$

JCS Chem Comm, 130 (1978)

$$C_7H_{15}\text{-}CH_2\text{-}CH\text{=}CH_2 \xrightarrow[\text{t-BuOOH}]{\text{SeO}_2} C_7H_{15}\text{-}\overset{O}{\overset{\|}{C}}\text{-}CH\text{=}CH_2 \qquad 61\%$$

JACS <u>99</u>, 5526 (1977)

1) PhLi

2) pyridinium
    chlorochromate

81%

JOC <u>42</u>, 682 (1977)

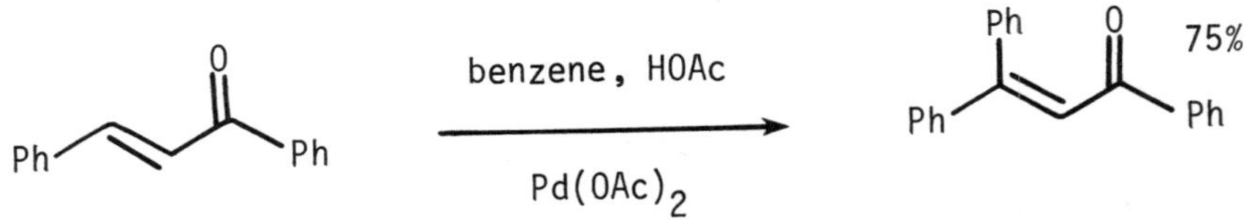

JOC **43**, 724 (1978)

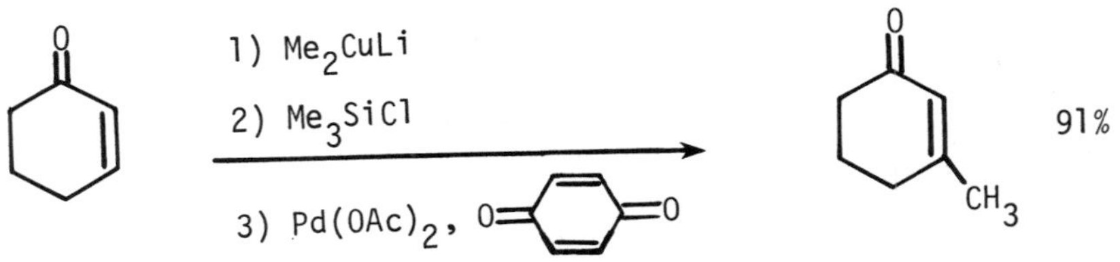

JOC **43**, 1011 (1978)

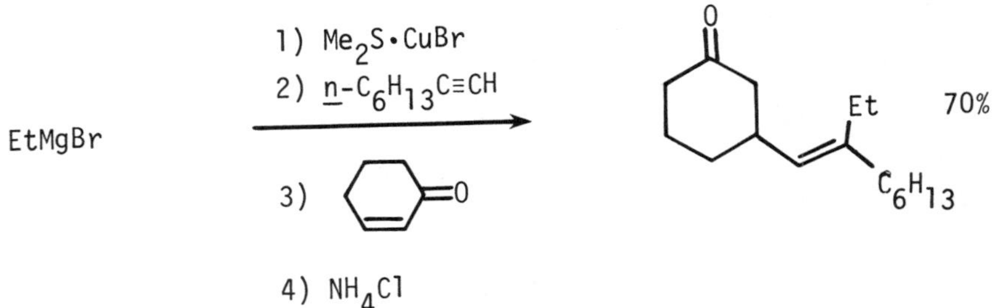

JACS **99**, 253 (1977)

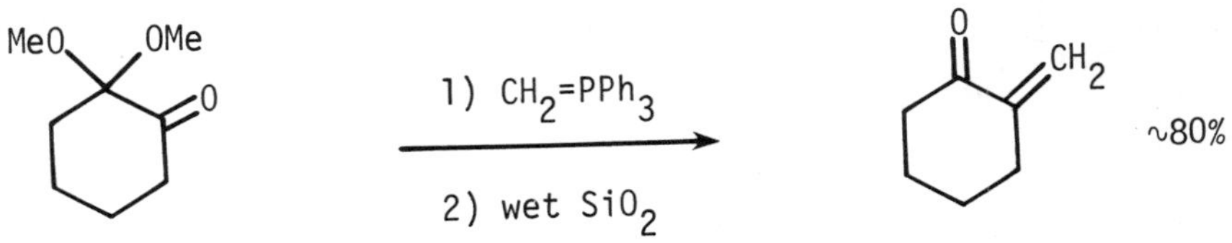

Tetr Lett, 3505 (1977)

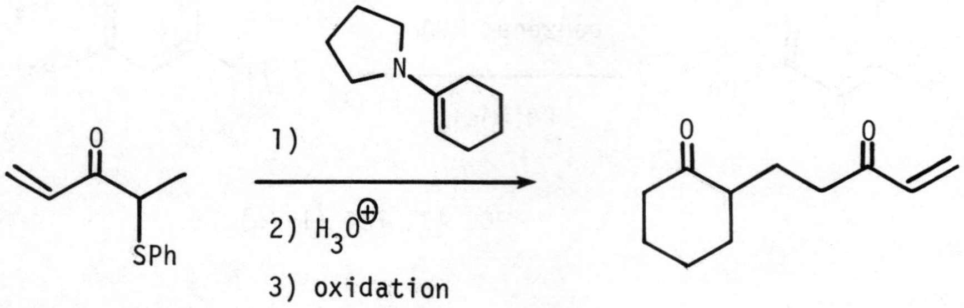

JCS Chem Comm, 821 (1978)

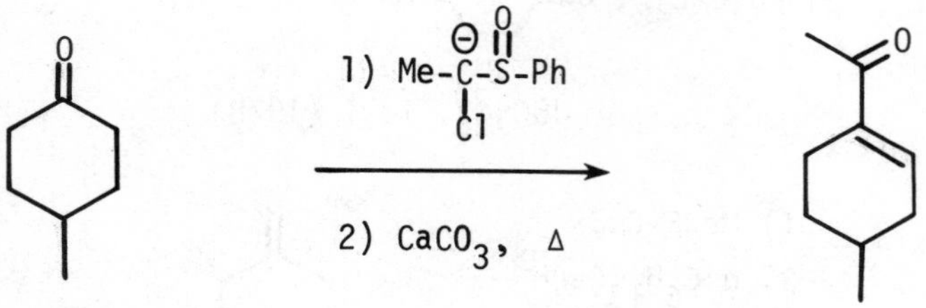

54%

JOC 44, 450 (1979)

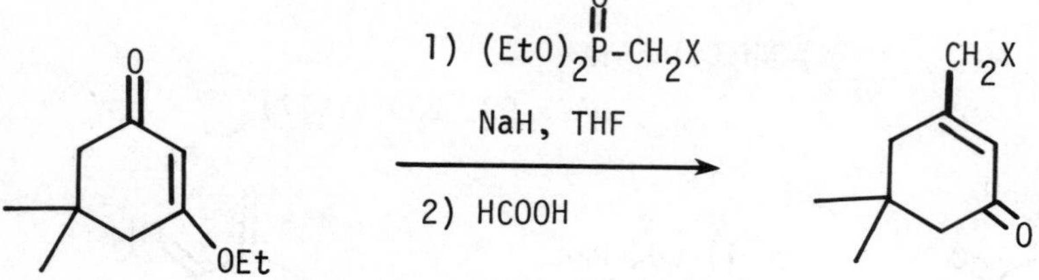

53-83%

X = -COOEt, -CN, $-\overset{\overset{\text{O}}{\|}}{\text{C}}$-R

JOC 43, 1817 (1978)

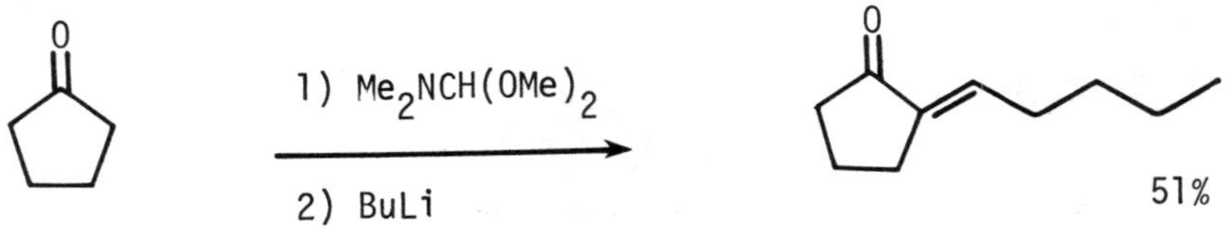

JOC <u>43</u>, 4248 (1978)

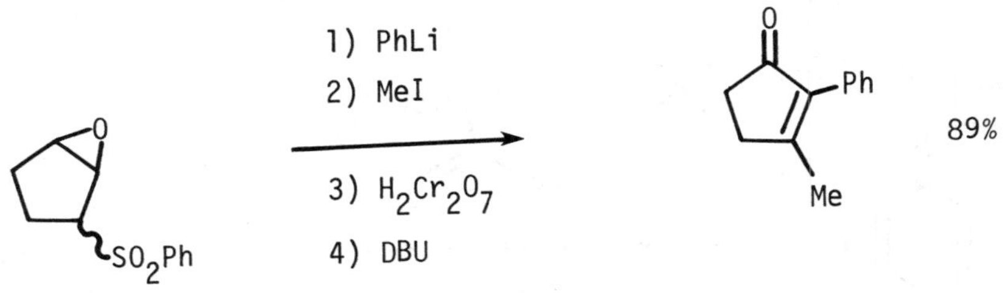

JACS <u>100</u>, 346 (1978)

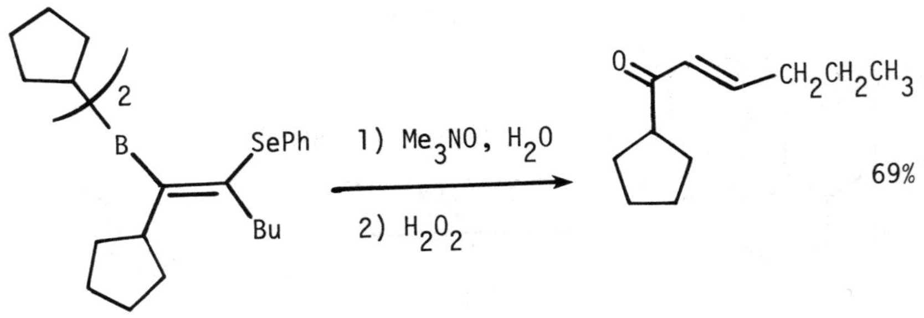

Can J Chem <u>56</u>, 2786 (1978)

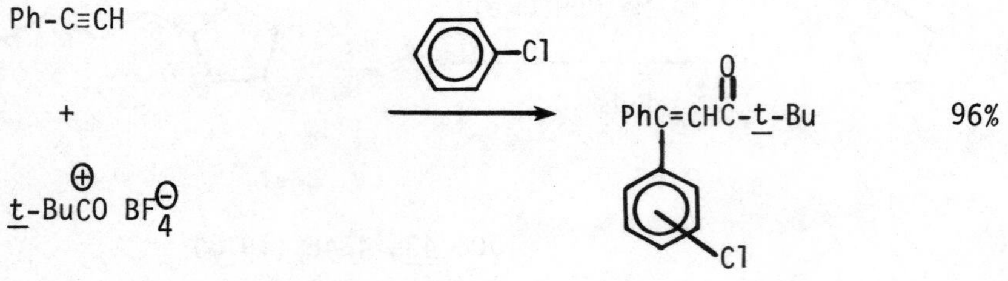

Ph-C≡CH

+

$\overset{\oplus}{\underline{t}\text{-BuCO}}$ BF$\overset{\ominus}{_4}$

$\text{PhC=CHC-}\underline{t}\text{-Bu}$ 96%

Synthesis, 324 (1977)

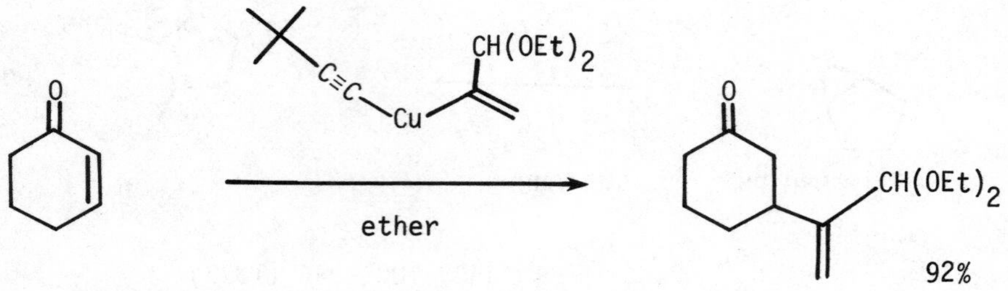

ether

92%

JOC 42, 1581 (1977)

1) Ph$_3$C⊕

2) H$_2$O

85%

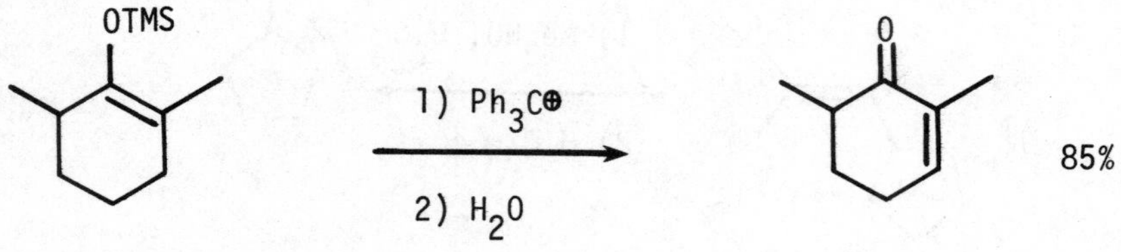

JOC 42, 3961 (1977)

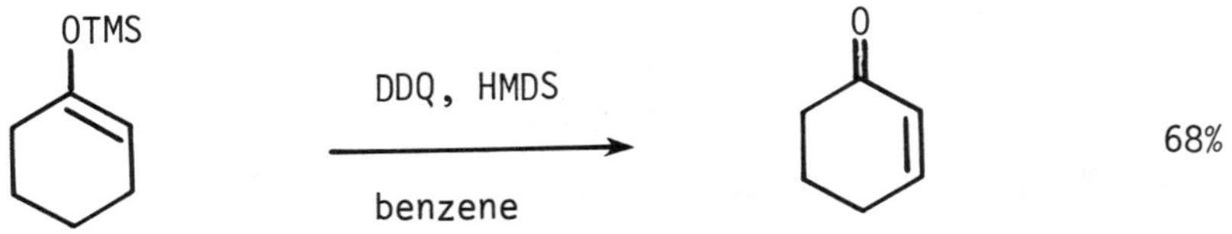

Tetr Lett, 3455 (1978)

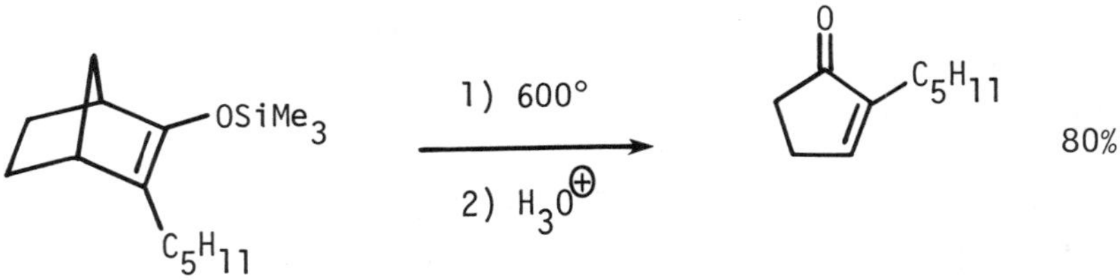

(from alkylation of norbornanone)

Tetr Lett, 3945 (1979)

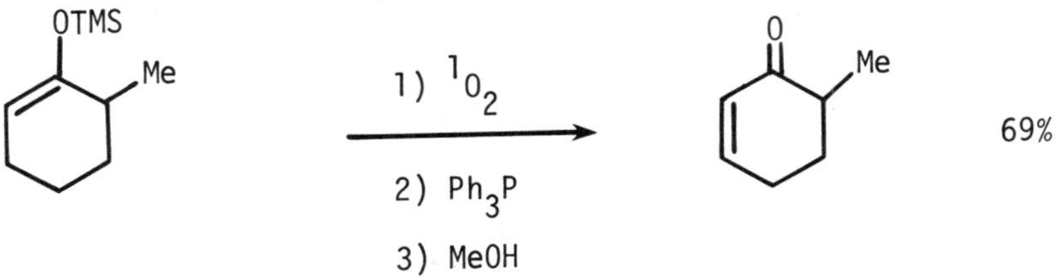

Angew Int Ed 16, 413 (1977)

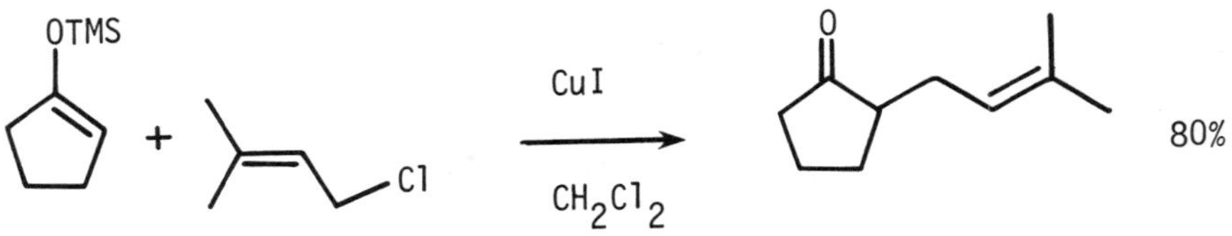

Tetr Lett, 4971 (1979)

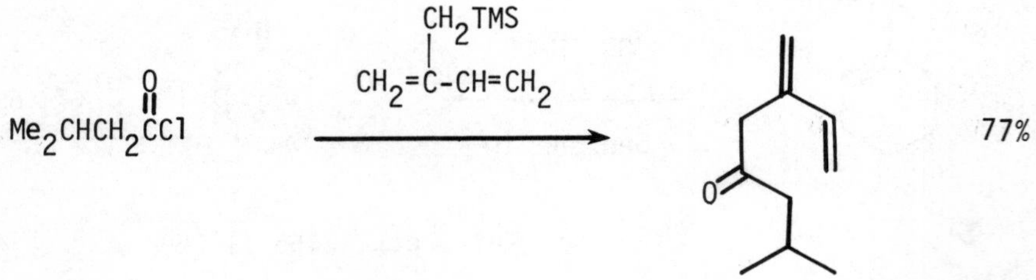

Tetr Lett, 429 (1979)

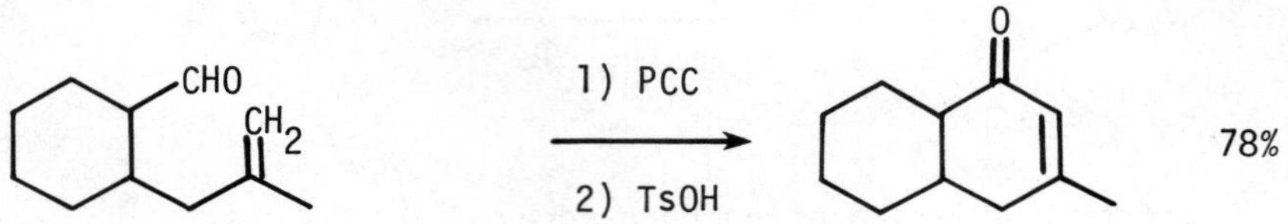

Tetr Lett, 2461 (1978)

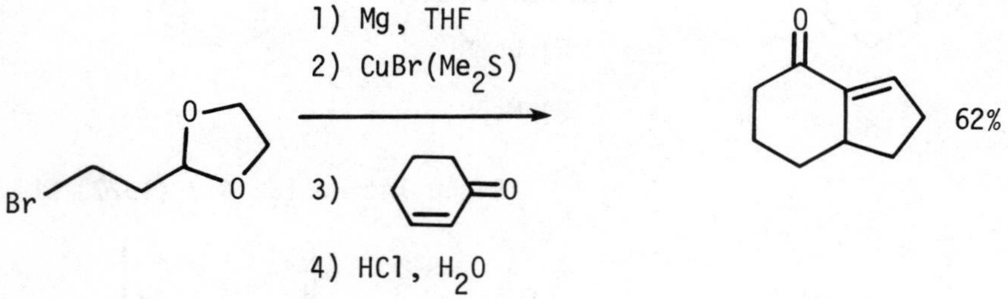

Tetr Lett, 4217 (1978)

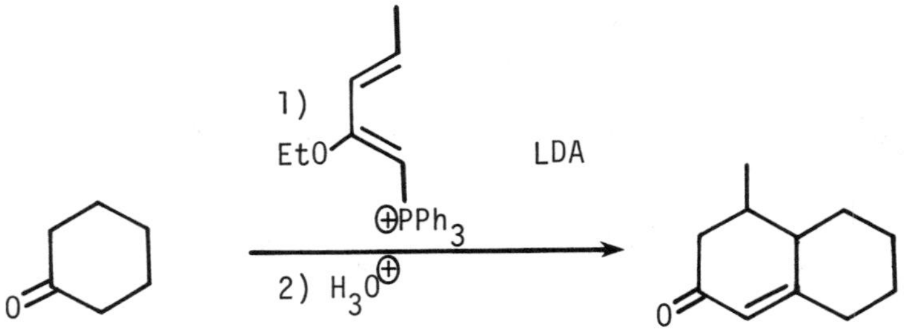

JOC 43, 4673 (1978)

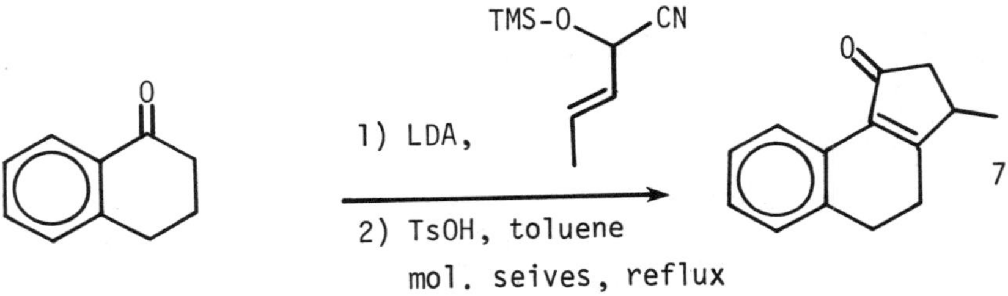

JOC 44, 462 (1979)

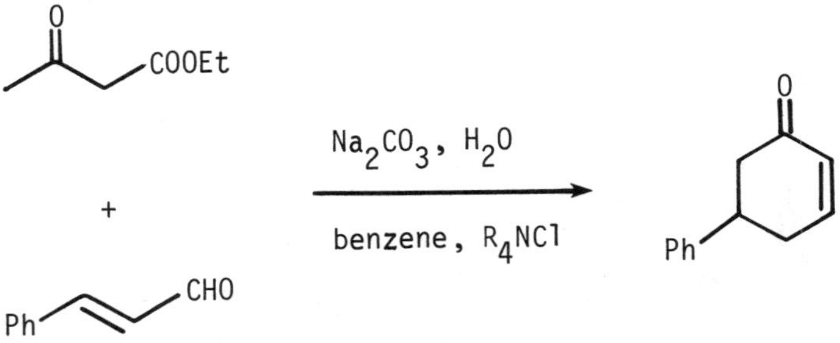

Synthesis, 107 (1979)

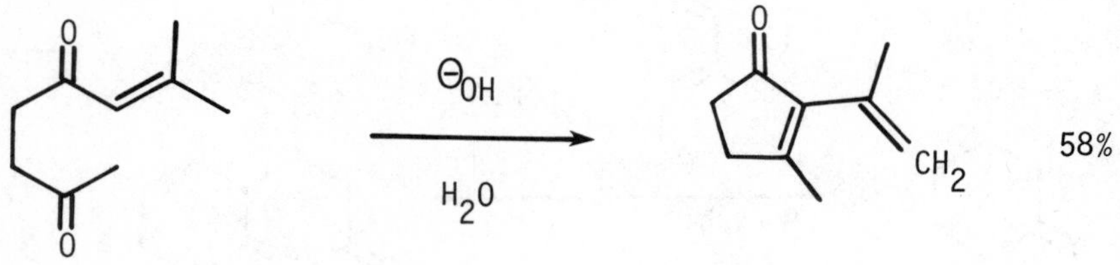

Synthesis, 187 (1979)

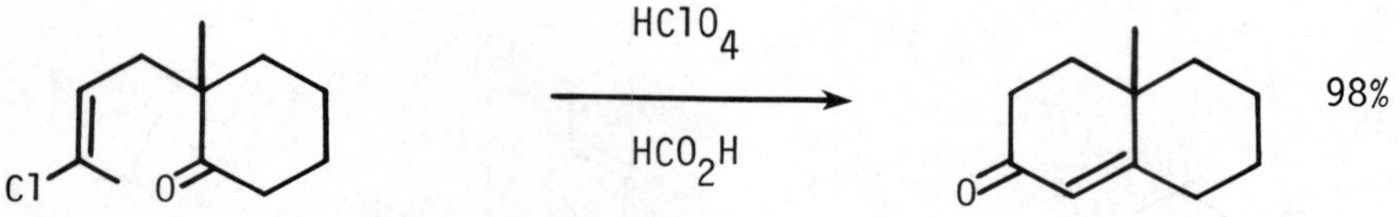

Bull Chem Soc Japan 52, 2978 (1979)

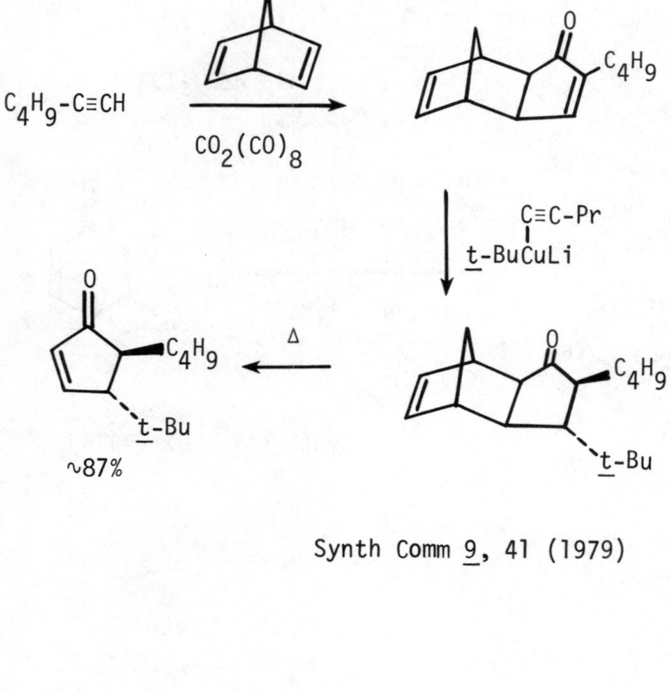

Synth Comm 9, 41 (1979)

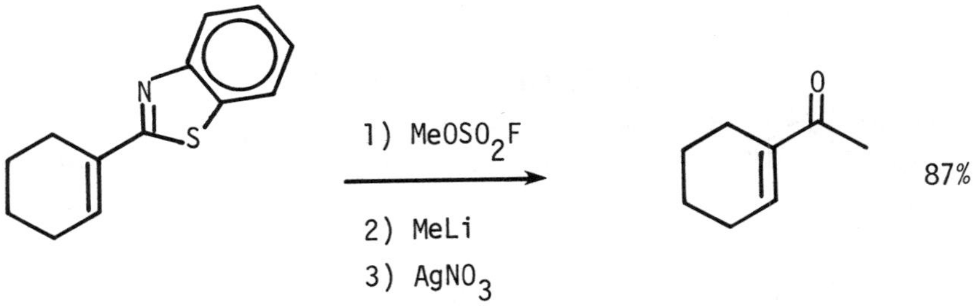

1) MeOSO$_2$F

2) MeLi

3) AgNO$_3$

87%

Tetr Lett, 5 (1978)

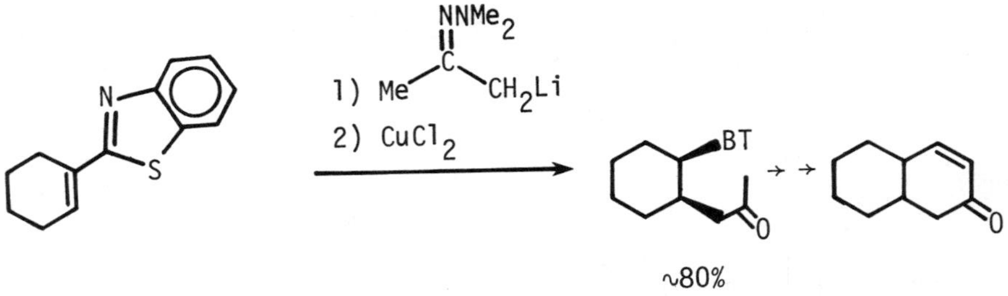

1) Me—C(=NNMe$_2$)—CH$_2$Li

2) CuCl$_2$

~80%

Tetr Lett, 13 (1978)

Na, NH$_3$

t-BuOH

88%

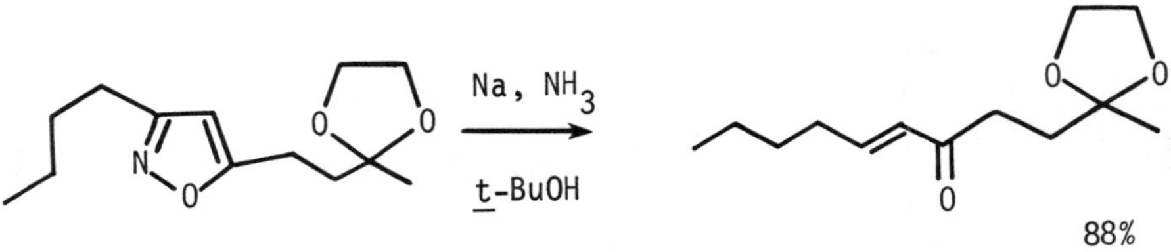

JOC **44**, 105 (1979)

Ph⌒⌒⌒ $\xrightarrow[\text{CO}_2(\text{CO})_8]{\text{CH}_3\text{I, CO}}$ Ph⌒⌒⌒⌒ 86%

Tetr Lett, 2665 (1979)

⌒⌒⌒⌒HgCl $\xrightarrow[\text{AlCl}_3]{\text{CH}_3\text{COCl}}$ ⌒⌒⌒⌒ 97%

JOC <u>43</u>, 710 (1978)

[CH₃CH₂CH₂CFe(CO)₄]⁻

$[\text{CH}_3\text{CH}_2\text{CH}_2\overset{\text{O}}{\overset{\|}{\text{C}}}\text{Fe(CO)}_4]^{\ominus}$

+

$\xrightarrow{}$

⌒⌒⌒⌒Ph   81%

Ph-CH-CH₂

$\text{Ph-}\overset{\text{O}}{\overset{\diagup\!\diagdown}{\text{CH-CH}_2}}$

Chem Lett, 1067 (1979)

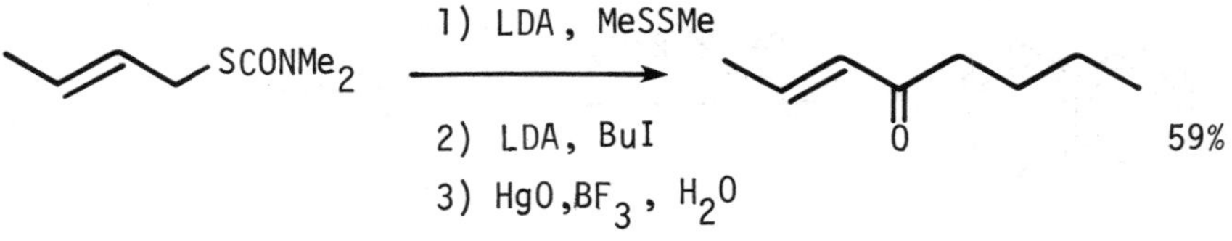
SCONMe2
1) LDA, MeSSMe
2) LDA, BuI
3) HgO,BF3, H2O
O
59%

Tetr Lett, 2895 (1978)

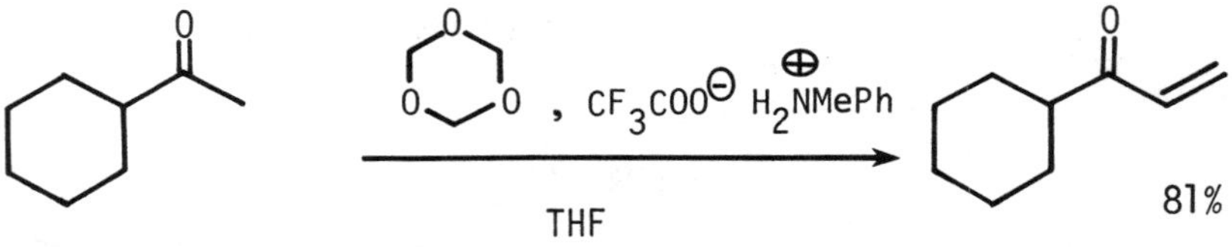
O
O O O
, CF3COO⊖ H2N⊕MePh
THF
81%

Tetr Lett, 2955 (1978)

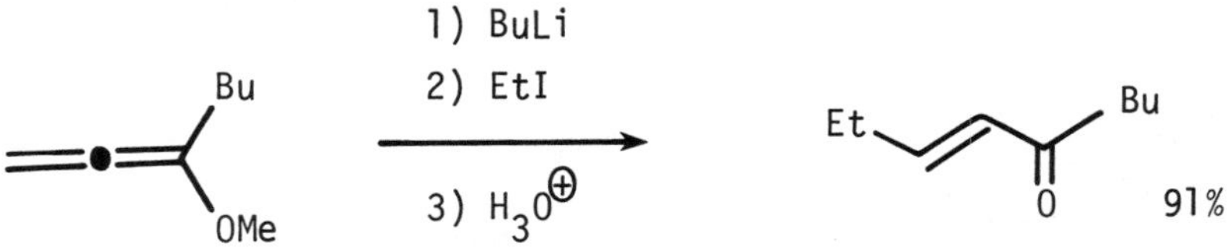
Bu
OMe
1) BuLi
2) EtI
3) H3O⊕
Et
Bu
O
91%

Tetr Lett, 1137 (1978)

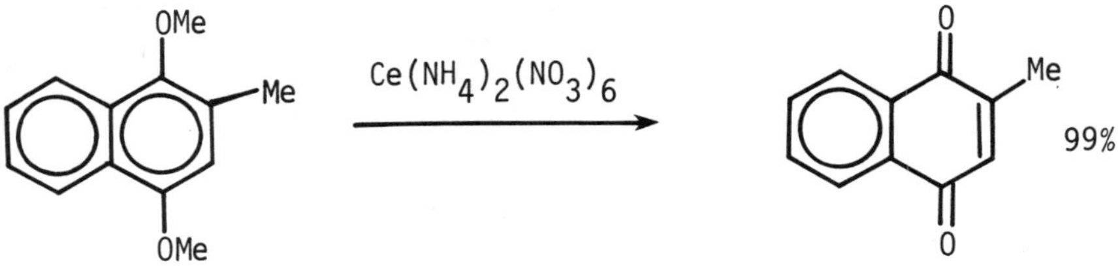
OMe
Me
OMe
Ce(NH4)2(NO3)6
O
Me
O
99%

Synthesis, 521 (1979)

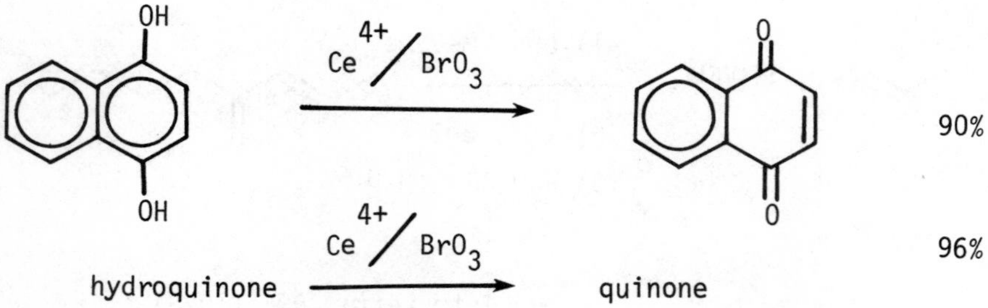

OH
OH
Ce 4+ / BrO3
O
O
90%
Ce 4+ / BrO3
hydroquinone
quinone
96%

Synth Comm 9, 237 (1979)

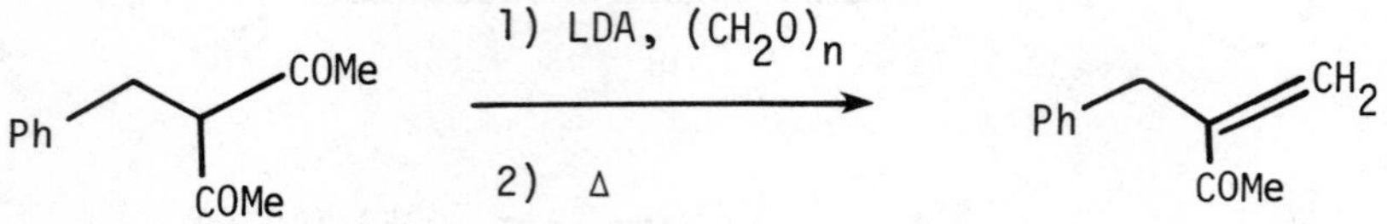

Ph
COMe
COMe
1) LDA, (CH2O)n
2) Δ
Ph
CH2
COMe

Tetr Lett, 3753 (1978)

HO
1) Pb(OAc)4
I2, hν
2) H2Cr2O7
O
O=
Pb(OAc)4
Cu(OAc)2
O
42%
57%

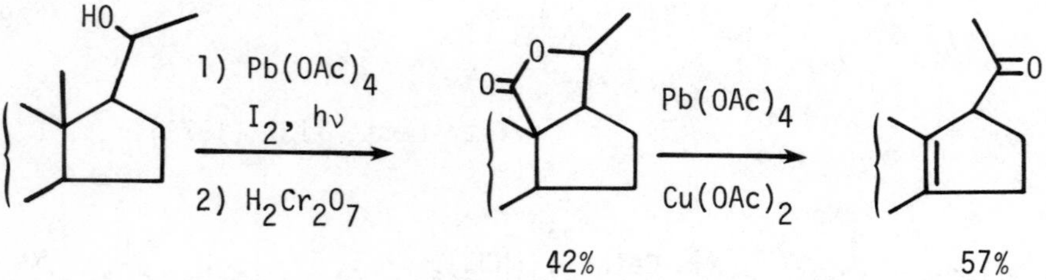

Tetr Lett, 3403 (1977)

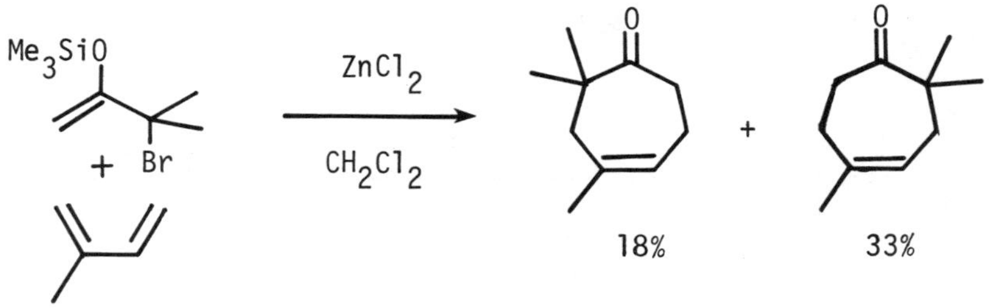

18%   33%

Angew Int Ed 18, 163 (1979)

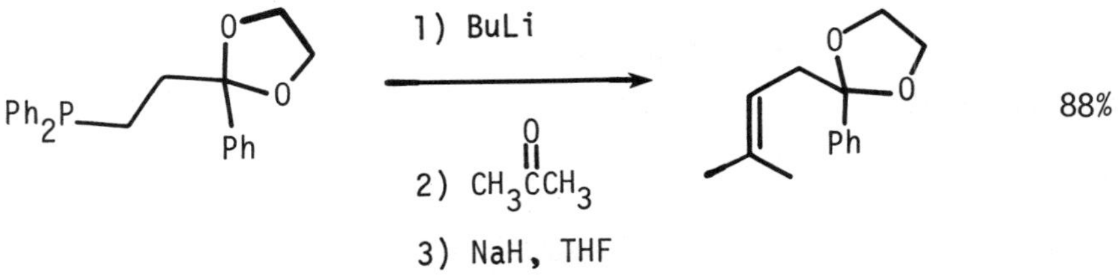

1) BuLi

2) $CH_3\overset{O}{\overset{\|}{C}}CH_3$

3) NaH, THF

88%

JCS Chem Comm, 988 (1978)

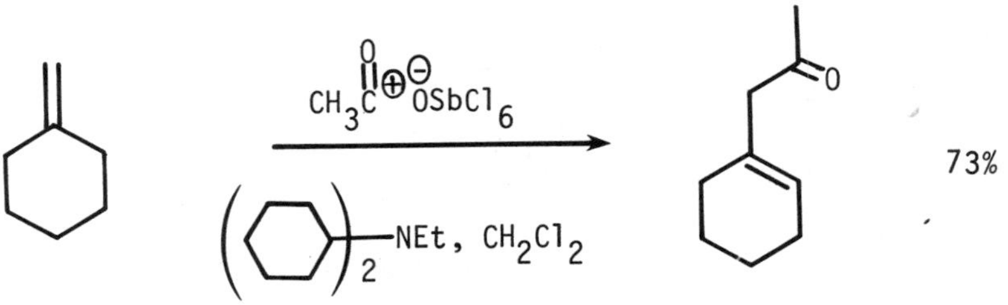

$CH_3\overset{O}{\overset{\|}{C}}{}^{\oplus}\overset{\ominus}{O}SbCl_6$

$\left(\text{cyclohexyl}\right)_2$—NEt, CH_2Cl_2

73%

JACS 99, 6008 (1977)

Section 375   Nitrile - Nitrile

No additional examples

Section 376   <u>Nitrile - Olefin</u>

$$\text{Bu}\diagup\!\!\diagdown\!\!\text{Br} \quad \xrightarrow[\text{crown ether, benzene}]{\text{KCN, PdL}_4} \quad \text{Bu}\diagup\!\!\diagdown\!\!\diagup\text{CN} \qquad 96\%$$

Tetr Lett, 4429 (1977)

$$\text{Ph-C}\equiv\text{C-CN} \quad \xrightarrow[\text{2) H}_3\text{O}^{\oplus}]{\text{1) Bu[CuBr]MgCl}} \quad \begin{array}{c}\text{Ph}\\ \diagdown \\ \text{Bu}\diagup\end{array}\!\!\text{C=CHCN} \qquad 96\%$$

Synthesis, 454 (1978)

$$\underline{n}\text{-Bu-C}\equiv\text{C-CN} \quad \xrightarrow[\text{2) H}_3\text{O}^{\oplus}]{\text{1) LiAlH}_4} \quad \begin{array}{cc}\text{H} & \text{CN}\\ \diagdown & \diagup\\ \text{C=C}\\ \diagup & \diagdown\\ \underline{n}\text{-Bu} & \text{H}\end{array} \qquad 90\%$$

Synthesis, 430 (1979)

$$\left[\begin{array}{cc}\text{Me} & \text{H}\\ \diagdown & \diagup\\ \text{C=C}\\ \diagup & \diagdown\\ \text{Bu} & \text{CuBr}\end{array}\right]\text{MgBr} \quad \xrightarrow[\text{THF}]{\text{ClCN}} \quad \begin{array}{cc}\text{Me} & \text{H}\\ \diagdown & \diagup\\ \text{C=C}\\ \diagup & \diagdown\\ \text{Bu} & \text{CN}\end{array} \qquad 97\%$$

Synthesis, 784 (1977)

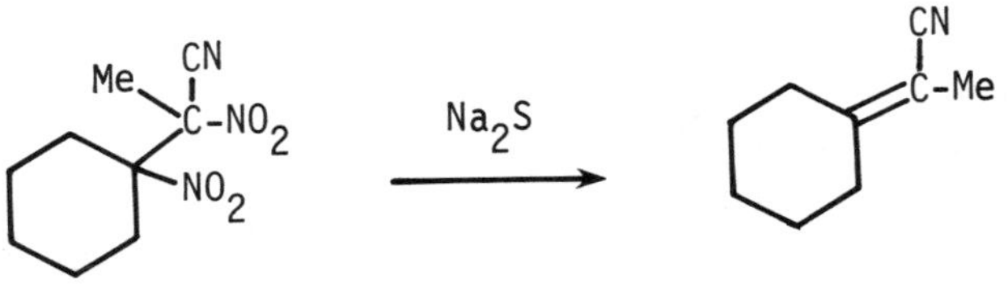

Tetr Lett, 763 (1978)

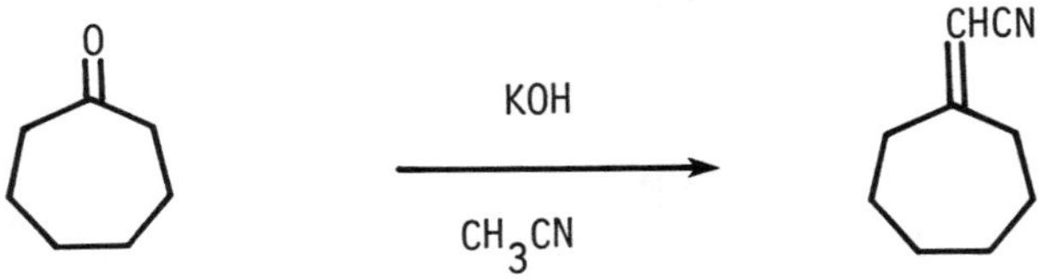

Synthesis, 629 (1977)

Ph-CHO $\xrightarrow{\text{KOH, CH}_3\text{CN}}$ PhCH=CHCN   82%

JOC 44, 4640 (1979)

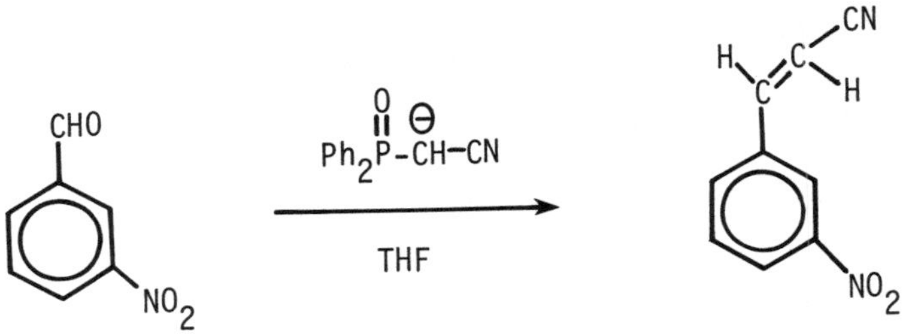

Synthesis, 126 (1977)

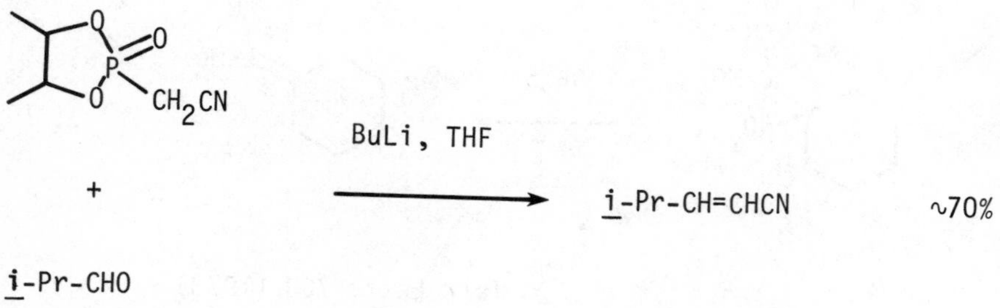

i-Pr-CH=CHCN   ~70%

BuLi, THF

i-Pr-CHO

Tetrahedron **34**, 997 (1978)

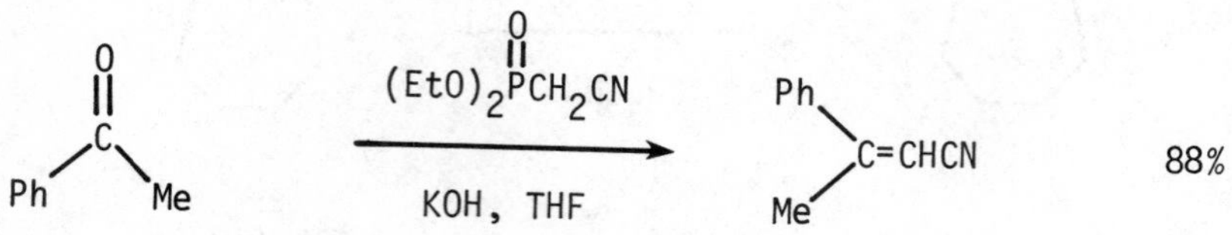

$(EtO)_2\overset{\overset{O}{\|}}{P}CH_2CN$

KOH, THF

88%

Synthesis, 884 (1979)

$EtO-\overset{\overset{S}{\|}}{C}-S-\overset{\overset{Me}{|}}{C}H-CN$

NaOH, H_2O/MeCN

R_4NCl

56%

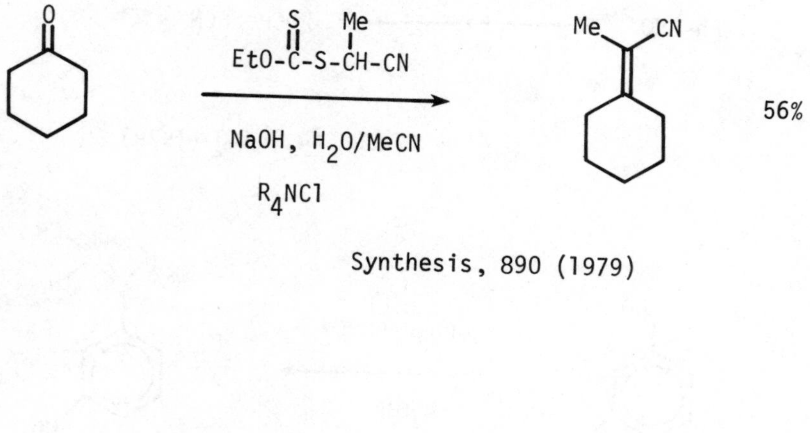

Synthesis, 890 (1979)

Section 377   <u>Olefin - Olefin</u>

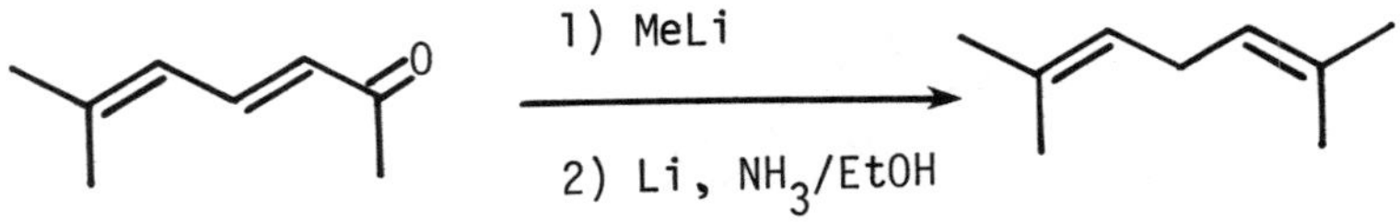

1) MeLi

2) Li, NH$_3$/EtOH

98%

JOC <u>44</u>, 1159 (1979)

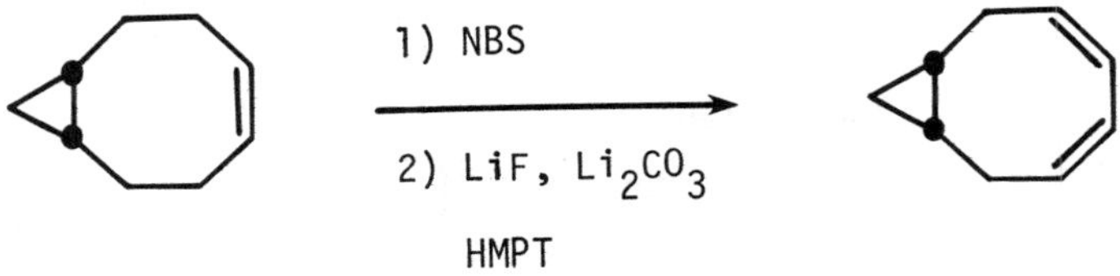

1) NBS

2) LiF, Li$_2$CO$_3$

HMPT

75%

Synthesis, 279 (1977)

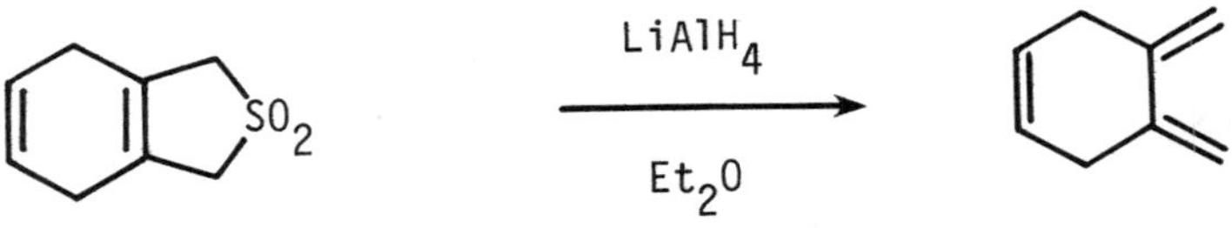

LiAlH$_4$

Et$_2$O

74%

Tetr Lett, 947 (1977)

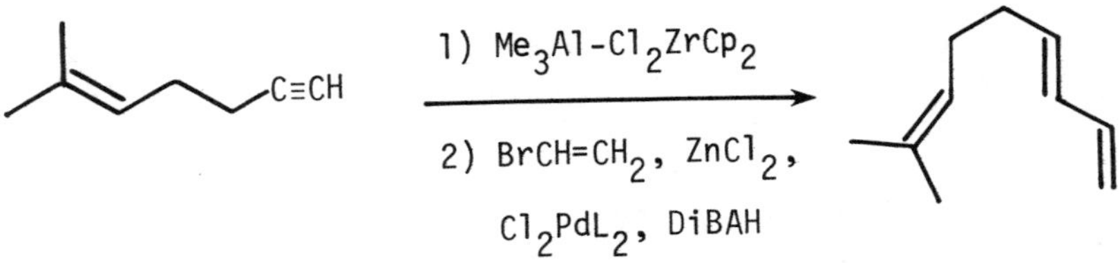

1) Me$_3$Al-Cl$_2$ZrCp$_2$

2) BrCH=CH$_2$, ZnCl$_2$,
   Cl$_2$PdL$_2$, DiBAH

70%

JACS <u>100</u>, 2256 (1978)

$C_7H_{15}-C\equiv CH$ 1) [Me, CH_2ZnBr]   →   [Me, C_7H_{15}]   72%

2) $H_3O^{\oplus}$

BSC France, 1173 (1976)

$Cl(CH_2)_3C\equiv CH$ 1) Sia_2BH → $\left(Cl \cdots \right)_2$ 70%

2) MeCu

JACS 99, 5652 (1977)

C_5H_{11} ... $ZrCp_2Cl$

+

I ... Bu

PdL_4 → C_5H_{11} ... Bu   91%

Tetr Lett, 1027 (1978)

$2\left[Bu \cdots SiF_5 \right] K_2$ $\xrightarrow{Ag(I)}$ Bu ... Bu   66%

Tetr Lett, 1137 (1979)

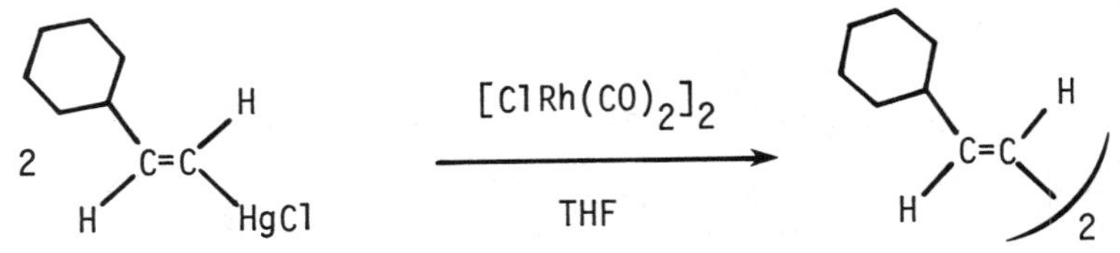

JOC <u>42</u>, 1680 (1977)

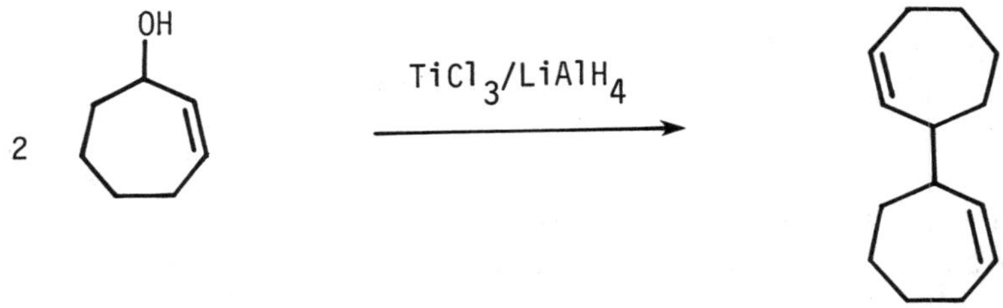

87%

JOC <u>43</u>, 3249 (1978)

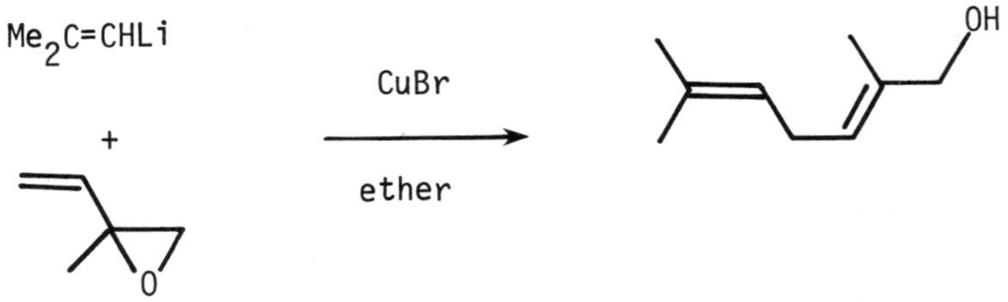

$Me_2C=CHLi$

$\xrightarrow[\text{ether}]{\text{CuBr}}$

94%

Synthesis, 528 (1978)

1) $CH_2=C(OLi)_2$

2) $MeO\underset{H}{\overset{OMe}{\underset{\displaystyle C}{|}}}NMe_2$

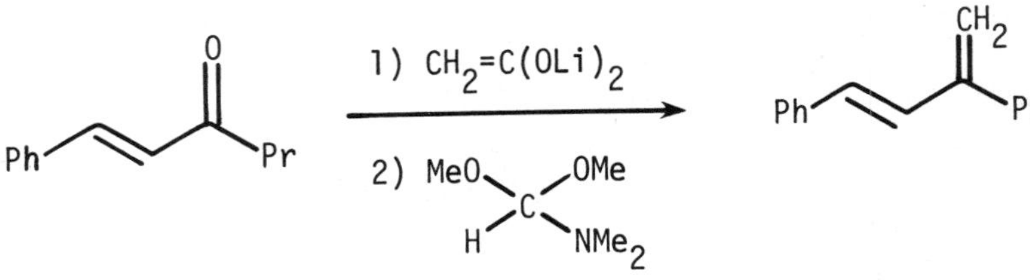

75%

Tetr Lett, 2953 (1978)

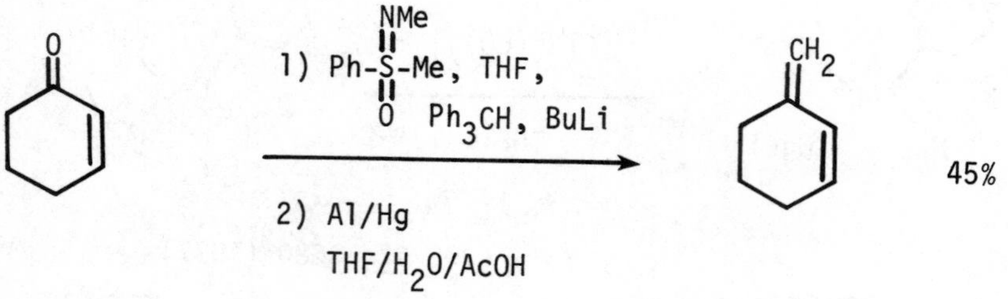
O
NMe
1) Ph-S-Me, THF,
O Ph3CH, BuLi
2) Al/Hg
THF/H2O/AcOH
CH2
45%

JACS 101, 3602 (1979)

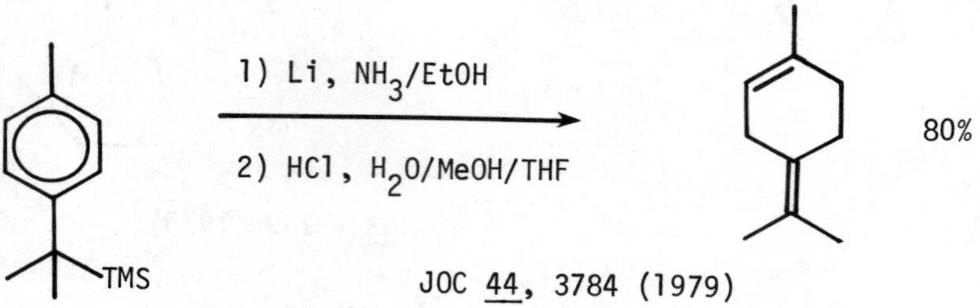
1) Li, NH3/EtOH
2) HCl, H2O/MeOH/THF
TMS
80%

JOC 44, 3784 (1979)

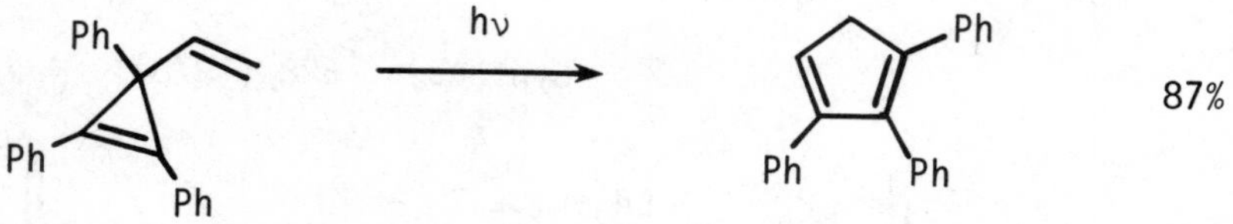
Ph
Ph
Ph
hν
Ph
Ph
Ph
87%

JACS 99, 2342 (1977)

Review: "Synthesis of Polyenes via Phosphonium Ylides"

Pure and Appl Chem <u>51</u>, 515 (1979)

Review: "The Methods of Synthesis and Properties of Conjugated Enallenic Hydrocarbons"

Russ Chem Rev <u>47</u>, 470 (1978)